BIOLOGY AND GEOLOGY OF CORAL REEFS

VOLUME I: Geology 1

CONTRIBUTORS

Dale E. Brandon

J. P. Chevalier

E. Heidecker

Harry S. Ladd

Alan R. Lloyd

W. G. H. Maxwell

John D. Milliman

D. R. Stoddart

F. W. Whitehouse

BIOLOGY AND GEOLOGY OF CORAL REEFS

EDITED BY

O. A. JONES

Department of Geology
University of Queensland
St. Lucia, Brisbane
Queensland, Australia

R. ENDEAN

Department of Zoology
University of Queensland
St. Lucia, Brisbane
Queensland, Australia

VOLUME I: Geology 1

ACADEMIC PRESS   New York San Francisco London   1973

A Subsidiary of Harcourt Brace Jovanovich, Publishers

ACADEMIC PRESS, INC.
111 Fifth Avenue, New York, New York 10003

United Kingdom Edition published by
ACADEMIC PRESS, INC. (LONDON) LTD.
24/28 Oval Road, London NW1

LIBRARY OF CONGRESS CATALOG CARD NUMBER: 72-84368

PRINTED IN THE UNITED STATES OF AMERICA

To the Great Barrier Reef Committee, its office-bearers and its many members who have worked unremittingly for fifty years to further our knowledge of the Great Barrier Reef; and to the memory of geologist Professor H. C. Richards, one of its founders, and biologist Professor E. J. Goddard, an early enthusiastic supporter

CONTENTS

1. Caribbean Coral Reefs

John D. Milliman

2. Coral Reefs of the Indian Ocean

D. R. Stoddart

3. Bikini and Eniwetok Atolls, Marshall Islands

Harry S. Ladd

4. Geomorphology and Geology of Coral Reefs in French Polynesia

J. P. Chevalier

5. Coral Reefs of New Caledonia

J. P. Chevalier

6. Coral Reefs of the New Guinea Region

F. W. Whitehouse*

7. Waters of the Great Barrier Reef Province

Dale E. Brandon

* Deceased.

8.   Geomorphology of Eastern Queensland in Relation to the Great Barrier Reef

W. G. H. Maxwell

9.   Structural and Tectonic Factors Influencing the Development of Recent Coral Reefs off Northeastern Queensland

E. Heidecker

10.   Sediments of the Great Barrier Reef Province

W. G. H. Maxwell

11. Foraminifera of the Great Barrier Reefs Bores

Alan R. Lloyd

LIST OF CONTRIBUTORS

Numbers in parentheses indicate the pages on which the authors' contributions begin.

DALE E. BRANDON, Esso Production Research Co., Houston, Texas (187)

J. P. CHEVALIER, Museum of Natural History, Paris, France (113, 143)

E. HEIDECKER, University of Queensland, St. Lucia, Brisbane, Queensland, Australia (273)

HARRY S. LADD, Smithsonian Institution, Washington, D.C. (93)

ALAN R. LLOYD,* French's Forest, New South Wales, Australia (347)

W. G. H. MAXWELL,† Department of Geology and Geophysics, University of Sydney, Sydney N. S. W., Australia (233, 299)

JOHN D. MILLIMAN, Woods Hole Oceanographic Institution, Woods Hole, Massachusetts (1)

D. R. STODDART, Department of Geography, University of Cambridge, Cambridge, England (51)

F. W. WHITEHOUSE,** Department of Geology, University of Queensland, St. Lucia, Brisbane, Queensland, Australia (169)

* Present address. P.O.B. 175, Euntwood, N.S.W., Australia.
† Present address: Appraisals Mining and Geological, Chatswood N.S.W., Australia.
** Deceased.

GENERAL PREFACE

This four-volume work (two volumes covering geological and two biological topics) originated from an article on The Great Barrier Reefs of Australia written by the editors and published in the November 1967 issue of *Science*. The prime aim of this treatise is to publish in one source as many as possible of the major advances made in the diverse facets of coral reef problems, advances scattered in a multitude of papers and published in a variety of journals.

Initially a one-volume work was projected, but the wealth of material available led to this four-volume treatise. Two contain chapters on aspects of geomorphology, tectonics, sedimentology, hydrology, and radiometric chronology relevant to coral reefs, and two accommodate articles on pertinent biological topics.

The task of organizing the contributions by some forty-one authors (situated in eight different countries) on forty-six different topics proved a formidable one. We wish to thank the contributors, all of whom have been as cooperative as their teaching and/or other commitments permitted.

We feel that the material presented in these volumes demonstrates that many major advances in our knowledge have been made in recent years. We realize that the treatment of the topics covered is not complete and that in many cases new problems have been brought to light. It is our hope that the volumes will prove a powerful stimulus to further work on all aspects of coral reefs.

The editors, of course, accept overall responsibility for all four volumes. Dr. Jones is mainly responsible for the editing of the two geology volumes; Dr. Endean for that of the two biology volumes.

O. A. JONES
R. ENDEAN

PREFACE TO VOLUME I: GEOLOGY 1

This first geology volume is restricted to chapters dealing with the reefs of particular areas; discussion of topics applicable to coral reefs throughout the world is deferred to the second geology volume. Important additions of our knowledge of coral reefs, particularly the spectacular results that stemmed in recent years from work in the Marshall Islands—seismic surveys and dredging at depths of 1460–3660 meters, as well as deep borings—are discussed. This work not only provided proof of Darwin's explanation of the origin of deep ocean atolls by progressive subsidence of their foundations but also yielded important conclusions on other aspects of the history of those reefs. Dr. Ladd summarizes this work in his chapter on Bikini and Eniwetok Atolls.

Ironically, the stimulus to the undertaking of these studies was the desirability of surveys prior and subsequent to the testing of nuclear explosive devices. The surveys were primarily to document the effects of the explosions, but incidentally, during a few years of study, they yielded far more definitive information on coral atolls than had the total of all previous work.

Similar considerations prompted surveys of Mururoa Atoll in the Tuamotu Archipelago, surveys which in turn produced much new data on the reefs, only some of which can be correlated with data on the atolls of the Marshall Islands. The reefs of French Polynesia, including Mururoa Atoll, as well as those of New Caledonia are described by Dr. Chevalier in Chapters 4 and 5.

As Stoddart (*Biol. Rev.* 44, 1969, p. 437) has pointed out and as is briefly discussed by Whitehouse in Chapter 6, Darwin's theory is not necessarily applicable to barrier reefs on continental shelves or surrounding "high" (continental) islands. Barrier reefs in particular need further study in regard to the nature and geological history of their foundations quite apart from work in other geological fields.

Proposals to establish an air base on one of the islands in the Indian Ocean initiated studies of the fauna of the islands involved, studies which were widened to include most of the reefs in the ocean. The more important features of these reefs are summarized by Dr. Stoddart in Chapter 2.

In Chapter 1 Dr. Milliman gives a summary of available information on the intensively studied reefs of the Caribbean Sea, including a valuable, complete bibliography of papers dealing with the area.

Dr. Whitehouse had unique opportunities to observe and record the surface features of the reefs in the New Guinea region in the course of various missions for the Allied Forces during World War II. He records and discusses his observations in Chapter 6.

Finally, the post-war period has seen a great upsurge in interest in the Great Barrier Reefs of Australia, leading to the application of techniques hitherto unused in this area and studies of aspects previously neglected. Thus Chapters 7–11 record the results of work in the fields of hydrology, sedimentology, and tectonics, as well as a new treatment of the geomorphology of the adjacent mainland coast and revised descriptions and correlations of the Foraminifera of several deep bores (including some drilled in search of oil) scattered over the whole length of the reefs.

We feel certain that the two geology volumes of this treatise will go a long way toward filling the need for an up-to-date summary of much of the available (geological) information on coral reefs. At the same time we emphasize that all aspects of coral reefs are not covered, that the volumes do not provide solutions to all coral reef problems, that there are still many gaps in our knowledge, and that there is still ample scope for further work. We hope these volumes will prove to be a stimulus to such work.

O. A. JONES

R. ENDEAN

BIOLOGY AND GEOLOGY OF CORAL REEFS

VOLUME I: Geology 1

1

CARIBBEAN CORAL REEFS[1]

John D. Milliman

I. Introduction

Most coral reefs in the Atlantic Ocean are limited to the western parts of the ocean. Although some hermatypic coral species are present in the shallow waters of the Gulf of Guinea, corals off West Africa never grow into well-defined hermatypic structures (Thiel, 1928; Chevalier, 1966). The paucity of corals in this area is the result of seasonal influxes of cooler waters by the Guinea Current and local up-wellings (Fig. 1), together with the large amount of terrigenous sedimentation by coastal rivers, which has effectively covered much of the potential substrate upon which coral planulae could settle. The high rates of sedimentation in the areas off the Orinoco and Amazon Rivers have limited reef development in the western Atlantic south of about 5°N. Some coral reefs have been reported off Brazil (Branner, 1904), and even one atoll, Atoll das Rocas (Ottmann, 1963), is known, but these reefs mostly consist of encrusted beachrock rather than constructional bioherms, and the major biological component is coralline algae,

[1] Contribution number 2927, from the Woods Hole Oceanographic Institution.

1

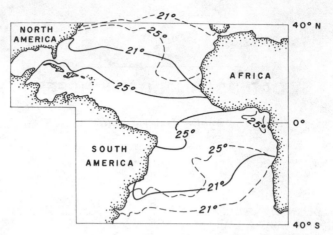

Fig. 1. Average winter (—) and summer (– –) surface isotherms (°C) in the Atlantic Ocean (after Sverdrup *et al.*, 1942). Areas with prolonged exposures to temperatures less than 22°C, or with short exposures less than 18°, at best will have a depauperate coral population. Periodic influxes of cold waters from the Guinea Current have helped to severely limit reef development off tropical Africa.

not coral (major coral genera are *Millepora, Porites,* and *Favia:* Greely, in Branner, 1904).

The northern limit of coral reefs in the western Atlantic is southern Florida; north of this area winter temperatures are too low for corals to produce viable reefs (Fig. 2 and 3), although hermatypic corals can survive as far north as Cape Hatteras (35°N) (Macintyre and Pilkey, 1969).[2] Work (1969) has stated that the northern limit of the tropical West Indian assemblage is St. Lucie, Florida (27°10′N); this boundary also approximates the northern limit of coral reef sediments on the neighboring continental shelf (Milliman, 1972a). Thus most coral reef development in the Atlantic Ocean is restricted to the Caribbean and adjacent areas in southern Florida and the Bahamas (Fig. 4). The Gulf of Mexico is basically an area of terrigenous sedimentation, but some scattered marginal reef growth is present near Veracruz, Mexico, together with many relict coral and algal mounds throughout much

[2] Bermuda (32°N) is an exception in that it contains a surprisingly large number of corals, but still lacks some of the more prolific West Indian corals (such as *Acropora*), and the reefs appear to be only thin encrustations over Pleistocene rock. Locally, however, reef accumulations have been significant. For example, Ginsburg *et al.* (1971) report reef accretion rates of greater than 1 meter per thousand years. A recent paper by Garrett and others (1972) describes the ecology and sediments of several lagoonal reefs from the Bermuda platform.

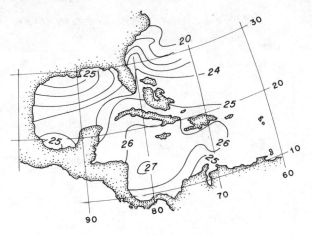

February

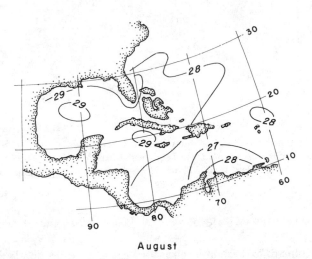

August

Fig. 2. Average winter (February) and summer (August) surface temperatures (°C) in the Caribbean (after Sverdrup *et al.*, 1942).

of the Gulf shelf (see Table V). The northern portion of this area, roughly that area north of a line passing through Progreso, Mexico and Havana, Cuba, can be considered to be marginal tropics in that winter temperatures commonly fall below 22°C (Fig. 3). Corals in this marginal belt seem to have slower growth rates [for example, Shinn (1966) reports a growth rate of *Acropora cervicornis* of 10 cm/year, while the same species grows at 26 cm/year in Jamaica; Lewis *et al.*, 1969)], and reefs

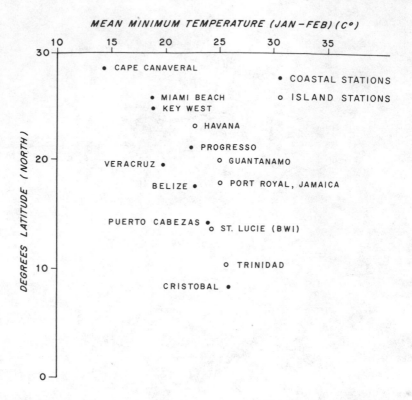

Fig. 3. Mean minimum winter temperature vs latitude in the Caribbean. (Data from Environmental Science and Services Administration, 1968.) Areas with mean minimum temperatures less than 21° or 22°C generally do not display optimal reef development.

generally are not as well developed as those in the central and southern Caribbean.

The Caribbean area is subject to northeasterly trade winds throughout most of the year, although during the winter months winds may shift to ESE. As a result of the nearly constant wind direction, currents and waves are predominantly from the east. Tides throughout most of the Caribbean are semidiurnal, although in the Gulf of Mexico tides are diurnal and in the southwestern Caribbean they are mixed, with a strong diurnal component. The tidal range in most Caribbean areas does not exceed 0.5–1 m.

Hurricanes are a common occurrence throughout much of the Caribbean, but most are restricted to two broad paths, from the Lesser Antilles up to the Bahamas and Florida, and along northern Central America

Fig. 4. Bathymetric chart of the Caribbean Sea and surrounding areas, showing the major morphological features.

into the Gulf of Mexico area. Hurricanes can result in the destruction of reef organisms (Stoddart, 1962b; Thomas *et al.*, 1961; Goreau, 1964), the alteration of both reef and island morphology (Vermeer, 1963; Stoddart, 1962b, 1963), and the large-scale erosion and transport of sediments (Ball *et al.*, 1967; Perkins and Enos, 1968).

II. Previous Workers

Most early studies on Caribbean coral reefs were limited to those areas immediately adjacent to the continental United States, notably Florida and the Bahamas. Perhaps the most noteworthy early studies were those by A. G. Mayor, T. W. Vaughan, and their co-workers at the Tortugas Laboratory, who investigated the interrelationships between environment, reef organisms, and sediments in the Florida Keys and the Bahamas. The death of Mayor in the early 1920's signaled a decline in Caribbean reef research, which lasted until after World War II. In the early 1950's, ecological and sedimentological parameters of the Bahamas and Florida were defined in a series of studies by N. D. Newell and his co-workers from Columbia University and the American Museum of Natural History, and by R. N. Ginsburg from the Shell Development Company. In more recent years, reef investigations have spread into

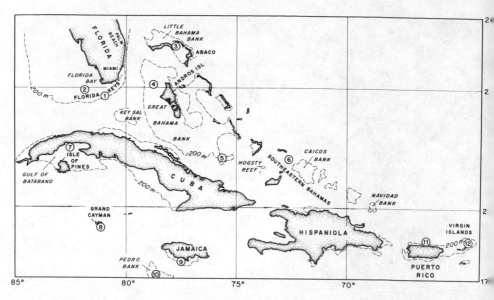

Fig. 5. Northern Caribbean reefs. Numbers refer to areas whose reefs have been investigated (see Table I).

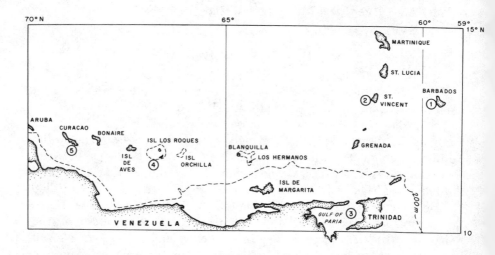

Fig. 6. Southeastern Caribbean reefs. Numbers refer to areas whose reefs have been investigated (see Table II).

Fig. 7. Southwestern Caribbean reefs. Numbers refer to areas whose reefs have been investigated (see Table III).

the more southerly parts of the Caribbean. Foremost were the studies by the late T. F. Goreau, who spent nearly 20 years studying the physiology of Caribbean corals (summarized in Goreau, 1963), as well as defining the ecology and geology of the reefs surrounding Jamaica. Other group studies include those by Texas A & M University at Alacran Reef and by McGill University in Barbados.

Fig. 8. Northwestern Caribbean reefs. Numbers refer to areas whose reefs have been investigated (see Table IV).

Perhaps the most distinctive difference between the studies in the Caribbean and those in the Pacific has been the direction of emphasis. Many of the more outstanding Pacific studies have been along biological lines (for example, the Palau studies and the Great Barrier Reef Expedition) or have been broad geologic–geomorphic studies (for example, the Bikini program). In the Caribbean, on the other hand, most recent studies have centered around the sedimentological aspects of reef and bank environments.

In such a short review it is not possible to discuss all the numerous investigations in Caribbean coral reefs. Therefore, the various reefs which have been studied are listed in Tables I–V and shown in Figs. 5–8. Note that most of the studies listed are relatively recent; references to earlier studies can be found in the bibliographies of the more recent papers.

TABLE I
STUDIES IN THE NORTHERN CARIBBEAN

Author[a]	Year	Morphology	Ecology	Sediments	History–structure
Florida Keys (1)					
Baars	1963	x		x	
Ball *et al.*	1967		x	x	
Banks	1959			x	
Brooks	1962			x	
Ginsburg	1956			x	
Ginsburg	1957			x	
Ginsburg and					
Lowenstam	1958		x	x	
Ginsburg	1964	x	x	x	
Ginsburg *et al.*	1954		x	x	
Hoffmeister and					
Multer	1964		x		
Hoffmeister and					
Multer	1965		x		
Hoffmeister *et al.*	1967		x		x
Jindrich	1969			x	
Jones	1963		x		
Kissling	1965		x		
Multer	1969	x	x	x	
Multer and					
Hoffmeister	1968			x	
Perkins and Enos	1968		x	x	
Shinn	1963	x	x		
Shinn	1966		x		
Siegel	1963			x	
Swinchatt	1965			x	
Taft and Harbaugh	1964			x	
Thomas *et al.*	1961		x		
Thorp	1936			x	
Vaughan	1918			x	
Voss and Voss	1955		x		
Florida Bay (2)					
Davis	1940		x		
Fleece	1962	x		x	
Ginsburg	1956			x	
Gorsline	1963		x	x	
Lloyd	1964		x		

TABLE I (*Continued*)

Author	Year	Morphology	Ecology	Sediments	History–structure
McCallum and					
Stockman	1964		x		
Müller and Müller	1967			x	
Multer	1969	x	x	x	
Scholl	1966		x	x	
Stockman *et al.*	1967		x	x	
Little Bahama Bank (3)					
Neumann *et al.*	1970		x		
Neumann and Land	1969		x	x	
Storr	1964	x	x	x	
Great Bahama Bank (4)					
Agassiz	1894	x			x
Broecker and					
Takahashi	1966		x	x	
Ball	1967a	x		x	
Ball	1967b				x
Bathurst	1967a			x	
Bathurst	1967b		x		
Cloud	1962		x	x	
Goldman	1926			x	
Kornicker	1958			x	
Kornicker	1964	x			x
Kornicker and Purdy	1957		x	x	
Lowenstam and					
Epstein	1957		x	x	
Monty	1967		x	x	
Multer	1969	x	x	x	
Newell	1955				x
Newell	1959		x		x
Newell and Rigby,	1957	x	x	x	
Newell *et al*	1951	x	x		
Newell *el al.*	1959	x	x		
Newell *et al.*	1960			x	
Purdy	1963			x	
Purdy and Imbrie	1964	x	x	x	
Scoffin	1970		x	x	
Shinn *et al.*	1965			x	
Shinn *et al.*	1969		x	x	
Squires	1958		x		
Taft *et al.*	1968			x	
Taft and Harbaugh	1964			x	
Thorp	1936			x	
Ragged Keys (5)					
Illing	1954	x		x	
Southeastern Bahamas (6)					
Agassiz	1894	x			x
Bock and Moore	1969		x		

TABLE I *(Continued)*

Author	Year	Morphology	Ecology	Sediments	History-structure
Doran	1955	x		x	
Milliman	1967a	x	x	x	
Milliman	1967b	x			x
Uchupi et al.	1972	x			x
Cuba (7)					
Bandy	1964		x		
Daetwyler and Kidwell	1960			x	
Duarte-Bello	1961		x		
Hoskin	1964		x		
Agassiz	1894	x			x
Cayman Islands (8)					
Doran	1954	x			
C. H. Moore and Billings	1971			x	
Roberts and Moore	1971			x	
Jamaica (9)					
Goreau	1959a	x			x
Goreau	1959b	x	x		
Goreau	1961	x			
Goreau	1963		x		
Goreau	1964		x		
Goreau and Burke	1966	x			x
Goreau and Graham	1967		x	x	
Goreau and Hartman	1963	x	x		
Goreau and Wells	1967		x		
Goreau et al.	1960		x		
Land and Goreau	1970			x	
Steers	1940		x	x	
Steers et al.	1940	x		x	
Zans	1958a	x			
Pedro Bank (10)					
Zans	1958b	x		x	
Puerto Rico (11)					
Almy and Carrion-Torres	1963		x		
Burkholder and Burkholder	1960		x		
Glynn	1963		x		
Kaye	1959	x		x	
Odum et al.	1959		x		
Randall	1965		x		
Virgin Islands (12)					
Clifton et al.	1970	x			
Kumpf and Randall	1961		x		
Vaughan	1916	x	x		

[a] The numbers in parentheses indicate locations given in Fig. 5.

TABLE II

STUDIES IN THE SOUTHEASTERN CARIBBEAN

Author[a]	Year	Morphology	Ecology	Sediments	History–structure
Barbados (1)					
Lewis	1960	x	x		
Lewis	1965		x		
Lewis et al.	1969		x		
Macintyre	1967a	x	x	x	
Macintyre	1967b	x	x		
Macintyre et al.	1968			x	
St. Vincent (2)					
Adams	1968	x	x		
Trinidad–Venezuela shelf (3)					
Koldewijn	1958	x		x	x
Nota	1958	x		x	x
Los Roques (4)					
Work	1969		x		
Netherlands Antilles (5)					
DeBuisonje and Zonneveld	1960	x			
Deffeyes et al.	1965			x	
Lucia	1968			x	
Roos	1964		x		
Zaneveld	1957		x		
Zaneveld	1958	x	x		

[a] The numbers in parentheses indicate locations given in Fig. 6.

TABLE III

STUDIES IN THE SOUTHWESTERN CARIBBEAN

Author[a]	Year	Morphology	Ecology	Sediments	History–structure
Nicaraguan Atolls (1–3)					
Bock and Moore	1969		x		
Milliman	1969a	x	x		
Milliman	1969b			x	
Milliman and Supko	1968	x			x

[a] The numbers in parentheses indicate locations given in Fig. 7.

TABLE IV

STUDIES IN THE NORTHWESTERN CARIBBEAN

Author[a]	Year	Morphology	Ecology	Sediments	History–structure
Campeche Bank–Alacran Reef (1)					
Folk	1962			x	
Folk	1967	x		x	
Folk and Robles	1964			x	
Fosberg	1962	x	x		
Hoskin	1963	x	x	x	
Hoskin	1966			x	
Hoskin	1968			x	
Kornicker *et al.*	1959	x	x		
Kronicker and Boyd	1962		x		
Logan	1969	x	x		x
Logan *et al.*	1969	x	x	x	x
Rice and Kornicker	1962		x		
Eastern Yucatan (2)					
Boyd *et al.*	1962	x	x		
Folk *et al.*	1962			x	
Ward *et al.*	1970			x	
British Honduras (3, 4)					
Ebanks	1967	x		x	
High	1969			x	
Matthews	1966			x	
Purdy and Matthews	1964				x
Pusey	1964			x	
Stoddart	1962a	x	x		x
Stoddart	1962b		x		
Stoddart	1963		x	x	
Stoddart	1964			x	
Stoddart and Cann	1965			x	
Vermeer	1963		x		
Wantland	1967		x	x	
Guatamala–Honduras shelf (5)					
Kornicker and Bryant	1969			x	

[a] Numbers in parentheses indicate locations shown in Fig. 8.

TABLE V
STUDIES IN THE GULF OF MEXICO

Author	Year	Morphology	Ecology	Sediments	History–structure
Dalrymple	1964			x	
Emery	1962		x	x	
Freeman	1962			x	
Gould and Stewart	1953			x	
Heilprin	1890		x		
Ludwick and Walton	1957	x		x	
Matthews	1963	x			x
D. R. Moore	1958	x	x		
Rusnak	1960	x		x	
F. G. W. Smith	1954		x		
Stetson	1953	x		x	

III. Modern Caribbean Coral Reefs

A. CARIBBEAN CORALS

A general discussion of modern Caribbean corals is difficult for several reasons. First, reefs have been studied in different manners and with different intensities. Thus, for instance, it would be difficult to compare quantitatively the estimates of various reef fauna made by Heilprin in 1890 with those made in the mid-twentieth century with the aid of underwater breathing apparatus. Similarly, corals identified during a short cruise may not necessarily represent the species present at that location. For example, after several years of residence in Jamaica, Goreau (1959b) listed 41 species of coral; eight years later he listed 62 species (Goreau and Wells, 1967), and it seems likely that further new species would have been found in later years. Second, different types of reefs should not necessarily contain the same species of coral, nor should they exhibit similar zonations. One would not expect to find the same types or numbers of species on a windward reef flat as on a reef leeward of a large island.

Even with these considerations, it is interesting to note that coral species are remarkably evenly distributed around the Caribbean. In almost all reefs six scleractinian genera (*Acropora, Montastrea, Porites, Diploria, Siderastrea,* and *Agaricia*) and one hydrozoan (*Millepora*) constitute over 90% of the total coral biomass. The corals *Montastrea annularis, Acropora palmata, A. cervicornis, Diploria clivosa, D. labyrin-*

thiformis, Porites porites, P. astroides, Siderastrea siderea, S. radians, and *Agaricia agaricites* generally constitute the dominant scleractinian species. In most reefs *Montastrea annularis* is the dominant coral (Goreau, 1959b; Lewis, 1960; Stoddart, 1962a). Milliman (1969a) estimated that *M. annularis* accounts for about 60% of the coral biomass at four southwestern Caribbean atolls. In the more northern Caribbean reefs, such as those in Florida, Alacran Reef, and the northern Bahamas, however, *A. palmata* may be the dominant coral (Newell *et al.*, 1951, 1959; Ginsburg, 1956; Shinn, 1963; Logan, 1969). This generalization, however, does not always apply; for instance *A. palmata* is by far the dominant coral at St. Croix (Virgin Islands) even though the climate is definitely more akin to the southern than northern Caribbean.

The paucity of hermatypic coral species in the Caribbean is well documented (Wells, 1957; Newell, 1959), but part of this scarcity has been counterbalanced by the numerous growth forms possible for many species. For example, in deeper waters *Acropora palmata* grows with its branches extended in an unoriented fashion (Plate I B). In shallower waters, the branches are oriented into the waves, perhaps to better withstand the incoming surf (Newell *et al.*, 1951 Shinn, 1963) (Plate I C), and in very shallow water, the coral can exhibit a massive growth form in which both the branches and trunk thicken (Plate I D). In extreme conditions, the coral may even become encrusting. Other reef corals, notably *Montastrea annularis* (Plates I A; II D; V B), also exhibit more than one growth form.

B. REEF TYPES AND THEIR DISTRIBUTION IN THE CARIBBEAN

Most coral reef workers recognize three basic reef types: fringing reefs, barrier reefs, and atolls. Fringing reefs occur around most Caribbean islands. On some of the higher islands in the Lesser Antilles, such as St. Vincent and Barbados, however, fringing reefs are restricted to the leeward sides, the turbidity of the windward waters being too great (the result of erosion of high-standing terrane) to support active reefs (Lewis, 1960; Adams, 1968). Many fringing reefs, such as those in the Bahamas and Florida, perhaps more accurately could be termed fringing–barrier reefs, since they are separated from land by as much as several kilometers and have lagoonal depths that can reach 10 m. The only true barrier reefs in the Caribbean are the one off British Honduras, which stretches over 200 km and has lagoonal depths that exceed 20 m (Matthews, 1966; Wantland, 1967), and the one extending some 10 km north of Providencia Island in the southwestern Caribbean.

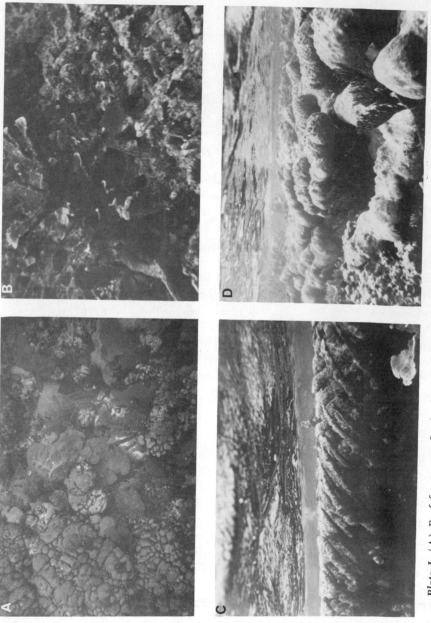

Plate 1. (A) Reef front, at a depth of about 15 m. Most coral growth is either hemispherical or foliose. Most corals are *Montastrea annularis.* (B) *Acropora palmata* on the reef front at a depth of 5 m. Note the unoriented nature of the branches. (C) Oriented *Acropora palmata* on the outer windward reef flat of Horsty Reef (D) *Mos*

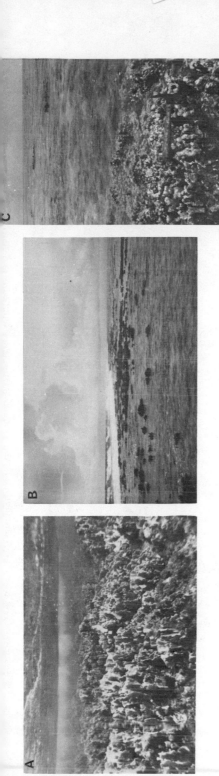

Plate II. (A) The *Millepora* zone on the outermost edge of the windward reef flat at Hogsty Reef. This zone is seldom emergent at low tide. (B) In contrast, the *Millepora* zone at Roncador Bank, in the southwestern Caribbean, extends above low tide level by as much as 40 cm. Incoming surf is able to bathe the exposed organisms with seawater. (C) The primary component organisms in the Roncador *Millepora* zone are the hydrozoan, *Millepora alcicornis*, encrusting red algae, and the soft coral *Palythoa mamillosa*; the green algae, *Halimeda*, may also be prominent. (D) The outer reef flat can also contain massive *Montastrea annularis* reefs, such as these at Hogsty Reef.

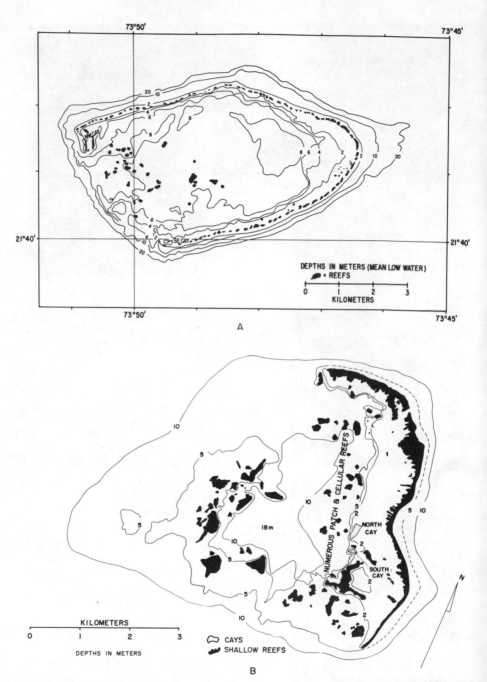

Fig. 9. Various Caribbean atolls. (A) Hogsty Reef is a small open atoll, with few patch reefs. (Based on British Soundings, 1920.) (B) Albuquerque Cays is small, but has an enclosed lagoon. (C) Roncador Bank lacks a prominant leeward reef. but has many lagoonal reefs. (D) Serrana Bank, which is one of the largest

18

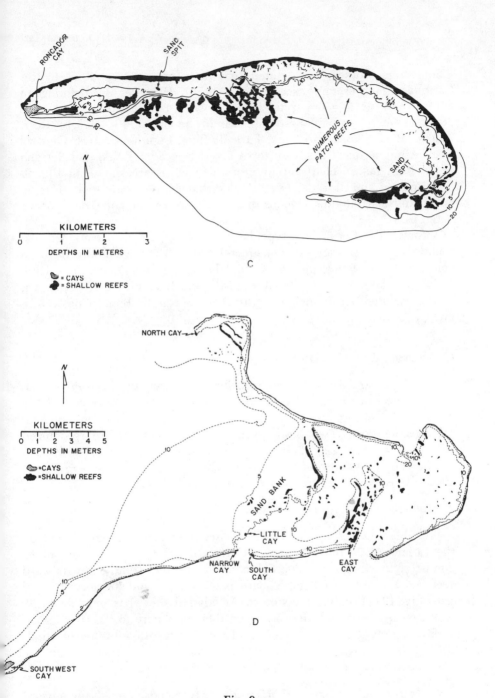

Fig. 9.

Caribbean atolls, has two sublagoons, separated from one another by a shallow sand bank. The eastern lagoon is enclosed and contains many patch reefs; the western lagoon is open and contains few reefs. (After Milliman, 1967b, 1969a.)

19

Some reef workers have stated that there are no true atolls in the Caribbean (for example, see Darwin, 1851). However, by defining an atoll strictly as a geomorphic form, Bryan (1953) listed 27 possible atolls. A more probable estimate would be about 10—Hogsty Reef, Alacran Reef, Chinchorro Bank, Turneffe Reef, Lighthouse Reef, Glover's Reef, Albuquerque Cays, Courtown Cays, Roncador Bank, and Serrana Bank. Bathymetric charts of four of these atolls are shown in Fig. 9. Los Roques and Isla de Aves off Venezuela are composed of reefs and islands surrounding lagoons, but they also contain subaerial igneous outcrops.

A fourth Caribbean reef type is the bank reef. These reefs are surrounded by deep water (often greater than 20 m), generally have no cohesive linear trend, and have no adjoining lagoon. Numerous bank reefs occur on the Yucatan Shelf and Misquito Bank off Central America, and on isolated banks, such as Pedro Bank, Serranilla Bank, Baja Nuevo, and Quita Sueno Bank.

C. MODERN REEF ZONATIONS

Reef types are defined by morphology. The zonations within a reef, however, are determined by the distributions and growth forms of the component reef organisms, which in turn are dependent on environmental considerations. Two basic forms of reef zonations are recognized in the Caribbean: (1) high energy reefs such as those on atolls and on the exposed windward portions of islands; and (2) low-energy reefs, such as those on the protected, leeward sides of islands.

1. High-Energy (Exposed) Reefs

Probably the best examples of exposed reef zonations are found on the various atolls throughout the Caribbean. The reefs on these atolls consist of four major physiographic zones: the reef front, the windward reef flat, the lagoon (together with patch reefs), and the leeward reef flat (Fig. 10). Fringing reefs on the windward side of islands obviously lack leeward reefs and, depending on their proximity to the island, may lack a lagoon. The morphology, ecology, and the distribution of corals and other organisms in these zones is generally the same in all the Caribbean reefs, although some specific differences are found.

a. Reef Front. The reef front (fore reef) extends from the windward edge of the reef flat to the lower limits of coral growth, probably in excess of 70 m (Goreau and Wells, 1967). The reef front area has been thought by many workers to be one in which faunal richness and divers-

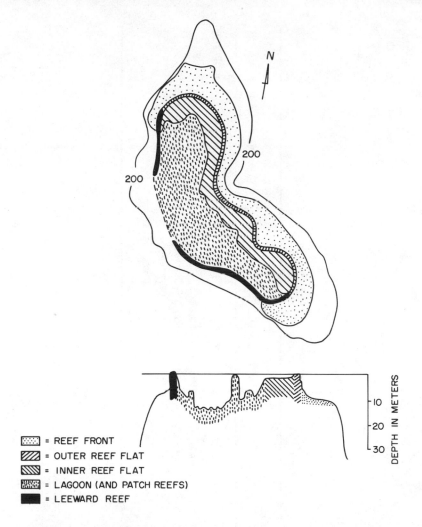

Fig. 10. Reef zonations at Courtown Cays.

ity quickly decreases with depth, but studies in Jamaica (Goreau, 1959b; Goreau and Hartman, 1963; Goreau and Wells, 1967; Goreau and Graham, 1967) have shown this zone to be prolific, perhaps the most prolific portion of the reef complex. Within this vertical interval, Logan (1969) has recognized five basic reef communities, each containing distinct faunal and floral components (Table VI). In shallower depths *Acropora palmata* and *A. cervicornis* are abundant, but decrease quickly in depths greater than about 15 m (Goreau, 1959b). *Montastrea, Porites,* and *Siderastrea* also are common corals, and tend to predominate below

TABLE VI

COMMUNITY ZONES ON THE YUCATAN SHELF[a]

Community	Depth (m)	Dominant species	Dominant growth habits
1. *Acropora palmata*	0–10	*Acropora palmata; Millepora* sp.; *Palythoa mammilosa; Porolithon* sp.	Branching; encrusting; hemispheroid
2. *Diploria–Montastrea–Porites*	5–25	*Diploria* sp.; *Montastrea* sp.; *Porites astroides*	Hemispheroid-encrusting
3. *Agaricia–Montastrea*	25–35	*Agaricia agaricites; Montastrea* sp.; *Solenastrea* sp.	Encrusting-hemispheroid
4. *Gypsina–Lithothamnium*	15–60	*Gypsina plana* (encrusting foraminifer); *Lithothamnium* sp.	Encrusting
5. *Lithophyllum–Lithoporella*	20–60	*Lithophyllum* sp.; *Lithoporella* sp.	Encrusting

[a] After Logan (1969).

25 m (Plate I A). Growth forms trend from massive in very shallow waters to hemispherical and foliose in mid-depths to crustose in deeper waters (Table VI). In depths greater than about 15 m the reef tends to be less cohesive and the corals often grow as individual heads, although these heads can attain considerable size. At still greater depths corals assume foliose and crustose growth patterns.

The uppermost part of the reef front often is punctuated by spurs (buttresses) and grooves oriented normal to the reef. *Acropora palmata* is especially common on the buttresses. Relief between the spurs and grooves can exceed 5 m, and the system can extend seaward to depths as great as 10 m and shoreward to the outer reef flat (Goreau, 1959b; Shinn, 1963; Stoddart, 1962a). Not all Caribbean reefs, however, possess a well-developed spur-and-groove system, and some reefs (for example, those at Hogsty Reef) appear to have none.

Many workers in Pacific reefs believe that the spurs and grooves are basically erosional in origin (Cloud, 1959; Wiens, 1962), and that this system acts as an effective baffle for the enormous energy expended by incoming surf (Munk and Sargent, 1954). Grooves off Andros Island are cut into Pleistocene bedrock (Newell *et al.*, 1951), indicating an erosional origin. However, in other parts of the Caribbean the spur-and-groove system may owe much of its relief to differential rates of coral growth. Goreau (1959b) suggested that the Jamaica spurs are the result of the preferred growth of *Montastrea annularis;* Shinn (1963) offered a similar origin for the buttresses found in South Florida, but suggested that the coral was *Acropora palmata.*

b. Windward Reef Flat. The outer rim of the windward reef flat often is the shallowest part of the coral reef complex. Depths range from 1 m in the northern parts of the Caribbean to less than 10 cm (at low tide) in some of the more southerly Caribbean reefs. Prominent organisms include the hydrozoan *Millepora alcicornis,* soft corals (*Palythoa* and *Zoanthus*), and encrusting red algae (Plate II A,C). This association is characteristic of the outer rim of many Caribbean coral reefs (Shinn, 1963; Kornicker and Boyd, 1962; Goreau, 1959b; Stoddart, 1962a), but in no place is it better developed than in the southwestern Caribbean atolls (Milliman, 1969a), where the so-called *Millepora* zone actually extends above low-tide level, forming an emergent community not unlike the leeward portions of the algal ridge found in Pacific reefs (Wells, 1954) [compare Plate II B with those in Tracey *et al.* (1948), Plate I, Fig. 2, and Emery *et al.* (1954), Plate 19, Fig. 2]. The obvious differences between the *Millepora* zone and Indo-Pacific algal ridges are the abundance of coral and the lack of massive coralline algae, especially *Porolithon,* in the Caribbean.[3]

The outer reef flat leeward of the *Millepora* zone has depths ranging from tens of centimeters to more than 1 m. Much of the bottom is devoid of living reef, although coarse coral gravel and encrusting coralline algae commonly cover much of the bottom (Plate III A). In other areas of the outer reef flat, however, many coral genera are found; many heads extend up to the surface, and in some instances, especially in warmer climates, extend above the surface at low tide. Incoming surf apparently is able to bathe the emergent corals so that they can survive subaerial emergence. The main species found on the outer reef flat include *Montastrea annularis* (especially at Jamaica) (Plate II D), *Diploria strigosa* (especially in the southwestern Caribbean atolls and Barbados), *Acropora palmata* (especially in the Bahamas and Florida), plus *Porites porites, Agaricia agaricites,* and *Siderastrea siderea.* The green alga, *Halimeda,* often growing in large bushes, is especially prominent in this portion of the reef flat.

In contrast to the outer reef flat, the inner reef flat is generally deep, 1–2 m (Plate III B), and is covered with rubble and sand as opposed to the bare rock and gravel found on the outer reef flat (plates III C,D and IV A). Heads of *Siderastrea* and *Porites* are locally prominent. In deeper reef flats, *Montastrea, Diploria,* and even thickets of *Acropora cervicornis* may be present. In some atolls and reefs large areas may be covered with marine grasses (*Thalassia, Halodule,* and *Syringodium*) and marine algae, such as *Halimeda* and noncalcareous brown algae.

c. Lagoon. In many barrier reefs and atolls, the transition from the windward reef flat to the lagoon is gradual. On other reefs, however, the transition is abrupt, the bottom often dropping from a depth of 1 or 2 m on the reef flat to 10 or 15 m in the lagoon, within a horizontal distance of 20–50 m (Plate IV D). This abrupt slope has been termed the sand cliff (Milliman, 1969a), and is prominent at Alacran Reef (Kornicker and Boyd, 1962), Glover's Reef (Stoddart, 1962a), and the southwestern Caribbean atolls (Milliman, 1969a).

The average depth for most Caribbean lagoons seems to be between 10 and 15 m, and in contrast to the Indo-Pacific atolls (Wiens, 1962), the average lagoon depth bears no relation to the size of the atoll. At most atolls, the lack of leeward peripheral reefs results in relatively

---

[3] P. Glynn (Volume II, Chapter 9) and W. H. Adey (in preparation) recently have found excellent examples of algal ridges in the high energy zones in the cays off Panama and St. Croix, Virgin Islands.

Plate 7II. (A) Outer windward reef coral debris. (B) The transition from the outer reef flat to the inner reef flat can be abrupt. For instance, in this picture, the transition is marked by the decrease in massive, shallow reefs, an increase in depth by more than 1 m, and the presence of sand-sized sediment. (C) In the northern Caribbean, such as Hogsty Reef pictured here, inner reef flats may contain relatively few viable coral reefs. (D) Inner reef debris often can be heavily encrusted with coralline algae.

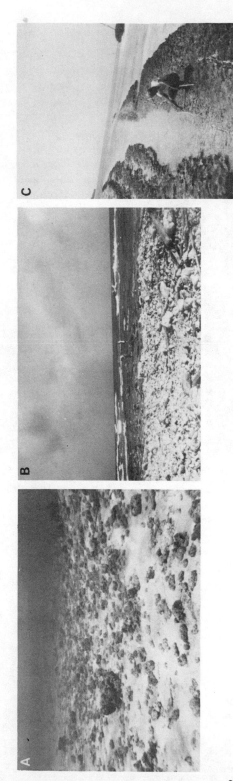

Plate IV. (A) Current energy decreases approaching the lagoon, and as a result, the sediment generally becomes finer. The gravel-sized material is composed mainly of massive and encrusting coralline algae. (B) Sand cays often begin as waves of sand and gravel slowly migrating across the reef flat. The rate of migration of such sand waves is strongly dependent upon the protection offered by beachrock and the stabilizing effect of terrestrial vegetation. (C) The leeward migration of sand cays is illustrated by the windward bands of beachrock which mark old stands of the cay. The windward shore of this cay is only barely visible on the extreme right portion of the picture. (D) The "sand cliff" marks the abrupt transition from the reef flat into the lagoons. Slopes can surpass 15°.

open lagoons, with ready access to the surrounding ocean. The shallow, open nature of the lagoons, plus the relatively small size of most Caribbean atolls, means that the circulation of water is much less complex than in larger and deeper Indo-Pacific atolls (von Arx, 1954). At Hogsty Reef and the atolls in the southwestern Caribbean, lagoonal currents generally are wind-driven, with local tidal reversals. In most cases residency times of lagoonal water masses are short, probably not exceeding 1–2 days. Possible exceptions may be the atolls of Chinchorro Bank and Lighthouse Reef, whose lagoons are mostly enclosed, and whose circulation therefore may be restricted.

Bank lagoons, such as those at Great Bahama Bank, and Batabano Bay and Florida Bay are relatively shallow, but because of restricted and sluggish circulation, waters can have extremely long residency times. The residency time of waters on Great Bahama Bank is estimated at 1–2 months (C. L. Smith, 1940; Broecker and Takahashi, 1966; Traganza, 1967). This restricted circulation, plus the high rate of evaporation results in supersaline waters. Reef fauna are found only at the extreme periphery of the Bahama Banks. In contrast, Florida Bay, whose circulation is restricted by the Florida Keys and shallow mud and mangrove banks, is exposed to seasonal fluctuations in salinities. Runoff from the neighboring Evergades in the winter causes the interior bay waters to be brackish, but during the summer months salinities can surpass 50‰ (McCallum and Stockman, 1964).

The floors of most Caribbean lagoons are covered by coral reefs, present as solitary mounds, low-lying patch reefs, or high-standing pinnacle reefs. In the "Back Reef" lagoon separating the Florida reefs and Florida Keys, lagoonal reefs occur in occasional patches, such as those at Misquito Bank; these reefs generally contain only a few types of massive stony corals, such as *Montastrea*, *Siderastrea*, and *Diploria*, together with large numbers of gorgonians. The restriction of coral growth in the Florida lagoons is probably the result of the relatively low current energy (and therefore a silty environment) together with low winter temperatures which may restrict the development of some coral genera.

In the lagoons of more southern Caribbean areas, reefs are plentiful and the number of coral genera and of growth forms more varied. *Montastrea annularis*, *Acropora cervicornis*, *Porites porites*, *Agaricia*, and *Diploria* are among the more common corals (Plate V A,B). Some lagoonal reefs, such as those at Alacran Reef and Albuquerque Cays, have coalesced into cellular-like reefs (Hoskin, 1963). The abundance of shallow pinnacle reefs has left some lagoons, such as those at Alacran Reef, Glover's Reef and parts of Roncador and Serrana Banks practi-

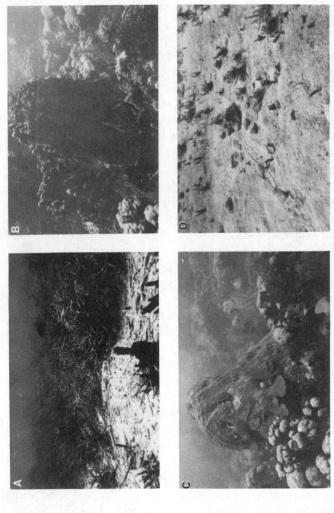

Plate V. (A) Thickets of *Acropora cervicornis* often are present throughout Caribbean coral reefs, especially on inner reef flats and in shallow lagoons (such as pictured here). A notable exception is the almost complete lack of this coral at Hogsty Reef (Milliman, 1967a). (B) A deep lagoonal patch reef, showing *M. annularis* in pinnacle growth form. (C) A deep lagoonal patch reef with foliose *M. annularis*. Note the predominance of gorgonians. (D) Grasses and green algae (*Halimeda* is pictured here) are critical components in lagoonal sedimentation. Grasses host numerous epiphytic and epifaunal organisms (Bock, 1967) and also can form excellent sediment-accreting baffles. Green algae are sources of aragonitic plates (in the case of *Halimeda*) and aragonitic muds. The browsing gastropod, *Strombus gigas*, provides a necessary source of protein for many Caribbean natives, as well as providing considerable carbonate to the lagoonal sediment.

28

cally unnavigable (Fig. 11). In deeper lagoons, gorgonians are plentiful, and may constitute an important sedimentary component (Cary,
1918) (Plate V C).

Most lagoons are floored by skeletal debris derived from both peripheral reefs and from neighboring lagoonal patch reefs. Common organisms
include green algae (*Penicillus, Rhipocephalus,* and *Halimeda*), infaunal
and epifaunal molluscs (the gastropod *Strombus* being the most visually
obvious), and asteroids and irregular echinoids (Plate V D). In some
lagoons marine grasses are plentiful; the trapping and accretion of sediment by their active root systems can be an important sedimentary agent
(Ginsburg and Lowenstam, 1958); at Alacran Reef *Thalassia* colonies
have accreted large sediment banks (Logan, 1969). In the southewestern
Caribbean atolls, however, marine grasses are virtually absent (Milliman,
1969a).

Blue-green algae also can form cohesive mats on the lagoon floor,
which provide sediment stability (Scoffin, 1970; Neumann *et al.*, 1970),
and also may provide both food for lagoonal organisms and a suitable
microenvironment for biogenetic alterations of sedimentary components
(Bathurst, 1967b). Burrowing animals, such as worms and crustaceans,
can produce complex structures that can provide environments for dia-
genetic alterations (Shinn, 1968).

d. Leeward Peripheral Reefs. Most Caribbean atolls have poorly developed leeward peripheral reefs. In some atolls, such as Courtown Cays
and Roncador Bank, leeward reefs are almost completely lacking, re-

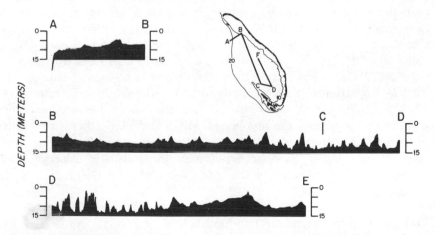

Fig. 11. Bathymetric profiles across Roncador Bank, illustrating the abundance of
lagoonal patch reefs.

sulting in open lagoons. Most leeward reefs are composed of anastomosing coral heads which seldom break the surface; the solid, continuous reef flat surface that is characteristic of the windward reef is absent. At several of the southwestern Caribbean atolls, the leeward reefs appear to be formed by coalescing patch reefs (Milliman, 1969a). The dominant coral genera are similar to those found on the windward reef, although *Millepora* and *Acropora palmata* generally are not as prevalent. In contrast, gorgonians often are more common than on windward reefs.

2. Low-Energy (Leeward Island) Reefs

Reefs located on the leeward sides of islands tend to be more poorly developed than those found on the exposed windward sides. For instance, because of the lower ambient current and wave energies in these environments, high-energy corals, such as *Acropora palmata,* are virtually absent over much of the leeward reef tract at both Barbados and St. Vincent (Lewis, 1960; Adams, 1968). *Montastrea annularis* is the most common coral at Barbados, while at St. Vincent, *Porites porites* is dominant. *Millepora alcicornis* is present at both reefs, located on the outer crest of the reef flat at Barbados and on the inner reef flat at St. Vincent.

D. REEF SEDIMENTS

Sedimentological studies in the Caribbean have achieved a far greater degree of sophistication than similar studies in the Pacific. For example, most present-day knowledge concerning the diagenesis of shallow-water carbonate sediments has been derived from investigations made in the Caribbean. The main reason for this difference in technique and knowledge is related to the proximity of many Caribbean reefs to American universities and laboratories. In terms of sedimentological studies, the Florida Keys is probably the best-studied carbonate area in the world (Table I). Numerous oil companies an universities have annual field trips to this area, and the number of M.Sc. and Ph.D. theses and other published and unpublished reports easily surpasses 100. Other sedimentological studies have been made in the Bahamas, Alacran Reef, British Honduras, Barbados, and several atolls in the southwestern Caribbean.

Caribbean carbonate sediments contain two types of components, skeletal derived fragments and nonskeletal particles (grains that do not have a skeletal origin, or whose origin, if skeletal, is not discernible). Carbonate component identification generally is limited to grains coarser than 125 μm (very fine sand). Although some petrographic studies have been made on finer sized material (for example, Matthews, 1966), knowledge concerning this fine material is much poorer than for the coarser sediments.

In the coarser fractions, most skeletal grains consist of fragments of coral, mollusc, foraminifera, coralline algae, and *Halimeda;* echinoids, bryozoans, and serpulids also provide some grains, but generally total less than a few percent of the sand and gravel fraction. In the fine sizes, sponge, holothurian and gorgonian spicules and green algal aragonitic needles may also be important. Nonskeletal fragments are generally composed of four different types: ooids, pellets, aggregates (grapestone), and cryptocrystalline particles (Illing, 1954; Purdy, 1963).

The composition of Caribbean carbonate sediments depends strongly upon the depositional environment (Fig. 12). Peripheral reefs generally produce a very coarse sediment, composed of gravel and coarse sands; coral and coralline algae, together with *Halimeda,* are the primary skeletal components (Table VII). The encrusting foraminifera, *Homotrema rubrum,* lives mainly on the fore reef and reef flat, and therefore its presence in the sediments can indicate either environment or transport (MacKenzie *et al.,* 1965). Nonskeletal fragments usually are absent or constitute a very small portion of the sediment. Lagoonward of the peripheral reefs, sediment size decreases as current and wave energy decrease. *Halimeda* and coralline algae are dominant components, although coral fragments washed in from the outer reefs can also be prominent.

In most lagoonal areas, epifaunal and infaunal organisms, especially molluscs and foraminifera, are more important sedimentary components than on the peripheral reefs (Fig. 12; Table VIII). In some lagoons, such as the eastern part of Serrana Bank, *Halimeda* contributes more than 50% of carbonate sediment. Coral and coralline algae are only abundant in those areas immediately surrounding patch reefs or peripheral reefs.

Lagoonal areas can contain a considerable amount of fine sediment (Fig. 12). The absolute content of fine-grained sediment depends upon the current and wave energy regime, which in turn is related to the number of patch reefs and grass banks and to the exchange of lagoonal water with the open sea. For instance, Hogsty Reef has a relatively open lagoon with few patch reefs; as a result, the sediment is generally coarse. In contrast, Roncador Bank lagoon contains many patch reefs which restrict circulation and trap sediment, resulting in a large amount of very fine sediment. Other areas with relatively low-energy environments, such as the "Back Reef" in the Florida Keys, also provide good sites for mud accumulation. These fine sediments probably represent a mixing of comminuted peripheral reef and lagoonal debris, together with the spicules and aragonitic sheaths of lagoonal organisms (Matthews, 1966).

Many of the shallow restricted banks and bays also contain an

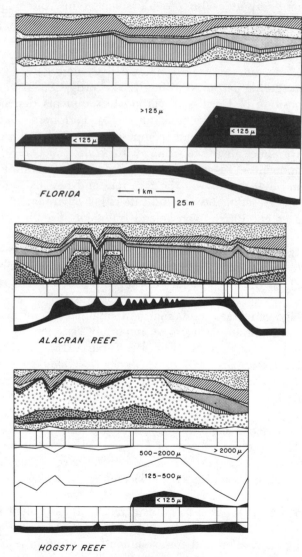

Fig. 12.

abundance of fine aragonitic needles (5–15 μm long). Three general theories regarding the origin of these aragonitic needles have been proposed: (1) bacterial action (Drew, 1914; Kellerman and Smith, 1914;

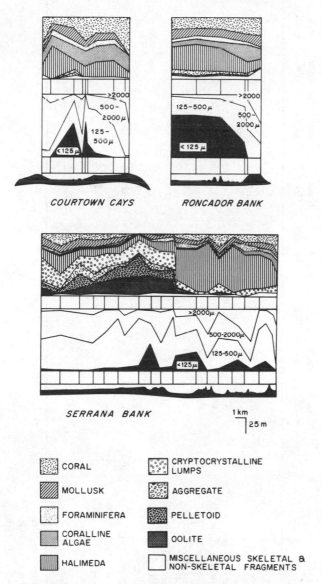

Fig. 12. Distribution of sediments across several Caribbean reefs and atolls. Data are from Ginsburg (1956), Hoskin (1963), and Milliman (1967b, 1969b).

TABLE VII

AVERAGE COMPOSITION OF PERIPHERAL REEF SEDIMENTS FROM VARIOUS CARIBBEAN REEFS AND ATOLLS

	Florida[a]	Alacran Reef[b]	Andros, Bahamas[c]	Abaco, Bahamas[d]	Ragged Islands, Bahamas[e]	Hogsty Reef[f]	Courtown Cays[g]	Albuquerque Cays[g]	Roncador Bank[g]	Serrana Bank[g]
Skeletal fragments										
Coral	20	26	24	27	12	27	35	30	25	33
Molluscs	12	7	6	11	18	22	10	9	15	5
Foraminifera	6	8	12	11	13	5	3	2	3	3
Coralline algae	10	11	33	10	39	19	21	21	24	24
Halimeda	30	40	17	16	39	2	28	32	17	17
Misc. skeletons	7	1	6	5	4	1		1	1	1
Nonskeletal fragments										
Oolite				4		1				
Pelletoids	← 10 →	2		4	← ? →	tr				
Aggregates		4		4						
Cryptocrystalline lumps				4		17		2		3
Misc. nonskeletal			2	15					11	10
Unknown	6					5				

[a] Ginsburg, 1956.
[b] Hoskin, 1963.
[c] Goldman, 1926.
[d] Storr, 1964.
[e] Illing, 1954.
[f] Milliman, 1967a.
[g] Milliman, 1969b.

TABLE VIII

AVERAGE COMPOSITION OF LAGOONAL SEDIMENTS FROM VARIOUS CARIBBEAN REEFS AND ATOLLS

	Florida[a]	Alacran Reef[b]	Hogsty Reef[c]	Courtown Cays[d]	Albuquerque Cays[d]	Roncador Bank[d]	Serrana Bank (east)[d]	Serrana Bank (west)[d]
Skeletal fragments								
Coral	7	15	4	28	20	22	9	8
Molluscs	18	7	10	12	12	12	7	10
Foraminifera	7	7	4	6	6	5	5	3
Coralline algae	3	5	1	21	14	12	4	1
Halimeda	38	23	1	28	31	37	61	13
Misc. skeletons	7	2	1	1	3		1	1
Nonskeletal fragments								
Oolite	←—— 12 ——→		1			tr	1	15
Peletoids		33	16				1	15
Aggregates		9	13					8
Cryptocrystalline lumps			46	2	12	6	4	21
Misc. nonskeletal						1		
Unknown	7		1	2	2		5	4

[a] Ginsburg, 1956.
[b] Hoskin, 1963.
[c] Milliman, 1967a.
[d] Milliman, 1969b.

N. R. Smith, 1926; Bavendamm, 1932); (2) inorganic precipitation (Black, 1933a; C. L. Smith, 1940; Cloud, 1962; Broecker and Takahashi, 1966; Traganza, 1967); and (3) algal sheaths of codiacean green algae (Lowenstam, 1955; Lowenstam and Epstein, 1957). Algal production in Florida Bay and on Little Bahama Bank appears to be sufficient to explain the entire amount of aragonite required for the aragonitic muds in these areas (Stockman et al., 1967; Neumann and Land, 1969). On the other hand, Cloud (1962) calculated that codiacean algal production leeward of Andros Island can account for no more than 25% of the aragonitic muds found in this area, and concluded that most of the aragonite mud is inorganically precipitated. Mineralogical and chemical analysis of Andros muds by the writer (Milliman, 1973) also support the contention that at least part of the Andros Island muds are inorganically precipitated.

In areas where current energy is relatively high but reefs are lacking or peripheral reef productivity is low, nonskeletal sedimentation may be important. The importance of inorganic precipitation in such areas may be the result of low skeletal productivity (and less uptake of $CaCO_3$ by hermatypic organisms) or the low rates of sediment accumulation (Milliman, 1969b). At Great Bahama Bank skeletal sediments are generally restricted to those areas adjacent to the peripheral reefs. The interior of the bank is almost completely covered with nonskeletal sediments (Table IX). Purdy (1963) defined four major nonskeletal facies on Great Bahana Bank: the oolite facies, the grapestone (aggregate) facies, pellet-mud and mud facies (Table IX). The location of these facies depends greatly upon ambient energy of the environment, the oolite facies being found at the periphery of the banks and the mud facies forming in the lowest energy environment, the interior of the bank leeward of Andros Island.

A similar type of nonskeletal sedimentation occurs in relatively high-energy atoll lagoons, where skeletally-derived sediment does not accumulate (Table IX; Fig. 12). At Serrana Bank, the western lagoon contains no patch reefs, and the peripheral reefs are too far distant to supply much sediment. At Hogsty Reef the peripheral reefs surround the lagoon, but they are relatively unproductive, so that little skeletal sediment reaches the lagoon. In both areas the sediment is dominated by aggregates and cryptocrystalline grains. At Serrana Bank ooids and pelletoids are also prominent.

E. CARBONATE CAYS

Three distinct types of carbonate cays are found in the Caribbean: mud cays, sand cays, and cays composed of, or surrounding, pre-Holo-

TABLE IX

AVERAGE COMPOSITION OF SEDIMENTS FROM VARIOUS CARIBBEAN BANKS AND BAYS

	Florida Bay[a]	Batabano Bay, Cuba[b]	Ragged Keys, Bahamas[c]	Great Bahama Bank[d]			
				Oolite facies	Grapestone facies	Pellet–mud	Mud
Skeletal fragments							
Coral	80	<5	tr	tr	tr		
Mollusc	11	10	6	1	4	5	12
Foraminifera	tr	<5	4	1	3	7	14
Coralline algae	tr	<5	1	tr	tr		
Halimeda	tr	0–10		2	3	2	9
Misc. skeletons	3		1	tr	1	1	1
Nonskeletal fragments							
Oolite				70	16	12	4
Pelletoids	3 (←→)	60 (←→)	6	7	5	57	33
Aggregates			27	8	36	10	12
Cryptocrystalline lumps			55	7	28	3	5
Misc. nonskeletal						1	2
Unknown	1			2	2	5	7

[a] Ginsburg, 1956.
[b] Daetwyler and Kidwell, 1960.
[c] Illing, 1954.
[d] Purdy, 1963.

cene bedrock. In the latter category are many of the cays in the Bahamas and offshore from British Honduras, which stand about 2 m high and are composed of Pleistocene reef rock, perhaps of Sangamon age. Other cays contain lithified or semi-lithified Pleistocene aeolian dunes that can attain elevations higher than 30 m. On some cays, such as Andros Island in the Bahamas, extensive mud flats have accumulated leeward of the older Pleistocene bedrock.

Mud cays are usually restricted to quiet shallow environments, and few rise more than 1 m above high tide. In Florida Bay, the presence of thriving mangrove forests have acted as efficient sediment traps, which have resulted in the horizontal expansion of many mud cays (Davis, 1940). Most of the sediment is silt and clay, although some discrete sand layers, perhaps derived from periodic hurricanes and other storms (Ball et al., 1967; Perkins and Enos, 1968), can be present.

Mud carbonate cays have two sedimentological characteristics that distinguish them from other carbonate environments. The first is the widespread occurrence of layered algal mats (Black, 1933b; Ginsburg et al., 1954; Monty, 1967; Gebelein and Hoffman, 1968; Shinn et al., 1969). Periodic wettings, algal growth, and subsequent sediment trapping are responsible for the layered appearance. Regular laminations and desiccation marks (the result of subaerial drying) are two characteristics that distinguish supratidal mats from those found in the intertidal and littoral mats. The second diagnostic sedimentological characteristic is the presence of dolomite. In some areas recent protodolomite is found forming in the supratidal mud flats. Dolomite crystals, which are mainly between 1 and 3 μm in size, are believed to be formed by the alteration of calcium carbonate through exposure to supersaline brines, either by the simple sinking of brines (Shinn et al., 1965, 1969; Bubb and Atwood, 1968; Atwood and Bubb, 1970) or by seepage refluxion (Deffeyes et al., 1965; Lucia, 1968).

Sand cays can be thought of as slowly migrating sand waves (Plate IV B,C). Sediment is derived from the neighboring reefs and thus corresponds closely in composition with the ambient reef sediments (Table X). The ultimate destiny of most sand cays on atoll reef flats is transport from the reef into the lagoon. The immediate stability of these cays depends upon the degree to which vegetation cover and beachrock can retard leeward migration. Stoddart (1962b, 1963) found that during Hurricane Hattie, cays in the British Honduras area without ground cover were completely destroyed, while those cays with heavy ground cover received relatively little damage.

One basic difference between Caribbean and Pacific sand cays is their

TABLE X

COMPOSITION OF REEF CAY SEDIMENTS FROM THE CARIBBEAN

	Windward beaches, Bahamas[a]	Leeward beaches, Bahamas[a]	Hogsty Reef[b]	Isla Perez[c]	Windward, Ambergris Cay[d]	Windward, cays, Courtown[e]	Leeward cays, Courtown[e]	Albuquerque[e]
Skeletal fragments								
Coral	4		31	25	9	20	22	19
Mollusc	14	2	20		18	10	6	6
Foraminifera	18	2	1	15	7	3	3	3
Coralline algae	27	tr	16		15	33	10	22
Halimeda	5	1	1	60	35	31	54	39
Misc. skeletons	3	tr	1		3	tr	tr	1
Nonskeletal fragments								
Oolite	16	38	tr					
Pelletoids	tr	3	5					
Aggregates	6	49	1					
Cryptocrystalline lumps			18		6			6
Cay Rock					4			5
Misc. nonskeletal	6	6			2			

[a] Illing, 1954.
[b] Milliman, 1967a.
[c] Folk and Robles, 1964.
[d] Ebanks, 1967.
[e] Milliman, 1969b.

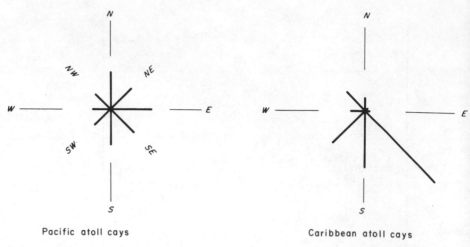

Pacific atoll cays   Caribbean atoll cays

Fig. 13. Location of sand cays on Caribbean and Indo-Pacific atolls.

position on the reef flats. Most Pacific cays are located on the windward
reef flats (Wiens, 1962), but most Caribbean cays are located more
leeward (Fig. 13). Folk (1967) hypothesized that leeward Caribbean
cays form by sand transport from the shallow lagoon and patch reefs,
whereas lagoon sand in the deeper Pacific lagoons is drowned before
it can reach the leeward margins. A more likely explanation is that
the lack of an algal ridge and the generally deeper reef flats result
in a greater exposure of Caribbean windward reefs to incoming waves.
Perhaps the sand cays cannot survive in such a higher energy environment, and they therefore must form in slightly quieter conditions.

IV. Evolution of Modern Caribbean Coral Reefs

During the late Mesozoic and early Tertiary the Atlantic and Pacific
Oceans were connected through what is now Central America; as a
result, reef genera in both oceans were closely related. Since the closing
of the Isthmus of Panama in the Miocene, however, the situation has
changed markedly. For example, the number of Atlantic coral genera
has decreased to less than 20, while Pacific genera have increased to
more than 80 (Vaughan, 1919; Wells, 1957). Scleractinian genera, such
as *Stylopora, Pocillopora, Goniostrea, Goniopora, Pavona,* and *Seriatopora* and the octocoral *Heliopora* are no longer present in the Caribbean,
although they are still prominent in the Pacific. There are three
species of *Acropora* and four species of *Porites* in the Atlantic,

and more than 150 and 80 species of these two corals in the Pacific (Wells, 1957). Numerous important reef genera, such as the octocoral *Tubipora,* many molluscs, and some coralline algae species (notably *Porolithon onkodes*) also are restricted to the Pacific (Ekman, 1953; Wells, 1957).

The decrease in Atlantic genera probably is the result of the general cooling of the world climate since the early Tertiary (Ekman, 1953; Emiliani, 1954). Being adjacent to North America, Pleistocene temperature fluctuations throughout much of the Caribbean were more severe than tropical parts of the Pacific (Emiliani and Flint, 1963). Such climatological conditions must have been deleterious to Caribbean reef growth. The increase in Pacific genera also is partly explained by the vast tropical expanse covered by reefs, thus allowing for more evolutionary divergency than in the smaller Caribbean.

During the last glaciation sea level fell more than 130 m below present-day sea level (Milliman and Emery, 1968), thus exposing and killing all existing shallow-water coral reefs. During the subsequent Holocene transgression, which began 15–19 thousand years ago (Curray, 1965; Milliman and Emery, 1968), erosional and constructional features formed at several horizons throughout much of the Caribbean. For instance —18 and —35 m terraces are present on many islands, banks, and atolls (Vaughan, 1919; Goreau, 1961; Goreau and Burke, 1966; Milliman, 1967b; Stanley and Swift, 1967). Submerged hermatypic reefs have been found on the shelf off Florida (Duane and Meisburger, 1969; Macintyre and Milliman, 1970), Yucatan (Logan *et al.,* 1969), and many Caribbean islands (Macintyre, 1967b, 1972). In addition, submerged oolitic dunes have been found off central Florida (Macintyre and Milliman, 1970) and Yucatan (Logan *et al.,* 1969). Carbon-14 dates indicate that these features were formed during the last transgression (Milliman and Emery, 1968), and suggest that reef and dune accretion was not able to keep pace with the rapid transgression.

Modern coral reefs have formed on shallow, preexisting platforms, such as the Pleistocene reefs upon which the modern reefs in Florida stand. The amount of Holocene reef accumulation has been small, averaging less than 1 m in Florida and the Bahamas, and less than 7 m in Jamaica (Goreau, 1961). More recent estimates however, suggest that these earlier figures may be too low and that accumulation rates are considerably higher (Land and Goreau, 1970; Ginsberg *et al.,* 1971). Logan (1969) has suggested that many of the emergent reefs on the Yucatan Shelf, such as Nuevo and Alacran Reefs, were able to keep pace with the transgression and in the process, accreted some 20–40 m of carbonate rock during the past 9000 years. However, it is equally possible that these Yucatan reefs may have formed on preexisting topographic expressions, and that the Holocene accretion has been considerably less than that estimated by Logan.

V. Caribbean versus Indo-Pacific Coral Reefs

Many workers have portrayed Caribbean coral reefs as pale images of their Indo-Pacific counterparts. Specifically, Caribbean reefs are said to be lacking in: (1) number of coral species (about 60 vs 700 in the Indo-Pacific); (2) the emergence of corals at low tide; (3) an algal ridge zone; and (4) general reef maturity (Wells, 1957; Newell, 1959; Yonge, 1963). Unfortunately, most of these comparisons have been made on the assumption that Florida–Bahama coral reefs represent typical Caribbean reefs. This assumption is probably invalid, since, as has been shown in other parts of this chapter, Florida and Bahama coral reefs may be significantly less prolific than those found in the more southern parts of the Caribbean.

In most respects, southern Caribbean coral reefs compare quite favorably with Indo-Pacific reefs (Table XI). The lack of coral species does not seem to prohibit prolific reef growth, partly because many Caribbean reef corals exhibit more than one growth form, which enables them to flourish in many different subenvironments. Year-round temperatures in the southern Caribbean are warm and incoming surf is sufficiently great for many reef corals to be able to survive subaerial emergence during low tides.

The major ecological difference between Caribbean and Indo-Pacific coral reefs seems to be the relative importance of massive and encrusting coralline algae. In the Caribbean, coralline algae encrust and coat many reefs and much reef debris, but massive encrustations generally are absent. Thus Caribbean reef flats often tend to be composed of discrete, anastomosing reef corals, in contrast to the solid, shallow algal pavements found in the Indo-Pacific. No viable algal ridge is present in the Caribbean, although the emergent *Millepora* zone may fill a similar ecological niche.[4] In turn, the green alga, *Halimeda,* seems far more prolific on Caribbean reef flats than in the Pacific. As a result Caribbean reef flat sediments usually contain large amounts of *Halimeda* plates, whereas Indo-Pacific reef flats than in the Pacific. As a result Caribbean reef flat sediments foraminifera.

[4] An important discovery of an algae ridge complex on Panamanian reefs has been reported by Glynn (1973) in Volume II of this book.

TABLE XI

Comparison of Indo-Pacific and (Southern) Caribbean Coral Reefs

Indo-Pacific	Caribbean
1. About 700 coral species	1. About 60 coral species
2. Foliose and encrusting corals often found in low energy conditions, while branching and massive corals found in higher energy conditions	2. Foliose and encrusting corals often found in low energy conditions, while branching and massive corals found in higher energy conditions
3. Many corals can survive prolonged subaerial exposures	3. Many corals can survive prolonged subaerial exposures
4. Algal ridge present on many outer windward reef flats	4. Emergent *Millepora* zones present on some reefs
5. Reef flat often paved with coralline algae, resulting in a relatively smooth and shallow reef flat	5. Coralline algae not so prevalent and thus the reef flats often are not as solid
6. Reef flat sediments dominated by coralline algae and benthonic foraminifera	6. Reef flat sediments dominated by coral, coralline algae, and *Halimeda* fragments
7. Sand cays mostly found on windward reef flats	7. Cays slightly more leeward
8. Lagoon depth often related to atoll diameter; large atolls can have lagoons deeper than 50 m	8. Lagoons generally shallower than 15 m
9. Most lagoonal sediments biogenic	9. Many bank and lagoonal sediments nonskeletal
10. Large lagoons can have distinctive depth-related sedimentary facies	10. Lagoonal facies usually similar to those found on peripheral reefs or related to patch reef sedimentation
11. Leeward reefs often well-developed	11. Leeward reefs usually poorly developed, resulting in rather open lagoons
12. More than 300 atolls and extensive barrier reefs	12. About 10 atolls and only 2 definite barrier reefs

Caribbean reefs generally are smaller, have shallower lagoons, and have less well-developed leeward reefs than their Indo-Pacific counterparts. As a result, Caribbean lagoonal sediments do not exhibit the depth-dependent facies found in many Indo-Pacific atolls (Wiens, 1962). Although some areas of nonskeletal sedimentation have been found in the Indo-Pacific (notably Shark Bay, Australia; Logan et al., 1970), nonskeletal sediments do not appear to be as extensive as on the shallow Caribbean banks and atolls.

One of the more obvious differences between Caribbean and Indo-Pacific reefs is the lack of extensive reef development in the Caribbean. The Indo-Pacific, for example, has more than 300 atolls, while the Caribbean has only about 10. Much of the difference in reef distribution is probably related to differences in regional tectonic histories. During much of the Tertiary, large portions of the Indo-Pacific experienced subsidence, with the result that many thick sequences of shallow-water reef structures were able to accumulate; the most obvious example is the Darwin Rise, upon which the Marshall, Gilbert, and Caroline Islands have formed. In contrast, much of the Caribbean (especially the Greater and Lesser Antilles) experienced uplift and volcanism during the late Tertiary. The only areas that have undergone extensive regional subsidence are the Florida–Bahama Plateau and the Yucatan peninsula, where considerable amounts of shallow-water carbonate deposits have accumulated.

Acknowledgments

Portions of this study were supported through the U.S. Office of Naval Research contract CO241 and the U.S. Geological Survey contract 12606. The writer thanks Dr. E. Uchupi and Dr. K. O. Emery for helpful comments on the manuscript.

References

For article titles, see the reference list at the end of this volume.

Adams, R. D. (1968). *J. Geol.* **76**, 587–595.
Agassiz, A. (1894). *Bull. Mus. Comp. Zool. Harvard Univ.* **26**, 1–203.
Almy, C. C., Jr., and Carrion-Torres, C. (1963). *Carib. J. Sci.* **3**, 133–162.
Atwood, D. K., and Bubb, J. N. (1970). *J. Geol.* **78**, 499–505.
Baars, D. L. (1963). *Shelf Carbonates Paradox Basin, Symp., 4th Field Conf., 1963,* Four Corners Geol. Soc. pp. 101–129.
Ball, M. M. (1967a). *J. Sediment. Petrol.* **37**, 556–571.
Ball, M. M. (1967b). *Trans., Gulf Coast Ass. Geol. Soc.* **17**, 265–267.
Ball, M. M., Shinn, E. A., and Stockman, K. W. (1967). *J. Geol.* **75**, 583–597.
Bandy, O. (1964). *Bull. Amer. Ass. Petrol. Geol.* **48**, 1666–1679.
Banks, J. E. (1959). *Bull. Amer. Ass. Petrol. Geol.* **43**, 2237–2243.
Bathurst, R. G. C. (1967a). *Mar. Geol.* **5**, 89–109.
Bathurst, R. G. C. (1967b). *J. Geol.* **75**, 736–738.

Bavendamm, W. (1932). *Arch. Mikrobiol.* 3, 205–276.
Black, M. (1933a). *Geol. Mag.* 70, 455–466.
Black, M. (1933b). *Phil. Trans. Roy. Soc. London, Ser. B* pp. 165–192.
Bock, W. D. (1967). Unpublished Ph.D. Thesis, University of Miami, Florida.
Bock, W. D., and Moore, D. R. (1969). *Carib. Geol. Congr., Prepr.*
Boyd, D. W., Kornicker, L. S., and Rezak, R. (1962). *Geol. Soc. Amer., Spec. Pap.* 73, 121–122 (abstr.).
Branner, J. C. (1904). *Bull. Mus. Comp. Zool., Harvard Univ.* 44, 1–285.
Broecker, W. S., and Takahashi, T. (1966). *J. Geophys. Res.* 71, 1575–1602.
Brooks, H. K. (1962). *Geol. Soc. Amer., Spec. Pap.* 73, 1–2 (abstr.).
Bryan, E. H., Jr. (1953). *Atoll Res. Bull.* No. 19, pp. 1–38.
Bubb, J. N., and Atwood, D. K. (1968). *Bull. Amer. Ass. Petrol. Geol.* 52, 552 (abstr.).
Burkholder, P. R., and Burkholder, L. M. (1960). *Amer. J. Bot.* 47, 866 872.
Cary, L. R. (1918). *Carnegie Inst. Wash., Dep. Mar. Biol. Pap., Publ.* No. 213, pp. 341–362.
Chevalier, J. P. (1966). *Bull. Inst. Fr. Afr. Noire, Ser. A* 28, 912–975.
Clifton, H. E., Mahnken, C. v. W., van Derwalker, J. C., and Waller, R. A. (1970). *Science* 168, 659–663.
Cloud, P. E., Jr. (1959). *U.S. Geol. Surv., Prof. Pap.* 280-K, 361–445.
Cloud, P. E., Jr. (1962). *U.S. Geol. Surv., Prof. Pap.* 350, 1–138.
Curray, J. R. (1965). *In* "The Quaternary of the United States" (H. E. Wright and D. G. Frey, eds.), pp. 723–735. Princeton Univ. Press, Princeton, New Jersey.
Daetwyler, C. C., and Kidwell, A. L. (1960). *World Petrol. Congr., Proc., 5th, 1959* Sect. 1, Pap. 1, pp. 1–21.
Dalrymple, D. W. (1964). Unpublished Ph.D. Thesis, Rice University, Houston, Texas.
Darwin, C. (1851). "The Structure and Distribution of Coral Reefs" (Reprinted by University of California Press, 1962).
Davis, J. H. (1940). *Pap. Tortugas Lab.* 32, 302–412.
DeBuisonje, P. H., and Zonneveld, J. I. S. (1960). *Natuurwetensch. Werkgroep Ned. Antillen* No. 11, pp. 121–144.
Deffeyes, K. W., Lucia, F. J., and Weyl, P. K. (1965). *Soc. Econ. Paleontol. Mineral., Spec. Publ.* 13, 71–88.
Doran, E. (1954). *Tex. J. Sci.* 6, 360–377.
Doran, E. (1955). *Tex., Univ., Dep. Geogr., Publ.* 5509, 1–38.
Drew, G. H. (1914). *Carnegie Inst. Wash. Publ.* 182, 7–45.
Duane, D. B., and Meisburger, E. P. (1969). *U.S. Army Corps Eng., Coastal Eng. Res. Cent., Tech. Memo* 29, 1–47.
Duarte-Bello, P. P. (1961). *Acuario Nac., Ser. Educ.* (Cuba) No. 2, 1–85.
Ebanks, W. J., Jr. (1967). Unpublished Ph.D. Thesis, Rice University, Houston, Texas.
Ekman, S. (1953). "Zoogeography of the Sea." Sidgwick & Jackson, London.
Emery, K. O. (1962). *Geofis. Int.* 3, 11–17.
Emery, K. O., Tracey, J. I., Jr., and Ladd, H. S. (1954). *U.S., Geol. Surv., Prof. Pap.* 260-A, 1–254.
Emiliani, C. (1954). *Science* 119, 853–855.
Emiliani, C., and Flint, R. F. (1963). *In* "The Sea" (M. N. Hill, ed.), Vol. 3, pp. 888–927. Wiley (Interscience), New York.
Environmental Science and Services Administration. (1968). "Surface Water Temper-

ature and Density," C & GS Publ. No. 31-1. U.S. Department of Commerce, 102 pp.

Fleece, J. B. (1962). "The Carbonate Geochemistry and Sedimentology of the Keys of Florida Bay, Florida," Contrib. No. 5. Sediment. Res. Lab., Dep. Geol., Florida State University, Tallahassee.

Folk, R. L. (1962). *Trans. N.Y. Acad. Sci.* [2] **25**, 222–244.

Folk, R. L. (1967). *J. Geol.* **75**, 412–437.

Folk, R. L., and Robles, R. (1964). *J. Geol.* **72**, 255–292.

Folk, R. L., Hayes, M. O., and Shoii, R. (1962). "Carbonate Sediments of Isla Mujeres, Quintana Roo, Mexico and Vicinity." Guide Book to Field Trip to Peninsula of Yucatan, New Orleans Geol. Soc., Louisiana.

Fosberg, F. R. (1962). *Atoll Res. Bull.* No. 93, pp. 1–25.

Freeman, T. (1962). *J. Sediment. Petrol.* **32**, 475–483.

Garrett, P., Patriquin, D., Smith, C. L., and Wilson, A. O. (1972). *J. Geol.*, **79**, 647–668.

Gebelein, C. D., and Hoffman, P. (1968). *Geol. Soc. Amer.*, *Ann. Meet.* p. 109 (abstr.).

Ginsburg, R. N. (1956). *Bull. Amer. Ass. Petrol. Geol.* **40**, 2384–2427.

Ginsburg, R. N. (1957). *Soc. Econ. Paleontol. Mineral. Spec. Publ.* **5**, 80–100.

Ginsburg, R. N., ed. (1964). "South Florida Carbonate Sediments Guidebook." Geol. Soc. Amer. Conv., Miami.

Ginsburg, R. N., and Lowenstam, H. A. (1958). *J. Geol.* **66**, 310–318.

Ginsburg, R. N., Isham, L. B., Bein, S. J., and Kuperberg, J. (1954). "Laminated Algal Sediments of South Florida and their Recognition in the Fossil Records," unpublished rep. No. 54.21. Marine Laboratory, University of Miami, Coral Gables, Florida.

Ginsburg, R. N., Marszalek, D. S. and Schneidermann, N., 1971. *J. Sediment. Petrol.* **41**, 472–482.

Glynn, P. W. (1963). *Ass. Isl. Mar. Lab.*, *5th Meet.* pp. 6–9.

Goldman, M. I. (1926). *Pap. Tortugas Lab.* **23**, 37–66.

Goreau, T. F. (1959a). *Proc. Int. Congr. Zool.*, *15th*, *1958* p. 250.

Goreau, T. F. (1959b). *Ecology* **40**, 67–89.

Goreau, T. F. (1961). "The Structure of the Jamaican Reef Communities, Geological Aspects." Dep. Biochem. Ecol., N.Y. Zool. Soc., New York.

Goreau, T. F. (1963). *Ann. N.Y. Acad. Sci.* **109**, 127–167.

Goreau, T. F. (1964). *Science* **145**, 383–386.

Goreau, T., and Burke, K. (1966). *Mar. Geol.* **4**, 207–225.

Goreau, T. F., and Graham, E. A. (1967). *Bull. Mar. Sci.* **17**, 432–441.

Goreau, T. F., and Hartman, W. D. (1963). *In* "Mechanics of Hard Tissue Destruction," Publ. No. 75, pp. 25–54. Amer. Ass. Advance. Sci., Washington, D.C.

Goreau, T. F., and Wells, J. W. (1967). *Bull. Mar. Sci.* **17**, 442–453.

Goreau, T. F., Llauger, V. T., Mas, E. L., and Seda, E. R. (1960). *Ass. Isl. Mar. Lab.*, *3rd Meet.* pp. 8–9 (abstr.).

Gorsline, D. S. (1963). *Shelf Carbonates Paradox Basin, Symp.*, *4th Field Conf.*, *1963, Four Corners Geol. Soc.* pp. 130–143.

Gould, H. H., and Stewart, R. H. (1953). *Soc. Econ. Paleontol. Mineral.*, *Spec. Publ.* **5**, 2–19.

Heilprin, A. (1890). *Proc. Acad. Natur. Sci.*, *Philadelphia* **42**, 303–316.

High, L. R., Jr. (1969). *J. Sediment. Petrol.* **39**, 235–245.

Hoffmeister, J. E., and Multer, H. G. (1964). *Geol. Soc. Amer., Bull.* **75**, 353–358.
Hoffmeister, J. E., and Multer, H. G. (1965). *Geol. Soc. Amer., Bull.* **76**, 845–852.
Hoffmeister, J. E., Stockman, K. W., and Multer, H. G. (1967). *Geol. Soc. Amer., Bull.* **78**, 175–190.
Hoskin, C. M. (1963). *Nat. Acad. Sci.—Nat. Res. Counc., Publ.* **1089**, 1–160.
Hoskin, C. M. (1964). *Bull. Amer. Ass. Petrol. Geol.* **48**, 1680–1704.
Hoskin, C. M. (1966). *J. Sediment. Petrol.* **36**, 1058–1074.
Hoskin, C. M. (1968). *Amer. Ass. Petrol. Geol., Bull.* **52**, 2170–2177.
Illing, L. V. (1954). *Bull. Amer. Ass. Petrol. Geol.* **38**, 1–95.
Jindrich, V. (1969). *J. Sediment. Petrol.* **39**, 531–553.
Jones, J. A. (1963). *Bull. Mar. Sci. Gulf Carib.* **13**, 282–307.
Kaye, C. A. (1959). *U.S., Geol. Surv., Prof. Pap.* **317-B**, 49–140.
Kellerman, K. F., and Smith, N. R. (1914). *J. Wash. Acad. Sci.* **4**, 400–402.
Kissling, D. L. (1965). *Bull. Mar. Sci.* **15**, 599 611.
Koldewijn, B. W. (1958). *Rep. Orinoco Shelf Exped.* **3**, 1–109.
Kornicker, L. S. (1958). *Trans., Gulf Coast Ass. Geol. Soc.* **8**, 167–170.
Kornicker, L. S. (1964). *Bull. Mar. Sci. Gulf Carib.* **14**, 168–171.
Kornicker, L. S., and Boyd, D. W. (1962). *Bull. Amer. Ass. Petrol. Geol.* **46**, 640–673.
Kornicker, L. S., and Bryant, W. R. (1969). *Amer. Ass. Petrol. Geol., Mem.* **11**, 244–257.
Kornicker, L. S., and Purdy, E. G. (1957). *J. Sediment. Petrol.* **27**, 126–128.
Kornicker, L. S., Bonet, F., Cann, J. R., and Hoskin, C. M. (1959). *Publ. Inst. Mar. Sci., Univ. Tex.* **6**, 1–22.
Kumpf, H. E., and Randall, H. A. (1961). *Bull. Mar. Sci.* **11**, 543–551.
Land, L. S., and Goreau, T. F. (1970). *J. Sediment. Petrol.* **40**, 457–462.
Lewis, J. B. (1960). *Can. J. Zool.* **38**, 1133–1145.
Lewis, J. B. (1965). *Can. J. Zool.* **43**, 1049–1074.
Lewis, J. B., Axelsen, F., Goodbody, I., Page, C., and Chislett, G. (1969). "Comparative Growth Rates of Some Reef Corals in the Caribbean," Mar. Sci. Manuscript, Rep. No. 10. McGill University, Montreal.
Lloyd, R. M. (1964). *J. Geol.* **72**, 84–111.
Logan, B. W. (1969). *Amer. Ass. Petrol. Geol., Mem.* **11**, 129–198.
Logan, B. W., Harding, J. L., Ahr, W. M., Williams, J. D., and Snead, R. G. (1969). *Amer. Ass. Petrol. Geol., Mem.* **11**, 1–128.
Logan, B. W., Davies, G. R., Read, J. F., and Cebulski, D. E. (1970). *Amer. Ass. Petrol. Geol., Mem.* **13**, 1–223.
Lowenstam, H. A. (1955). *J. Sediment. Petrol.* **25**, 270–272.
Lowenstam, H. A., and Epstein, S. (1957). *J. Geol.* **65**, 364–375.
Lucia, F. J. (1968). *J. Sediment. Petrol.* **38**, 845–858.
Ludwick, J. C., and Walton, W. A. (1957). *Bull. Amer. Ass. Petrol. Geol.* **41**, 2054–2101.
McCallum, J. S., and Stockman, K. W. (1964). *In* "South Florida Carbonate Sediments Guidebook" (R. N. Ginsburg, ed.), pp. 11–15. Geol. Soc. Amer. Conv., Miami.
Macintyre, I. G. (1967a). Unpublished Ph.D. Thesis, McGill University, Montreal.
Macintyre, I. G. (1967b). *Can. J. Earth Sci.* **4**, 461–474.
Macintyre, I. G. (1972). *Bull. Amer. Assoc. Petrol. Geol.* **56**, 720–738.
Macintyre, I. G., and Milliman, J. D. (1970). *Geol. Soc. Amer., Bull.* **81**, 2577–2598.

Macintyre, I. G., and Pilkey, O. H. (1969). *Science* **166**, 374–375.
Macintyre, I. G., Mountjoy, E. W., and D'Anglejan, B. F. D. (1968). *J. Sediment. Petrol.* **38**, 660–664.
MacKenzie, F. T., Kulm, L. D., Cooley, R. L., and Barnhart, J. T. (1965). *J. Sediment. Petrol.* **35**, 265–272.
Matthews, R. K. (1963). *Trans., Gulf Coast Ass. Geol. Soc.* **13**, 49–58.
Matthews, R. K. (1966). *J. Sediment. Petrol.* **36**, 428–454.
Milliman, J. D. (1967a). *J. Sediment. Petrol.* **37**, 658–676.
Milliman, J. D. (1967b). *Bull. Mar. Sci.* **17**, 519–543.
Milliman, J. D. (1969a). *Atoll Res. Bull.* No. 129, pp. 1–41.
Milliman, J. D. (1969b). *Trans., Gulf Coast Ass. Geol. Sci.* **19**, 195–206.
Milliman, J. D. (1972a). *U.S. Geol. Surv., Prof. Pap.* **529-J**, 40.
Milliman, J. D. (1973). "Marine Carbonates." Springer–Verlag, Heidelberg (in press).
Milliman, J. D., and Emery, K. O. (1968). *Science* **162**, 1121–1123.
Milliman, J. D., and Supko, P. R. (1968). *Geol. Mijnbouw* **47**, 102–105.
Monty, C. L. V. (1967). *Ann. Soc. Geol. Belg.* **90**, 55–100.
Moore, C. H., Jr., and Billings, G. K. (1971). *Carbonate Cements, Johns Hopkins Univ. Stud. Geol.* **19**, 40–45.
Moore, D. R. (1958). *Publ. Inst. Mar. Sci., Univ. Tex.* **5**, 151–155.
Müller, G., and Müller, J. (1967). *Neues Jahrb. Mineral., Abh.* **106**, 257–286.
Multer, H. G. (1969). "Field Guide to Some Carbonate Rock Environments." Florida Keys and Western Bahamas, *Miami Geol. Soc.*
Multer, H. G., and Hoffmeister, J. E. (1968). *Geol. Soc. Amer., Bull.* **79**, 183–192.
Munk, W. H., and Sargent, M. C. (1954). *U.S. Geol. Surv., Prof. Pap.* **260-C**, 275–280.
Neumann, A. C., and Land, L. S. (1969). *Geol. Soc. Amer., Spec. Pap.* **121**, 219 (abstr.).
Neumann, A. C., Gebelein, C. D., and Scoffin, T. P. (1970). *J. Sediment. Petrol.* **40**, 274–297.
Newell, N. D. (1955). *Geol. Soc. Amer., Spec. Pap.* **62**, 303–316.
Newell, N. D. (1959). *Natur. Hist., N.Y.* **68**, 226–235.
Newell, N. D., and Rigby, J. K. (1957). *Soc. Econ. Paleontol. Mineral., Spec. Publ.* **5**, 13–72.
Newell, N. D., Rigby, J. K., Whiteman, A. J., and Bradley, J. S. (1951). *Bull. Amer. Mus. Natur. Hist.* **97**, 1–29.
Newell, N. D., Imbrie, J., Purdy, E. G., and Thurber, D. L. (1959). *Bull. Amer. Mus. Natur. Hist.* **117**, 177–228.
Newell, N. D., Purdy, E. G., and Imbrie, J. (1960). *J. Geol.* **68**, 481–497.
Nota, D. J. G. (1958). *Meded. Landbouwhogesch., Wageningen* **58**, 1–98.
Odum, H. T., Burkholder, P. R., and Rivero, J. (1959). *Publ. Inst. Mar. Sci., Univ. Tex.* **6**, 159–170.
Ottmann, F. (1963). *Rev. Geogr. Phys.* **5**, 101–107.
Perkins, R. D., and Enos, P. (1968). *J. Geol.* **76**, 710–717.
Purdy, E. G. (1963). *J. Geol.* **71**, 334–353 and 472–497.
Purdy, E. G., and Imbrie, J. (1964). "Carbonate Sediments, Great Bahama Bank Guidebook." Geol. Soc. Amer. Conv., Miami.
Purdy, E. G., and Matthews, R. K. (1964). *Geol. Soc. Amer., Spec. Pap.* **82**, 157 (abstr.).

Pusey, W. C., III. (1964). Unpublished Ph.D. Thesis, Rice University, Houston, Texas.

Randall, J. E. (1965). *Ecology* **46**, 255–260.

Rice, W. H., and Kornicker, L. S. (1962). *Publ. Inst. Mar. Sci., Univ. Tex.* **8**, 366–403.

Roberts, H. H., and Moore, C. H., Jr. (1971). *Carbonate Cements, Johns Hopkins Univ. Stud. Geol.* **19**.

Roos, P. J. (1964). *Stud. Fauna Curacao, Natuurwetensch. Werkgroep Nede. Antillen* **20**, 1–51.

Rusnak, G. A. (1960). *Sediment. Petrol.* **30**, 471–480.

Scholl, D. W. (1966). *In* "The Encyclopedia of Oceanography" (R. W. Fairbridge, ed.), pp. 282–288. Van Nostrand-Reinhold, Princeton, New Jersey.

Scoffin, T. P. (1970). *J. Sediment. Petrol.* **40**, 249–273.

Shinn, E. A. (1963). *J. Sediment. Petrol.* **33**, 291–303.

Shinn, E. A. (1966). *J. Paleontol.* **40**, 233–240.

Shinn, E. A. (1968). *J. Paleontol.* **42**, 879–794.

Shinn, E. A., Ginsburg, R. N., and Lloyd, R. M. (1965). *Soc. Econ. Paleontol. Mineral., Spec. Publ.* **13**, 112–123.

Shinn, E. A., Lloyd, R. M., and Ginsburg, R. N. (1969). *J. Sediment. Petrol.* **39**, 1202–1228.

Siegel, F. R. (1963). *J. Sediment. Petrol.* **31**, 336–342.

Smith, C. L. (1940). *J. Mar. Res.* **3**, 147–189.

Smith, F. G. W. (1954). *U.S., Fish Wildl. Serv., Fish. Bull.* **55**, 291–295.

Smith, N. R. (1926). *Carnegie Inst. Wash. Publ.* **344**, 69–72.

Squires, D. F. (1958). *Bull. Amer. Mus. Natur. Hist.* **115**, 219–262.

Stanley, D. J., and Swift, D. J. P. (1967). *Science* **157**, 677–681.

Steers, J. A. (1940). *Geogr. Rev.* **30**, 279–296.

Steers, J. A., Chapman, V. J., and Lofthouse, J. A. (1940). *Geogr. J.* **96**, 305–328.

Stetson, H. C. (1953). *Pap. Phys. Oceanogr. Meteorol.* **12**, 1–45.

Stockman, K. W., Ginsburg, R. N., and Shinn, E. A. (1967). *J. Sediment. Petrol.* **37**, 633–648.

Stoddart, D. R. (1962a). *Atoll Res. Bull.* No. 87, pp. 1–147.

Stoddart, D. R. (1962b). *Nature* (*London*) **196**, 512–515.

Stoddart, D. R. (1963). *Atoll Res. Bull.* No. 95, pp. 1–142.

Stoddart, D. R. (1964). *Atoll Res. Bull.* No. 104, pp. 1–16.

Stoddart, D. R., and Cann, J. R. (1965), *J. Sediment. Petrol.* **35**, 243–247.

Storr, J. F. (1964). *Geol. Soc. Amer., Spec. Pap.* **79**, 1–98.

Sverdrup, H. V., Johnson, M. W., and Fleming, R. H. (1942). "The Oceans, Their Physics, Chemistry and General Biology." Prentice-Hall, Englewood Cliffs, New Jersey.

Swinchatt, J. P. (1965). *J. Sediment. Petrol.* **35**, 71–90.

Taft, W. H., and Harbaugh, J. W. (1964). *Stanford Univ. Publ., Univ. Ser., Geol. Sci.* **8**, 1–133.

Taft, W. H., Arrington, F., Haimovitz, A., MacDonald, C., and Woolheater, C. (1968). *Bull. Mar. Sci.* **18**, 762–828.

Thiel, M. E. (1928). *In* "Meersfauna Westafrika" (D. D. Michaelsen, ed.), Vol. 3, pp. 253–350.

Thomas, L. P., Moore, D. R., and Work, R. C. (1961). *Bull. Mar. Sci. Gulf Carib.* **11**, 101–107.

Thorp, E. M. (1936). *Pap. Tortugas Lab.* **29**, 37–119.

Tracey, J. I., Jr., Ladd, H. S., and Hoffmeister, J. E. (1948). *Geol. Soc. Amer.*, *Bull.* **59**, 861–878.

Traganza, E. D. (1967). *Bull. Mar. Sci.* **17**, 348–366.

Uchupi, E., Milliman, J. D., Luyendyk, B. P., Bowin, C. O., and Emery, K. O. (1972). *Bull. Amer. Ass. Petrol. Geol.* **35**, 687–704.

Vaughan, T. W. (1916). *J. Wash. Acad. Sci.* **6**, 53–66.

Vaughan, T. W. (1918). *Carnegie Inst. Wash. Publ.* **213**, 235–288.

Vaughan, T. W. (1919). *U.S., Nat. Mus., Bull.* **103**, 189–524.

Vermeer, D. E. (1959). "The Cays of British Honduras." Dep. Geogr., University of California, Berkeley.

Vermeer, D. E. (1963). *Z. Geomorphol.* **7**, 332–354.

von Arx, W. S. (1954). *U.S. Geol. Surv., Prof. Pap.* **260-B**, 265–273.

Voss, G. L., and Voss, N. A. (1955). *Bull. Mar. Sci.* **5**, 203–229.

Wantland, K. F. (1967). Unpub. Ph.D. Thesis, Rice University, Houston, Texas.

Ward, W. C., Folk, R. L., and Wilson, J. L. (1970). *J. Sediment. Petrol.* **40**, 548–555.

Wells, J. W. (1954). *U.S. Geol. Surv., Prof. Pap.* **260-I**, 385–486.

Wells, J. W. (1957). *Geol. Soc. Amer., Mem.* **67**, 609–631.

Wiens, H. J. (1962). "Atoll Environment and Ecology." Yale Univ. Press, New Haven, Connecticut.

Work, R. C. (1969). *Bull. Mar. Sci.* **19**, 614–711.

Yonge, C. M. (1963). *Advan. Mar. Biol.* **1**, 209–260.

Zaneveld, J. S. (1957). *Rep. Inter-Isl. Mar. Biol. Conf., Inst. Mar. Biol., Univ. Puerto Rico* pp. 18–19.

Zaneveld, J. S. (1958). *Blumea, Suppl.* **4**, 206–219.

Zans, V. A. (1958a). *Geonotes* **1**, 18–25.

Zans, V. A. (1958b). *Jam., Geol. Surv. Dep., Bull.* No. 3, pp. 1–47.

2

CORAL REEFS OF THE INDIAN OCEAN

D. R. Stoddart

I. Introduction

This paper summarizes present knowledge of the coral reefs of the Indian Ocean: their origin and structure, surface morphology, zonation, and associated communities (Fig. 1). It is shown that in general features

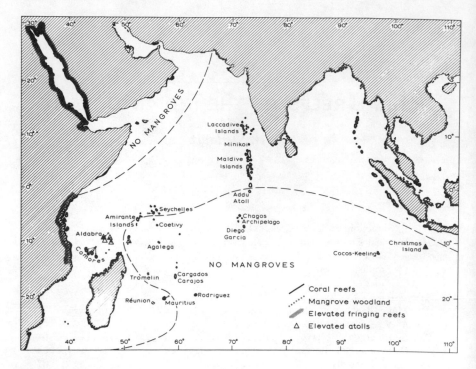

Fig. 1. Distribution of major reef types in the Indian Ocean. [From Stoddart (1969a).]

and diversity Indian Ocean reefs resemble those described since 1945 from the Pacific Ocean, though Indian Ocean studies are so far few in number and geographically scattered.

Darwin (1842) himself was among the first to study Indian Ocean reefs, at Cocos-Keeling and Mauritius in 1836, and his theory of subsidence accounts satisfactorily for the gross geological relations of reef limestones and their foundations in the Indian Ocean. The paucity of modern reef growth, however, particularly over large areas of older reef limestones, makes it necessary to pay particular attention to relative movements of land and sea level during the Pleistocene. As will become apparent, to explain reef variation in the Indian Ocean it is at least as important to consider cases where modern reefs do not grow as it is to consider those where they do. Thus we need to consider not only the effects of present-day environmental factors, but also to what extent the reefs have been affected by environmental changes during Pleistocene and earlier times. Reefs as geological structures are markedly resistant to erosional changes over time scales of 10^3–10^5 years (Stoddart,

1969b), and many modern Indian Ocean reef communities only veneer much older reef structures.

The general reviews of Indian Ocean reefs published by Sewell (1935) and Gardiner (1936) have been revised by Stoddart (1969a, 1971). Much recent information has been brought together in symposia organized by the Marine Biological Association of India (Mandapam, 1969; Cochin, 1971), the Royal Society (London, 1969, see Westoll and Stoddart, 1971), and the Zoological Society (London, 1970, see Stoddart and Yonge, 1971). Bibliographies on reefs and associated biota in the Indian Ocean are provided by Yentsch (1962), Alagarswami *et al.* (1968), and Sriramachandra Murty *et al.* (1969).

Apart from the early work in the Red Sea of Ehrenberg (1834), Klunzinger (1877–1879), Walther (1888), and von Marenzeller (1907), on the East African coast by Ortmann (1892) and Werth (1901), and at Cocos-Keeling and Mauritius by Darwin (1842), the reefs of the Indian Ocean were made known largely by two men: J. Stanley Gardiner and R. B. Seymour Sewell. Gardiner's first expedition in 1899–1900 took him from Minikoi through the Maldive Islands to Addu Atoll (Gardiner, 1903–1906); his second, the Percy Sladen Trust Expedition, in 1905 and 1908, covered the Chagos Archipelago, Seychelles, Amirantes, Coetivy, Cargados Carajos, Farquhar, Providence, St Pierre, and Mauritius (Gardiner, 1907–1936). Gardiner's collections covered both marine and terrestrial fauna and flora, and for many reefs his results form the only or the latest data available. Sewell worked in the Maldive, Andaman, and Nicobar Islands, and led the John Murray Expedition of 1933–1934. Both Gardiner (1931, 1936) and Sewell (1935) published summary accounts of the reefs they had studied. Alexander Agassiz (1903) also cruised through the Maldives in 1901–1902, but his voluminous account is of little value. Voeltzkow's (1897–1905, 1903, 1904, 1905, 1906, 1917) major contributions on the Malagasy region mainly concerned reef structure, geology, and terrestrial biota.

Other reef studies before 1945 were generally unrelated to Gardiner's and Sewell's work in the western and central Indian Ocean. Cocos-Keeling was studied by Forbes (1879), Guppy (1889), Wood-Jones (1910), and Gibson-Hill (1948, 1950), all workers with Malaysian rather than Indian Ocean experience. Collections were made at Christmas Island in 1897–1898 (Andrews, 1900), the Amirantes in 1882 (Coppinger, 1884), Aldabra in 1908–1909 (Fryer, 1911), and in the Laccadives (Oldham, 1895). Apart from Sewell's work (1932) on the Indian coast, that of Ortmann (1889) and Walther (1891) in Ceylon, of Coutière (1898) and Gravier (1911) in the Gulf of Aden, and of Crossland (1902, 1904,

1907) in East Africa and the Red Sea, there were almost no studies of mainland reefs around the Indian Ocean.

With notable exceptions such as Guppy, Crossland, and Sewell, these earlier workers were less concerned with the structure of reef communities and their physiographic expression than with the taxonomy and biogeography of the reef biota. Hence much of the data accumulated contributed little to the development of coral reef studies which took place after 1945. Apart from the work on Cocos, Addu, Goifurfehendu, Salomon, Minikoi, and Aldabra, it is often difficult to construct coherent accounts of the structure and zonation of Indian Ocean reefs from this earlier literature.

Recent Indian Ocean studies have been carried out largely by French and British workers. Guilcher's (1955) work on Banc Farsan, Red Sea, was followed by geomorphic studies in northwest Malagasy (Guilcher, 1956, 1958; Guilcher et al., 1958) and at Mayotte, Comoro Islands (Guilcher et al., 1965). Guilcher's geomorphic and sedimentological work has been supplemented in Madagascar and at Europa by that of Battistini (1964, 1966a; Berthois and Battistini, 1969), who has extended it to the East African coast (Battistini, 1966b, 1969). Extensive ecological work has been carried out at the Tuléar marine station in southwest Madagascar (M. Pichon, 1964; Vasseur, 1964; Blanc et al., 1966), and reconnaissance studies have begun in Mauritius (M. Pichon, 1967), Réunion (M. Pichon, 1971; Montaggioni, 1970; Faure and Montaggioni, 1970), and Rodriguez (Faure and Montaggioni, 1971). British workers have studied marine geomorphology and ecology at Addu Atoll, Maldive Islands (Stoddart, 1966; P. S. Davies et al., 1971), Diego Garcia Atoll, Chagos Archipelago (Stoddart and Taylor, 1971), Aldabra Atoll and neighboring islands north of Madagascar (Stoddart, 1967, 1970; Stoddart et al., 1971; Taylor, 1971a; Barnes et al., 1971). and the fringing reefs of Mahé, Seychelles (Taylor, 1968; Lewis, 1968, 1969; Taylor and Lewis, 1970; Braithwaite, 1971; Rosen, 1971a). The Royal Society of London, which initiated work on Aldabra in 1967, has now established a research station there for both land and marine studies.

There have also been brief visits to the Maldives and other reefs by the Yale Seychelles Expedition (Kohn, 1964, 1967, 1968), by the "Xarifa" (Hass, 1965; Scheer, 1969), and vessels of the International Indian Ocean Expedition (Hackett, 1969; Rice, 1971; Garth, 1971); to Cocos-Keeling, Mauritius, and the Seychelles by members of the Coastal Studies Institute, Baton Rouge; and to parts of the Red Sea by Israeli expeditions. Little work has been published on African reefs (Boshoff, in Macnae and Kalk, 1958; Talbot, 1965; Brander *et al.*, 1971); and that on the reefs of western Australia (Fairbridge, 1948a), Houtman's Abrolhos (Teichert, 1947; Fairbridge, 1948b), and the Sahul Shelf (Teichert and Fairbridge, 1948) remains largely unrelated to Indian Ocean studies.

In spite of this recent work (cf. Stoddart and Yonge, 1971), large gaps remain: most of the Maldives and Laccadives have not been visited since Gardiner and Sewell; there are no modern accounts of the reefs of the Amirantes, Tromelin, Agalega, Cargados Carajos, Coetivy, and other western Indian Ocean islands (though see Baker, 1963), or of Christmas Island, the Andamans and Nicobars; the Chagos Archipelago remains largely unknown; and apart from Pillai's work (1967; Nair and Pillai, 1969), no further studies have been made of the reefs of the Asian mainland.

This paper deals with the reefs of the Indian Ocean proper: only incidental mention can be made of reefs in the Red Sea and the Persian Gulf. Work is being carried out in the Red Sea by French, German, English, and Israeli workers, particularly at Eilat and at other marine stations at Ghardaqa, Port Sudan, and Jedda. In addition to purely geological studies, reef communities have been described from the southern Red Sea by Wainwright (1965), Goreau (1964), Nesteroff (1955), Einsele *et al.* (1967), Klausewitz (1967), Scheer (1964, 1967), and Rossi (1954), and from the central and northern Red Sea by Mergner (1967, 1971), Friedman (1968), Loya and Slobodkin (1971), and Ormond and Campbell (1971). Coral reefs are less well developed in the Persian Gulf (Harrison, 1911), where environmental conditions are less favorable. Most recent information has been obtained from studies of geomorphology and sediments (Emery, 1956; Evans, 1966; Sugden, 1963; Kendall and Skipwith, 1969; Kinsman, 1964).

II. Physical Environment

Reefs are adapted to and themselves modify their physical environment, and it is therefore important to distinguish between environmental factors which show wide regional differences, and those which may show high but local variability dependent on reef morphology itself. Thus while both temperature and salinity distributions show broad gradients over the ocean, in the area covered by Fig. 1 they nowhere constitute factors limiting reef growth, except possibly in areas of upwelling such as the Somali coast: extreme values are found (e.g., salinities as high as 48‰ in the Persian Gulf and as low as 26‰ at Mandapam) but these are purely local phenomena without regional significance. This section briefly summarizes available knowledge on wind, tropical storms, tides, and rainfall as regional reef environmental factors.

A. Wind

Wind is important primarily through the action of wind waves and swell. Because of the smaller size and monsoonal circulation of the Indian Ocean, adjustments of reefs to the trade wind circulation patterns is less marked than in the open Pacific. North of the equator the monsoonal circulation is dominant, with seasonal reversal and often inequality of winds. The Seychelles and the Chagos in the equatorial zone experience the southeast trades for part of the year and calms for the rest, and only the reefs south of 10°S are dominated by the trades. North of the equator one would not therefore expect the marked variation between windward and leeward reefs characteristic of Pacific trade wind reefs, while further south, as on Aldabra, Assumption, and Farquhar, such a distinction can be made. Sewell (1936a,b), for example, in the Maldives, has contrasted Addu Atoll, outside the limits of monsoon reversals, with Goifurfehendu Atoll further north, where both northeast and southeast monsoons are felt.

Figure 2 illustrates these patterns by showing the number of months in each year in which the frequency of winds greater than Beaufort force 4 (7 m sec^{-1}) exceeds 70%. The core of the trade wind belt in the southern hemisphere is clearly marked (Agalega has 11 months), and the seasonal extension toward the west and northwest is also apparent. The areas of the Arabian Sea and the Bay of Bengal are characterized by moderately high winds which are purely seasonal and occur during the northern summer monsoon. Within the areas covered by the trades the wind blows with great constancy and speeds of 6–9 m sec^{-1}. Figure 2 is clearly highly approximate: it suggests, for example, that Aldabra is affected by the trades for only 5 months of the year, which is an underestimate, and other modifications could be suggested. Nevertheless it clearly distinguishes the seasonal monsoonal, the equatorial calm, and the permanent trade wind areas, and thus serves as a useful substitute for a map of sea surface conditions.

One important reservation must be made: wind-generated waves, as swell, travel outside their area of generation, and morphologically significant sea conditions are not restricted to the areas shown in Fig. 2. Swell from the southeast is certainly important on reefs in the equatorial zone, as at Addu Atoll, and swell from the southern westerlies also reaches some reefs, particularly those in southwest Madagascar,

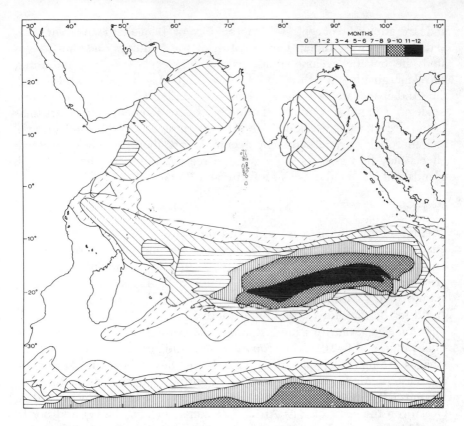

Fig. 2. Constancy of wind in the Indian Ocean. The map shows the number of months in the year in which the frequency of winds greater than Beaufort Force 4 (~7 m sec⁻¹) exceeds 70%. [Data from *Monthly meteorological charts of the Indian Ocean* (M.O. 470, 1949) with the sanction of the Controller, Her Majesty's Stationery Office, and of the Hydrographer of the Navy.]

sheltered from the trades, which, in spite of their leeward location, nevertheless show many features characteristic of windward reefs. These swellward reefs contrast with the more fully protected reefs of northwest Madagascar, where neither the southern swell nor the easterly trades are felt. Swell from the westerlies is probably also significant on reefs in the eastern part of the ocean.

B. Tropical Storms

Equally important in reef physiography is the distribution of tropical cyclones. These are almost absent from the zone between 6°N and 5°S, and are strongly concentrated in the Bay of Bengal, the Arabian Sea, and the southwest Indian Ocean. Most of the Maldives and the Chagos atolls lie outside cyclone areas, as do the Seychelles, the Amirantes, and (marginally) Aldabra and Coetivy. The Laccadives and Minikoi, the Andamans and Nicobars, Comoros, Madagascar, the Mascarenes, Agalega, Cargados Carajos, Tromelin, Cocos-Keeling, and Christmas Island are affected. Cyclones are an important control of coral growth, reef morphology, and sediment accumulation, though their destructive effects may be less marked on isolated reefs than on the shores of high islands and mainland coasts (McIntire and Walker, 1964).

C. TIDES

Tidal information is reasonably full for mainland coasts, but is inadequate for most Indian Ocean islands; the following information is based on predictions in *Admiralty Tide Tables 1970*. Most Indian Ocean tides are semidiurnal or mixed, mainly semidiurnal. Tides with a range at springs of 0.3–1.0 m, which may be classified as microtidal, are experienced in the Laccadives, Maldives, south India and Ceylon, Cocos-Keeling, Agalega, Mauritius, and Réunion. Mesotidal areas, with a spring tidal range of 1.0–2.0 m, are found in the Andamans, the Nicobars, the Chagos Archipelago, Christmas Island, the Seychelles, the Amirantes, and eastern Madagascar. A macrotidal springs range of more than 3.0 m, unusually large for reef areas, is experienced in the southwest Indian Ocean: on the coast of East Africa (Zanzibar, Pemba, Tutia) and northwest Madagascar, but also on the islands of the Comoros and the Aldabra group. Such extreme tides on atolls have no counterpart in other oceans. High tidal ranges are also found over small areas elsewhere in the Indian Ocean, particularly in the Mergui Archipelago (more than 5 m) and in the Gulf of Cambay (8.78 m). High tidal ranges are important in reef areas because reef corals normally only grow up to neap tide levels, and because desiccation and exposure to the atmosphere limit the growth of corals, algae, and other organisms on intertidal flats.

III. Origin of the Reefs

The distribution and size of Indian Ocean coral reefs must be viewed in the perspective of the development of the Indian Ocean itself. The bottom topography, especially of the western Indian Ocean, is now reasonably well known (Heezen and Tharp, 1966; Fisher *et al.*, 1967). Most of the ocean floor lies at depths of 4–5 km. Rising from it are linear ridges (the Chagos–Laccadive Ridge; the Ninetyeast Ridge), arcuate plateaus and ridges (Mid-Ocean Ridge; Mascarene Plateau), and isolated islands (e.g., Aldabra, Mascarenes, Cocos-Keeling). The Chagos–Laccadive Ridge, the Mascarene Plateau, and the isolated islands are, in addition to continental coasts, the main sites of reef limestone development; no sea-level islands are present on the Mid-Ocean and Ninetyeast Ridges.

From geological and paleomagnetic data on the surrounding continents it is apparent that the Indian Ocean began to form in Mesozoic time (McElhinny, 1970). It is likely that its main lineaments existed by the early Tertiary, when the foundations of the present reefs were formed and when much reef growth probably began. Basalts dredged from the ocean floor have been dated radiometrically, and associated sediments paleontologically, at late Cretaceous to Oligocene in age, and magnetic lineations suggest continued widening of the ocean during the later Tertiary (Fisher *et al.*, 1967; Le Pichon and Heirtzler, 1968). Subaerial volcanoes in the Mascarene and Comoro Islands are Pliocene to Recent in age.

The structure of the main island groups may be related to these developments. The Mascarene Plateau is of diverse origin. The Seychelles Bank in the north is an isolated granitic "microcontinent" of Upper pre-Cambrian age with Tertiary intrusions (Baker and Miller, 1963). Similar granites are inferred on geophysical grounds to underlie the Mascarene Plateau between the Seychelles Bank and Saya de Malha (Matthews and Davies, 1966). Saya de Malha and Nazareth Banks are thought to consist of coral-capped volcanic rocks, while Cargados Carajos bank is underlain by either granitic or volcanic rocks at shallow depth and is capped with coral (Shor and Pollard, 1963; Francis and Shor, 1966).

After the Mascarene Plateau, the most extensive reef limestone province in the Indian Ocean is formed by the Chagos, Maldive, and Laccadive Archipelagoes. This interrupted, linear, aseismic ridge, 3000 km long, appears from seismic refraction and gravity anomalies to consist of coral limestone capping massive volcanics (Glennie, 1936; Francis and Shor, 1966). It has been suggested that the volcanism responsible for this ridge moved northward over time to culminate in the outpourings of the Deccan Traps, which are largely Paleocene–Eocene in age (Wellman and McElhinny, 1970); if so, the southernmost volcanics, in the Chagos, could be as old as Cretaceous, and the Chagos–Laccadive Ridge could mark the track taken by the Indian subcontinent as it moved northward away from Africa (Francis and Shor, 1966).

There is less information on the history of isolated islands which now exist only as surface reefs. The bathymetry of many (e.g., Aldabra, Cosmoledo, Europa, Tromelin, Cocos-Keeling), rising from depths of 4–5 km, suggests a volcanic basement, and this is partially confirmed by dredging of basalts from their slopes (e.g., on Providence). From the structure of the ocean floor itself, it can be inferred that these volcanoes are unlikely to be older than Eocene. The volcanoes which are still above sea level, such as the Mascarenes and Comoros, are undoubtedly younger than those now drowned and have absolute ages ranging from Pliocene to Quaternary; two, Réunion and Grand Comore, are still active. Corals and limestone with *Lepidocyclina* from the summit of a guyot at 0.7 km depth near Cocos-Keeling indicate a Miocene age for sediments on the summit of a cone 3.3 km high (Niino and Oshite, 1966): more investigations are needed of guyots elsewhere in the Indian Ocean, e.g., in the Mozambique Channel.

No deep drilling has yet been carried out on open-ocean reefs in the Indian Ocean, and data on reef thicknesses are hence limited. Geophysical data indicate thicknesses of reef limestones of 1 km plus 0.2–0.5 km lagoonal carbonate sediments on Saya de Malha (Shor and Pollard, 1963), of less than 1 km in the Amirantes (Matthews and Davies, 1966), of 0.6–1.7 km under 1 km of water near Great Chagos Bank (Francis and Shor, 1966), and of less than 0.5 km on Seychelles Bank (D. Davies and Francis, 1964). These thicknesses compare with 1.25 km on Eniwetok Atoll in the Pacific, where paleontological evidence from cores shows that reef growth began in the Eocene, and with 0.4 km on Mururoa and 0.2 km on Midway, other cored atolls in the Pacific.

Reef histories are undoubtedly more complex in nonoceanic areas peripheral to the Indian Ocean basin. These include particularly the Red Sea and Gulf of Aden, which have characteristics of embryonic widening oceans and extensive fringing and elevated reefs (Laughton, 1966); Madagascar, particularly the northwest coast (Battistini, 1965); and the islands and coastlands of the tectonically active Andaman Sea (Rodolfo, 1969; Weeks et al., 1967).

We may conclude that the reefs of the Indian Ocean have been built almost entirely during Tertiary and Quaternary times; that the thicker reef sequences cap older, mainly volcanic, structures, forming the great atoll and submerged reef-bank chains; and that fringing reefs and barriers have formed on younger volcanoes, such as the Mascarenes and Comoros in Pliocene and Pleistocene times. The main Indian Ocean reefs had reached their approximate distribution, extent, and form by the end of the Tertiary, and their present surface features record the erosional and depositional consequences of Pleistocene and Holocene sea-level fluctuations.

IV. Major Reef Types

Indian Ocean reefs include sea-level atoll, fringing and barrier reefs, elevated reefs, and reef platforms now submerged. Reefs are poorly developed or absent on the continental coasts of Somalia, Arabia, the Indian subcontinent, and Malaya, but are well developed in parts of the Red Sea, the coasts of Kenya and Tanzania, western Madagascar, and parts of Sumatra and northwest Australia (Fig. 1).

A. SEA-LEVEL REEFS

1. Fringing Reefs

Reefs of continental coasts are mostly fringing reefs, a type particularly well developed in East Africa and the Red Sea. Fringing reefs are also found round the granitic Seychelles, where they have been studied in detail, the volcanic Comoros and Mascarenes, and the sedimentary Andamans and Nicobars. Windward fringing reefs generally have straight outlines with well-marked zonation; leeward fringes, e.g., in Victoria Harbour, Mahé, and Ramanetaka Bay, Madagascar, may be irregular in outline and lack pronounced zonation.

2. Barrier Reefs

Barrier reefs are not well developed in the Indian Ocean. There is a discontinuous submerged possible barrier in northwest Madagascar (Guilcher *et al.*, 1958), and a largely submerged barrier extends from Tuticorin to Jaffna, Gulf of Mannar, India. There is a fine barrier, in part a double feature, surrounding Mayotte, Comoros (Guilcher *et al.*, 1965), and in places the Mauritius fringing reefs have some barrier characteristics (e.g., at Mahébourg). The absence of barriers round other volcanic islands must reflect their youth, and their absence in the Seychelles the stability of this bank.

3. Atolls

After continental fringing reefs, atolls form the dominant reef type of the Indian Ocean. They comprise the Laccadive, Maldive, and Chagos Archipelagoes, and a number of isolated reefs, notably Cocos, Farquhar, and Cosmoledo. Suvadiva Atoll, Maldives (length 70 km, breadth 53 km, area 2240 km²) is one of the largest atolls in the world.

The atolls vary widely in morphology. Many of the Maldive atolls, in the area of monsoon reversals, are markedly symmetric in contrast to asymmetric trade-wind atolls such as Farquhar, which have linear or broadly convex windward reefs and general arcuate shape; reef response to unidirectional winds in the case of Cocos-Keeling was used by Wood-Jones (1910) in a general theory of atoll development. Similar asymmetry is shown by atolls affected by swell rather than wind-waves, notably Addu in the southern Maldives. In the northern Maldives the atoll rim, instead of being a continuous linear reef, breaks down into a series of small ring-shaped reefs called *faros*. Similar ring-shaped reefs, termed *mini-atolls* by Scheer (1969), are also found in the lagoons of many Maldive atolls. Faros have nowhere been studied in detail, though Guilcher (1971) has drawn attention to their changing orientation and the variation in the direction of reef openings in them in response to differences in wind direction and especially protection by other reefs (see especially the faros of Tiladummati, Miladummadulu, and North Malosmadulu). Faros probably form by reef extensions to leeward in the way envisaged for atoll development at Cocos-Keeling by Wood-Jones.

Lagoon depths vary with atoll size and also location. Scheer (1969) has shown a relationship between maximum lagoon depth and atoll equivalent diameter $(2[A/\pi]^{1/2}$, where A is atoll area) which extends from reef patches, through mini-atolls and faros, to small, medium, and large atolls in the Maldives (Fig. 3). The relationship is less good for the larger atolls, where it is dominated by a general latitudinal gradient in maximum lagoon depth, from 80–90 m near the equator to 30–40 m in the northern Maldives. Biewald (1964), using less adequate chart data, has found that other reef groups conform to a latitudinal depth trend. No obvious explanation exists for this phenomenon. Nor is it invariate, for the lagoons of Chagos atolls are comparatively shallow (e.g., 31 m at Diego Garcia and Salomon). In the Maldive atolls there also exists a second latitudinal trend, in the continuity of the peripheral reef round each atoll lagoon. This varies from 95% (Addu) in the south to only 40% (Tiladummati–Miladummadulu) in the north. This may result from temperature gradients controlling the rate of reef growth away from the equator. Because of these variations in maximum depth and in the surrounding reefs, atoll lagoons vary considerably in their degree of "basining," i.e., flatness of floor and steepness of sides.

Lagoons vary widely in the occurrence of reef knolls and patches.

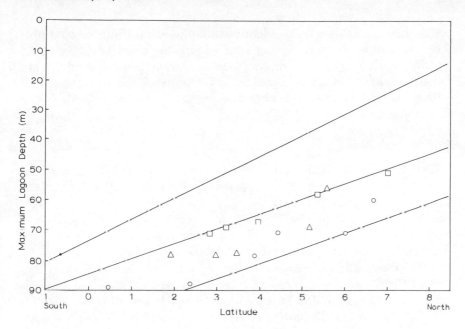

Fig. 3. Latitudinal trends in lagoon depths in atolls of different size in the Maldive Islands. Equivalent diameter of atoll: (●) 5–15 km; (□) 20–35 km; (△) 35–45 km; (○) >45 km. [From Scheer (1969).]

Diego Garcia has a highly irregular floor but virtually no patch reefs reaching the surface, while the lagoon of Farquhar is full of surface knolls and ridges. There is no general explanation of these differences.

4. Patch or Table Reefs

Some open-ocean reefs lack the typical atoll form, and are best described as linear patch or table reefs (e.g., Coetivy, Agalega, Cargados Carajos, Tromelin). None has been examined in the field, but air photographs show that many, particularly Providence, have well-developed lagoonal meshing in Maxwell's (1968) terminology. Some of the smaller Maldives reefs, such as Fua Mulaku, also fall in this category.

5. Sinuous Irregular Reefs with Lagoonlets

Guilcher (1960; Guilcher *et al.*, 1958, pp. 156–162) has described "reefs with many small lagoons" from protected near-shore areas in northwest Madagascar. These have a sinuous irregular outline with irregular, steep-walled, flat-floored lagoonlets in otherwise flat reef tops. These lagoonlets are either closed or partly open. Where sedimentation is not active they are as deep as the floor outside the reefs. Guilcher interprets them as resulting from the coalescence of growing reefs, but some could equally result from karst erosion of reefs during low sea-level stands. Such reefs have been recognized in other parts of the world, and fall within Maxwell's classes of closed and open mesh reefs with prong development.

B. Emerged Reefs

Elevated coral reefs are widely distributed on the coasts of East Africa (Werth, 1901; Ortmann, 1892; Crossland, 1902, 1904; Alexander 1968; Temple, 1970) and the Red Sea (Walther, 1888; Macfadyen, 1930; Guilcher, 1955, Schäfer, 1967). They are known from Christmas Island in the eastern Indian Ocean (Andrews, 1900; Trueman, 1965) and from Aldabra and the neighboring islands of Assumption, Astove, Cosmoledo, and St. Pierre in the western Indian Ocean (Baker, 1963; Stoddart, 1967, 1970). They are also found, though less well developed, on Madagascar and islands of the Mozambique Channel, and in the Seychelles and the Mascarenes. The extent of elevation varies from 360 m at Christmas Island to 10 m or less in the southwest Indian Ocean and along the African coast. The degree of preservation varies from that of entire elevated atolls such as Aldabra, with little-altered reef faunas, to the altered limestones of former fringing reefs on Mauritius and neighboring Ile aux Aigrettes, and to the small inconspicuous residual fragments of former fringing reefs on Mahé, Seychelles. Most of the elevated reefs of continental coasts are fringing reefs, while those of the open ocean are atolls or platform reefs.

In addition to these clearly elevated reefs, both Gardiner and Sewell have claimed elevation of varying amounts throughout the Chagos and the Maldives, largely on the basis of the existence of patches of reef-derived sediments, in some cases with corals said to be in the position of growth, slightly above present sea level. Reexamination of Gardiner's examples at Addu and Diego Garcia Atolls suggests that most such outcrops are of lithified island sediments not indicating any sea-level change, and this may also be true of Gardiner's classic site at Minikoi Atoll (Gardiner, 1901).

C. Submerged Reefs

Platforms of reef limestone which fail to reach present sea level are among the most remarkable reef features of the Indian Ocean. The total area of such banks in the western Indian Ocean (Fig. 4) is approxi-

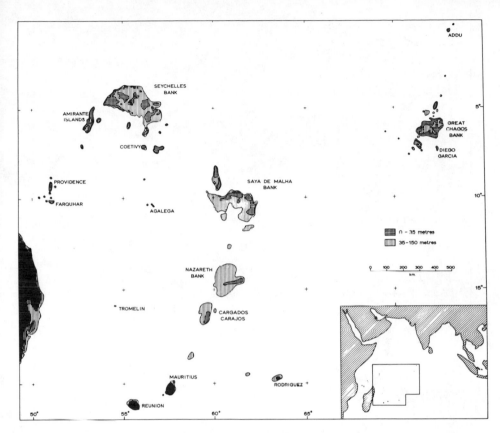

Fig. 4. Submerged reefs of the western Indian Ocean.

mately 140,000 km², compared with 155,000 km² for the Bahama Banks. Some of the banks are of great size: the Seychelles Bank covers 43,000 km², Saya de Malha 40,000 km², Nazareth Bank 26,000 km², and the banks of the Chagos Archipelago 13,500 km². The banks have general maximum depths of 33–90 m, and most have a shallower rim round at least part of their periphery, at 8–10 m depth. Though such banks are quite widespread in reef seas, Davis (1918) was the last to investigate them systematically; thus maximum depths and rim depths of banks in the China Sea are 30–110 and 7–26 m, respectively, and the mean depths of 13 banks in the Caribbean range from 20 to 42 m. Great Chagos Bank (Fig. 5), measuring 150 × 100 km, is one of the best known bathymetrically; the central basin is about 70 m deep; it has a rim 1.6 km wide at 10–18 m, and a very pronounced terrace within

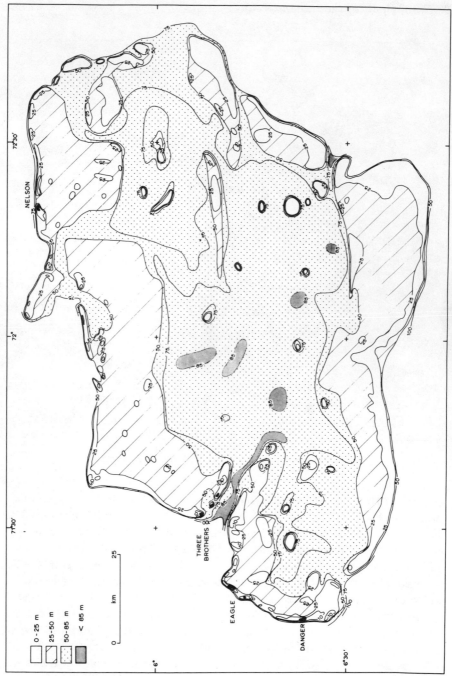

Fig. 5. Great Chagos Bank: bathymetry.

this rim, 8–20 km wide and 30 m deep. The Seychelles Bank is the only one of these banks on which sediments and bottom communities have been investigated in any detail (Matthews and Davies, 1966; Lewis and Taylor, 1966), and on this, as on the others, modern coral growth appears to be insignificant.

V. Coral Diversity

A. REGIONAL DIVERSITY

Until recently, knowledge of Indian Ocean coral faunas was limited to reasonably full studies of the Maldives, the Red Sea, and Cocos-Keeling, with partial accounts for a number of scattered localities. Earlier collections have not been completely described, largely because of taxonomic difficulties. Wells (1954) brought these data together in his map of generic diversity, which showed a westward decrease across the Indian Ocean from the center of coral diversity in the East Indies and West Pacific region.

Large collections have now been made in south India by Pillai (1971), at Addu in the Maldives (Wells and Davies, 1966), Mahé in the Seychelles (Rosen, *in litt.*), Aldabra (Barnes *et al.*, 1971), Tuléar in Madagascar (Pichon, 1964), and at several localities by Scheer (1971). A revision of the Chagos coral fauna has also been prepared (Rosen, 1971c). Most of the new collections have been made by underwater divers, and are thus more representative of total faunas than earlier collections.

Rosen (1971b) has collated this information in a preliminary revision of Indian Ocean coral distributions. In Fig. 6 he plots generic diversity contours based on known records. He finds a simple relationship between diversity and water temperature, which he has used to predict, in Fig. 7, the probable actual diversity, the difference between the two maps giving an indication of the extent of undercollecting, particularly in deeper waters. It is likely that the southern Maldives and the Seychelles form an area with more than 60 genera, and that most of the active reefs of the Indian Ocean have more than 50 genera. These figures are substantially higher than those used by Wells in 1954, and reflect a diversity comparable to that of the best-known West Pacific reefs. Major gaps remain, particularly in the area between the high-diversity zones of the West Pacific and the western Indian Ocean: very few records are available from the Indian Ocean shores of Sumatra and Java, from the Andamans, Nicobars, and Burma coast, or from Christmas

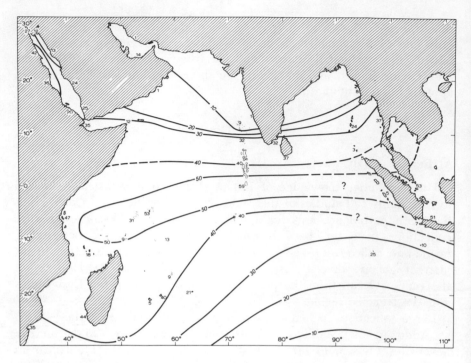

Fig. 6. Generic diversity of corals on Indian Ocean reefs: actual records. Contours show number of coral genera. [From Rosen (1971b).]

Island. Thus it is still uncertain whether or not the Seychelles–Maldives high-diversity area is continuous with that in the East Indies. We also need more information from the northern limits of the reefs, in Gujarat and the Persian Gulf, and from most of the remoter islands. The main pattern, however, has been established by Rosen; and further revision, particularly at the species level, will require much further systematic work.

Lack of reef coral diversity is therefore probably not a limiting factor in reef development on open-ocean reefs, though diversity gradients may be significant controls on continental shores. The southernmost coral reef on the African coast is at Inhaca, Mozambique (Macnae and Kalk, 1958), though scattered hermatypic corals occur as far south as Port Elizabeth in lat. 34°S (Day, 1969).

There is little information on the distribution of the non-Scleractinian corals. The Alcyonarian coral *Heliopora* is widespread in the Indian Ocean, though apparently absent in Houtman's Abrolhos (Thorpe, 1928), and is in places an important reef-builder, as is the hydrozoan

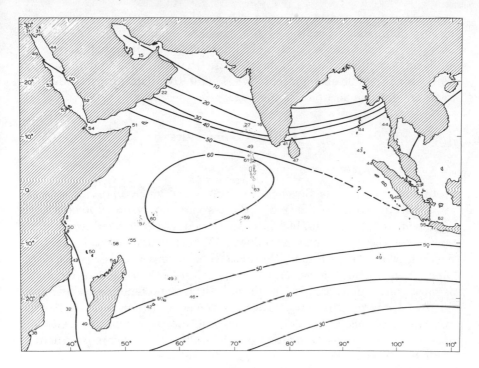

Fig. 7. Generic diversity of corals on Indian Ocean reefs: predicted levels. Contours show number of coral genera. [From Rosen (1971b).]

Millepora. The Alcyonarian soft corals *Sarcophyton* and *Lobophytum* are unimportant in the Maldives, Chagos, and western Indian Ocean reefs, but are numerous in the Nicobars and Andamans and are present in the Seychelles, perhaps indicating an important difference in the ecology of isolated reefs and of reefs adjacent to high islands.

B. LOCAL DIVERSITY

Largely because of technical problems, there is remarkably little information on coral abundance at any taxonomic level on individual reefs. Scheer (1967, 1969) has presented data on species diversity in quadrats of different sizes, from 0.1 to 100 m², at Rasdu Atoll and elsewhere in the Maldives. He found about 25 species in 10 m² at Rasdu, a figure comparable with that found at Addu by P. S. Davies *et al.* (1971). This diversity appears high by comparison with published data for reefs in the Torres Straits and at Arno Atoll in the Marshalls, but low when compared with Low Isles, Great Barrier Reef. The data are not, however, strictly comparable because of the range of quadrat sizes used in the surveys, nor do we have enough information on the spatial patterns of local variability on reefs to be sure that the quantitative surveys so far made are at all representative. Qualitative data on coral distributions are given in Sections VI and VII.

VI. Reef Communities

A. CORAL COMMUNITIES

Taylor (1968) in the Seychelles distinguished an *Acropora* community on exposed seaward reefs and a *Porites* community on protected reefs in channels. Rosen (1971a) has developed a generalized community model for the Mahé reefs as follows: "Where water movement is most vigorous, i.e., the surf zone, the most conspicuous coral genus is *Pocillopora*; where there is water movement, but not surf action, *Acropora* is most conspicuous; and where there is little or no tangible water movement, *Porites* predominates." He thus traces a sequence of coral communities on reefs, in response to changing energy conditions, vertically downward, transversely across, and laterally along the reef structure, with the three main communities appearing in the same order in each direction.

The *Pocillopora* assemblage is found on the edges of windward reefs, where robust massive and encrusting corals occupy about 15% of a surface largely covered by calcareous algae. The main corals are *Pocillopora* species of the *danae-meandrina* group, *Acropora humilis,* and *Millepora platyphylla,* with *Stylophora mordax* on the margins of surge channels.

The *Acropora* assemblage, typically developed round Mahé on less exposed reefs in Victoria Harbour, forms banks of living corals, mainly delicate ramose *Acropora* species (*A. formosa, A. irregularis*), with such associated corals as fungiids.

The *Porites* assemblage forms steep or vertical reef fronts in channels or hollows on the reefs, in protected situations. The reef structure is formed by massive corals such as *Porites lutea* and *Porites solida,* in colonies up to 5 m high and 3–4 m in diameter, with *Favia favus, Favia pallida, Goniastrea retiformis, Favites halicora, Platygyra lamellina,* and *Leptoria phrygia.* Delicate corals such as *Echinopora lamellosa* grow on this framework.

Rosen's scheme forms a useful device for organizing observational data, though it is probably too generalized to account for all features of community composition and distribution. In its ideal form, *Acropora* and *Porites* communities occur successively landward and with increasing depth from the reef edge *Pocillopora* community. In the Seychelles, however, an *Acropora* community is not found on reef flats, where the main corals are *Millepora exaesa* and *Montipora tuberculosa*. M. Pichon (1967) found an *Acropora* community on Mauritius reef flats, but also a *Pavona* community. P. S. Davies *et al.* (1971) on lagoon reefs at Addu found an *Acropora formosa* community both on the reef flats and on the reef slope, as suggested by Rosen, but many other corals are also present.

Though the *Acropora* community appears absent from the Mahé reef flats, however, a modified *Porites* community (with *Heliopora coerulea* and *Millepora exaesa*) is present. The *Porites* community is difficult to recognize on the protected reefs (channels and pools) described by M. Pichon (1964) from southwest Madagascar. Here the upper 2 m forms a zone of ramose corals (*Acropora pharaonis, Pocillopora damicornis, Millepora dichotoma, Stylophora* sp.), followed by a zone from 2 to 10 m with foliaceous and ramose corals (*Montipora foliosa, Pavona danai, Pachyseris* sp., *Millepora intricata*), and below 10 m, massive corals (species of *Platygyra, Leptoria, Porites, Goniopora, Favites,* and *Diploastrea*).

A major problem in applying Rosen's model to reefs is that (1) the reef structure itself affects water movement, but (2) the distribution of coral communities is at present often discordant with the limestone structures on which they are growing. Hence, while there is a general distinction between *Acropora* communities in agitated and *Porites* communities in quiet water, as recognized for example by Talbot (1965) in Tanzania and by Sewell (1922) in the Nicobar Islands, coral communities show great diversity and are affected by controls other than water movement alone.

B. Calcareous Algae

Reef-edge ridges of encrusting melobesioid algae are less prominent in the Indian Ocean than in the central and western Pacific. Both Crossland (1902, pp. 497–498) and Gardiner (1906, p. 456) commented on the sparse development of encrusting algae on western Indian Ocean reefs, but the latter was wrong in reporting their absence in the Seychelles. Though ridges are generally not prominent, encrusting reef-edge algae are widespread, though Setchell (1930) was misleading in his overemphasis of the role of algae in the Maldives and Chagos.

Low algal ridges with surge channels are found locally on Addu,

Peros Banhos, and Diego Garcia Atolls, comparable in extent to Pacific examples, and Taylor (1968) reports similar ridges from exposed reefs on Mahé, Seychelles. The atolls where ridges have so far been described are all in the equatorial belt marginal to the trades. More data are needed from areas of constant surf on trade-wind reefs, though observations on Aldabra, Farquhar, and Cosmoledo have so far not revealed extensive algal ridges.

The best-known and best-developed algal ridge features are found on the Grand Récif and the Songeritelo barrier reef, southwest Madagascar, protected from the prevailing winds but open to southerly swell. These include groove–spur systems, surge channels, creeks, tunnels and caves, enclosed pools and irregular reentrants in the reef front, all forming intricate environments with sharp environmental gradients. Corals are not important colonizers except for *Acropora* and *Millepora* species on the edges of channels, *Tubastrea aurea* and *Dendrophyllia elegans* on the walls, and ahermatypic corals (*Culicia* and *Phyllangia*) and hydrocorals (*Stylaster* and *Distichopora*) in deeper holes. The main colonizers are sponges, especially boring clionid sponges, foraminiferans, hydroids, and algae. In the groove–spur system the spurs are 50–100 m long and 5–10 m wide, and the grooves are 3 m wide and up to 5 m deep (Vasseur, 1964; Vacelet and Vasseur, 1965).

C. Sea-Grass Communities

In spite of their importance in sediment stabilization, little is known of the distribution of marine angiosperms on Indian Ocean reefs (Aleem, 1966), and den Hartog's (1970) recent monograph is based on collections made before the present expansion of scientific work. Rich sea-grass communities of the genera *Cymodocea, Thalassodendron, Diplanthera, Syringodium, Halophila, Thalassia,* and *Enhalus* are known from the East African coast, Madagascar, the Seychelles, and Aldabra. Extensive beds are found in south Indian and elsewhere on the Asian coast. Seagrasses are, however, poorly represented on central Indian Ocean reefs. Only two species have been reported from the Maldives, both for the first time in 1966. Gardiner (1907) referred to the absence of sea-grasses in the western Indian Ocean, but his statement needs revision and in the case of Diego Garcia is certainly incorrect. At least three genera are important on Mauritius reefs (M. Pichon, 1967).

Taylor and Lewis (1970) have discussed the sediments and faunas of marine sea-grass communities on Mahé, Seychelles, and M. Pichon (1964) those of Tuléar, Madagascar. Taylor notes that *Thalassodendron ciliatum* (= *Cymodocea ciliata*) is apparently absent from reef flats around granitic islands in the Seychelles, where it is replaced by *Thalassia hemprichii; T. ciliatum* is, however, common on the coral island of Coetivy, as well as on the East African coast, and the possibility of a sediment or substrate control on the distribution of sea-grasses needs further study.

Round Mahé the sea-grass communities form a zone 100–300 m wide on the inner parts of wider reef flats, covered by 2–5 m of water at highest tides. *Thalassia hemprichii* is the main species, together with *Syringodium isoetifolium, Halophila ovata, Enhalus acoroides,* and *Cymodocea rotundata.* The underlying sediments are mainly poorly sorted coral and algal coarse sands and gravels, subject to constant re-working by organisms. Holothurians and other echinoderms are abundant, with *Holothuria atra* reaching 320/100 m². There are patches of *Pavona frondifera, Psammocora contigua, Porites lutea,* and *Pocillopora damicornis,* but corals are not numerous.

The sea-grass communities at Tuléar resemble those of Aldabra in diversity, with 10 species of marine phanerogams in 6 genera. M. Pichon (1964) distinguishes an open grassland and a dense grassland. The former comprises *Thalassia, Cymodocea,* and *Diplanthera* with scattered massive corals such as *Psammocora gonagra, Porites somaliensis, Goniastrea retiformis, Platygyra daedalea, Favia,* and *Turbinaria.* Other characteristic organisms are the echinoderms *Culcita, Diadema,* and *Tripneustes,* the anemone *Stoichactis,* and the crab *Calappa.* In the dense turf, corals are less common. *Turbinaria ornata* grows in holes through the turf, and the characteristic echinoderms are *Linckia, Oreaster,* and *Acanthaster.*

Chassé (1962) and Price (1971) have discussed the possibility that there is a successional sequence involving marine phanerogams, algae, and either mangrove or salt-marsh communities.

D. MANGROVES

Mangrove communities are widespread on continental coasts round the Indian Ocean and Madagascar, and are particularly well developed in East Africa estuaries and on the mainland and high island coasts east of the Ganges delta (Macnae, 1968). The limits of mangrove communities shown in Fig. 1 refer to high mangrove woodland of geological significance: mangroves are found outside these limits, for example, in the Red Sea (Kassas and Zahran, 1967), the Persian Gulf (Kendall and Skipwith, 1969), and the Makran coast (Snead, 1967), but generally as scattered stunted trees. Even on the Indian coast, as near Bombay and Mandapam, mangrove communities may be replaced by salt-marsh communities.

Away from the continental coasts, mangroves are generally less well represented, though nine species were present in the now largely destroyed mangrove forests of the Seychelles (Sauer, 1967). Mangroves are also well developed on Aldabra, and in the Andamans and Nicobars. No mangroves are recorded from the Laccadives, five species from the Maldives (forming "great groves" on some atolls, according to Gardiner, but certainly unimportant on Addu), and two from the Chagos, where they are very uncommon, or, as on Diego Garcia, nonexistent. Two species are known from Mauritius (Sauer, 1961). Mangroves are absent on Cocos-Keeling, Christmas, and most of the smaller western Indian Ocean islands, including the Amirantes. In this respect the central oceanic atolls compare with the more remote Pacific groups, such as the Tuamotus. The mangrove communities have so far been only studied in detail on the continental coasts, Madagascar (Dérijard, 1965; Weiss, 1966), and on Aldabra (Macnae, 1971), and more work is needed of their role in promoting sedimentation on reefs and in chemically weathering reef limestones.

E. OTHER COMMUNITIES

Increasing work is being carried out on the faunas of mobile substrates, particularly of reef flat and beach sands and gravels (Thomassin, 1969; M. Pichon, 1967), and on patterns of intertidal zonation, particularly on hard substrates. Important zonation studies in reef intertidal environments are those of Taylor (1968, 1971b) on Mahé and Aldabra; Hodgkin and Michel (1963) and Baissac et al. (1962) on Mauritius; Plante (1964) at Tuléar; Hodgkin et al. (1959) in west Australia; Macnae and Kalk (1958) in Mozambique; and Lawson (1969) in Kenya and Tanzania.

VII. Reef Zonation

In the absence of quantitative data, reefs are best compared using descriptive zonation schemes based on physiographic and biological distribution patterns. The method is best suited to reefs showing pronounced zonation parallel to the reef edge, i.e., windward reefs. Leeward reefs often show no pronounced zonation either in morphology or biota, and are often discontinuous and irregular with deep enclosed pools. Zonation studies in the Indian Ocean are almost entirely restricted to reefs adjacent to continental or high-island coasts, e.g., Mauritius, Seychelles, Madagascar. The atoll reefs of the open ocean are still poorly known, though it is possible to infer from earlier accounts that the reefs of Cocos-Keeling, the Chagos, and the Maldives are broadly similar to the reefs here described from Addu Atoll. Only four areas are described here, from areas with spring tidal ranges of 0.5–3 m. The zonal categories used are not necessarily equivalent between reefs, though the similarities are apparent. The reefs described may be compared with others recently described from the Mascarenes (Montaggioni, 1970; Faure and Montaggioni, 1970, 1971), and those of Aldabra, East Africa, the Red Sea, and India described at the 1970 London Symposium (Stoddart and Yonge, 1971).

A. TULÉAR, MADAGASCAR

The Grand Récif of southwest Madagascar has been described in detail by M. Pichon (1964), Blanc *et al.* (1966), and Thomassin (1969). The barrier reef itself is 1.5–2 km wide. Although it is located on a lee shore, it is nevertheless exposed to southerly swell and has many of the features, such as an algal ridge and groove–spur system, characteristic of windward reefs. The tidal range at springs is 3 m and at neaps 1.5 m. The zonation on the barrier is shown in Fig. 8, and the following zones are recognized:

1. Outer slope with corals. This slopes downward from about 20 m, and has not been studied in detail.

2. Outer reef wall, a vertical wall 10–18 m high with its upper edge slightly below the level of low water springs. Three coral zones are found. In the upper zone much of the surface is covered with melobesioid algae and *Millepora*, and corals are represented by species of *Acropora*, *Stylophora*, and *Pocillopora*. In the middle zone, between 4 and 13 m, coral cover is more extensive and diverse, with foliaceous, encrusting, and some massive species. Genera represented include *Pachyseris*, *Echinopora*, *Platygyra*, *Leptoria*, *Montipora*, *Hydnophora*, *Merulina*, *Leptoseris*, *Favia*, *Acropora*, and *Diploastrea*. The zoanthid *Palythoa* and the soft coral *Lobophytum* are also present. In the lower zone, below 13 m, there are large colonies of massive corals, including *Diploastrea heliopora*, *Porites somaliensis*, species of *Favia*, *Favites*, and *Lobophyllia*, and other species from the zones above. Such a vertical outer slope has not been described from other Indian Ocean reefs.

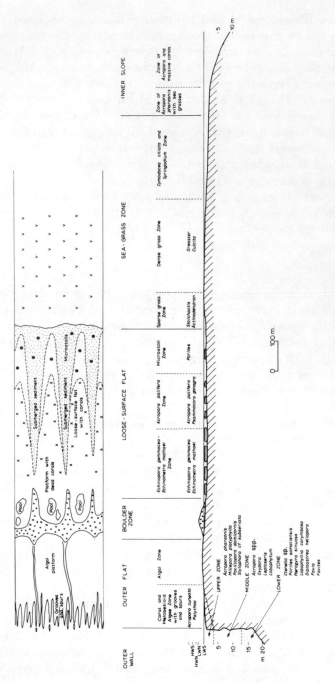

Fig. 8. Zonation of the Grand Récif, Tuléar, Madagascar. [After Pichon (1964) and Caye and Thomassin (1967).]

3. Outer flat, 100–150 m wide. The seaward half of the outer flat comprises the groove-and-spur zone, coated with melobesioid algae, and with encrusting corals, soft corals and sponges growing in the grooves and tunnels formed by algal growth. This zone has been described in detail by Vasseur (1964). Corals include *Acropora cuneata* and *Goniastrea*, with *Palythoa* and *Lobophytum*. The inner part of the outer flat consists of an algal pavement with few corals.

4. The boulder zone is 40–80 m wide and stands 0.6–1.5 m above the general level of the flat.

5. The "loose-surface flat" (*platier friable*) is about 0.5 km wide and carries up to 1 m of water even at low tide. It consists of three zones of corals, all sharply beveled off close to the level of low water springs: (a) a zone of *Echinopora gemmacea*, with *Goniastrea retiformis*, *Pocillopora*, *Stylophora*, *Pavona*, and *Leptoria*; (b) a zone of *Acropora palifera*, with *Psammocora gonagra*; and (c) a zone of microatolls of *Porites*, with *Turbinaria ornata*.

6. Sea-grass zone. The inner reef flat, about 0.7 km wide, in part drying at low water springs, is covered with marine phanerogams. Ten species are represented; the most dense beds are formed by *Thalassia*, with *Cymodocea* and *Syringodium* as the depth increases toward the lagoon.

7. Inner slope. This is dominated by species of *Acropora*, mainly *A. pharaonis*, with massive corals becoming more important with increasing water depth. These include species of *Favia*, *Goniastrea*, *Lobophyllia*, and *Porites*.

B. MAHÉ, SEYCHELLES

The fringing reefs on the east side of Mahé, one of the granitic Seychelles, vary in width from 100 to nearly 2000 m. They lie on the windward side of the island, but in the north they are protected by outlying islands. The more exposed reefs have weakly developed algal ridges, but groove–spur systems are less well developed than at Tuléar. Tidal range is approximately 2 m at springs and 1 m at neaps. The reefs have been studied in detail by Taylor (1968) and Lewis (1968, 1969). The following zones may be distinguished (Fig. 9):

1. Outer slope. Coral growth is active over the upper 6 m, and corals decrease rapidly below 10 m in depth. An *Acropora* community is dominant on exposed coasts, with *Acropora pharaonis*, *A. divaricata*, *A. irregularis*, *A. palifera*, and *A. humilis*, with *Seriatopora caliendrum* and *Fungia fungites*. Massive corals are more common lower on the

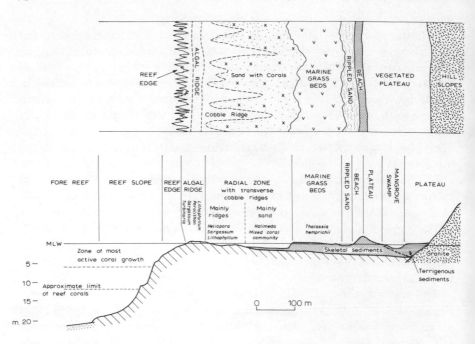

Fig. 9. Zonation of a windward reef, Mahé, Seychelles, generalized from the reef at Anse aux Pins. [After Lewis (1968, 1969), Taylor (1968), and Lewis and Taylor (1966).]

slope, represented by species of *Porites*, *Favia*, and *Leptoria*. Lewis (1968) recognized a series of buttresses from echograms at 0.25–0.5 km from the reef edge; these features are about 8 m high and about 200 m apart, and are clearly not analogous to normal groove–spur features.

2. Reef edge. This forms a zone 20–40 m wide with a gentle slope to seaward. Low encrusting corals are common, and include *Millepora dichotoma*, *Pocillopora danae*, *P. meandrina*, *Acropora digitifera*, *Goniastrea pectinata*, together with *Millepora dichotoma*, *Tubipora musica*, *Heliopora coerulea*, *Lobophytum*, *Sarcophyton*, *Xenia*, and *Palythoa*. Much of the surface is covered with encrusting calcareous algae.

3. Algal ridge. This is 10–20 m wide, rises up to 0.5 m above the general level of the flat and dries at low-water springs. It is intersected by surge channels, some of which are roofed over by growth. The coral fauna is diverse but individual corals are small; they include *Stylocoeniella armata*, *Hydnophora microconos*, *Goniastrea pectinata*, *Montipora*, *Cyphastrea*, *Porites*, and *Acanthastrea*. Note that there is no boulder zone on the Mahé reefs, and virtually no reef blocks.

4. Radial zone. This name is given to a zone where ridges normal to the reef edge and merging seaward with the algal ridge alternate with sand-filled troughs. The ridges, 2–3 m high, narrow toward the shore, and the troughs 1–2 m deep, widen and deepen toward the shore. The ridges consist of coral and algal cobbles or of reef-platform material; *Sargassum* and *Turbinaria* cover a large proportion of the surface, and the green alga *Halimeda opuntia* is also common. Large coral colonies are found in the sand-floored troughs. They include large branching colonies of *Acropora pharaonis, Porites divaricata*, and *Porites nigrescens; Pocillopora damicornis; Heliopora coerulea;* and in deeper water massive colonies of *Porites, Goniastrea, Favia*, and *Platygyra*.

5. Marine grass beds. These are dominated by *Thalassia hemprichii*. They vary from 100 to 300 m wide, and are absent on narrower sectors of the reefs. The margins of the beds are characteristically irregular, especially to seaward, where they are bounded by a step 15–20 cm high. Much of this zone dries at low-water springs, but is covered with 2–5 m of water at high tide.

6. Rippled sand zone. This is a narrow belt of strongly rippled mobile sand, 5–10 m wide, between the grass beds and the beach.

7. Beach. Round Mahé this is 15–25 m wide and 2.5–3.5 m high, mainly consisting of carbonate sands with admixture of quartz sands from the granites. Taylor (1968) describes it as the site of a *Donax–Ocypode* community, characterized by the bivalves *Donax cuneatus, D. fabra*, and *Atactodea glabrata* and the ghost crab *Ocypode ceratophthalma*. The area immediately landward from the beach, generally 2–3 m high, is locally known as "plateau," and between it and the upland slopes there is often a swampy area.

C. MAURITIUS

The Mauritius reefs are very variable, being best developed in the southeast, where they are described by M. Pichon (1967). Their most unusual distinguishing characteristic is the narrowness of the reef flat, which may be only 20 m wide, bounding a wide shallow lagoon a few hundred meters wide. Spring tidal range is 0.5 m. There are relict patches of raised reefs formed during the last interglacial period at intervals around the coast. The Mauritius reefs may be compared with the reefs of neighboring Réunion (Montaggioni, 1970, Faure and Montaggioni, 1970) and Rodriguez (Faure and Montaggioni, 1971); they show considerable differences both in physiography and species composition. Pichon in Mauritius distinguishes the following zones (Fig. 10):

1. Outer slope, below about 5 m; this has not been studied in detail.

2. Groove–spur zone, with grooves 1.5–3 m deep, with numerous corymbose and encrusting corals, including *Pocillopora, Millepora platy-phylla, Montipora, Hydnophora, Oxypora, Echinopora,* and *Leptoria,* and massive *Favia, Favites, Alveopora,* and *Porites.* The soft corals *Sarcophyton* and *Lobophytum* are also present.

3. Reef flat, 20 m or more wide. This is encrusted with *Porolithon onkodes,* but a number of corals are represented, chiefly *Platygyra, Leptoria, Pocillopora, Acropora, Favites,* and *Favia.*

4. Boulder zone on the inner slope of the reef flat. *Sargassum* sp. and *Turbinaria ornata* are the main benthic organisms in this zone, which is a few tens of meters wide.

5. *Acropora* zone. This forms the outer part of the shallow lagoon, and is of variable width. The main coral species is *Acropora ?pharaonis,* with *Galaxea fascicularis, Montipora foliosa,* and *Fungia.*

6. *Pavona* zone, dominated by a foliaceous *Pavona, P.* cf. *divaricata,* with locally up to 100% cover, is also variable in extent. In deeper water *P. decussata* becomes more important. Tongues of sand normal to the shore become wider seawards, and both *Turbinaria* and *Sargassum* colonize these and dead coral heads.

7. Sea-grass zone, occupies the inner part of the shallow lagoon, with a width generally of 100–200 m. *Syringodium isoetifolium* is dominant in shallower areas, and *Cymodocea ciliata* in deeper water; near the shore *Diplanthera* becomes more common. There are scattered corals in the grass beds, mainly *Millepora, Pocillopora, Porites,* and *Pavona.*

8. Beach and near-shore sand.

D. Addu Atoll, Maldive Islands

Addu Atoll is the only open-ocean reef so far studies in detail, by Scheer (1969) and by P. S. Davies *et al.* (1971). Though close to the equator, its reefs are affected by swell from the south, and there is a well-marked distinction between windward and leeward reefs. The reefs here described are a windward seaward reef (Fig. 11) and a lagoon reef, also on the south side of the atoll (Fig. 12). Tidal range is 1 m at springs.

1. Seaward Reef

The seaward reef on the south side of the atoll is exceptionally wide, reaching 1500–2000 m, and for most of its width it is a smooth planed-

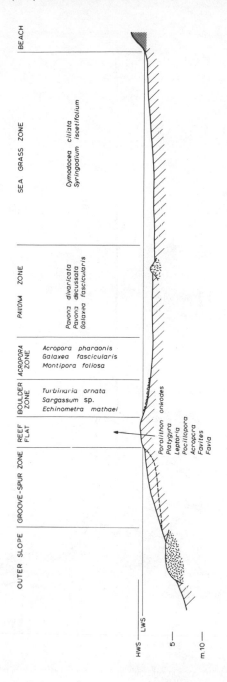

Fig. 10. Zonation of a windward reef, near Mahébourg, Mauritius. [After Pichon (1967).]

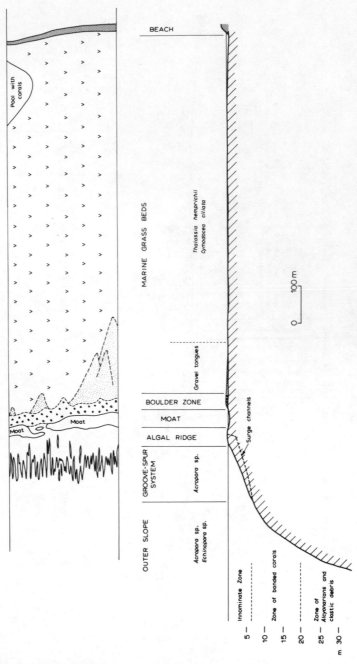

Fig. 11. Zonation of a seaward reef, Addu Atoll, Maldive Islands. [After P. S. Davies *et al.* (1971).]

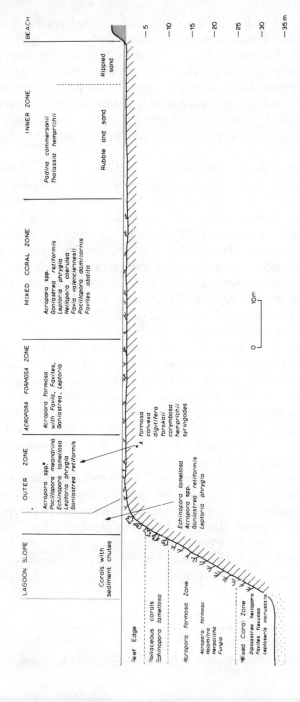

Fig. 12. Zonation of a lagoon reef, Addu Atoll, Maldive Islands. [After P. S. Davies *et al.* (1971).]

rock surface, drying at low spring tides, and covered with a thin sheet of sediment. The following zones are apparent:

1. Outer slope. Because of the surf this is known only imperfectly, from observations at depths of 12–45 m. At 25 m there is no trace of a groove–spur system, but corals are arranged in lineations normal to the reef edge. Branching and massively palmate *Acropora* species are the main corals, with foliaceous *Echinopora*. Alcyonarians become important in deeper water, where much of the floor is covered with mobile sand and gravel.

2. Groove–spur zone, 100–300 m wide on the south side of the atoll, and extending to a depth of about 12 m. Palmate *Acropora* species grow on the spurs, but little else is known of this zone.

3. Algal ridge, 30–50 m wide, and covered with encrusting *Porolithon*. The ridge generally rises about 0.5 m above the reef flat immediately landward of it. In some places surge channels cut back into the ridge surface from the groove–spur zone for distances of 25–35 m, and are up to 2 m deep at their heads; elsewhere the ridge is undissected. Small corymbose colonies of *Acropora, Pocillopora,* and *Millepora* grow on the sides of the channels, and benthic algae grow luxuriantly near their lips and heads (*Codium, Halimeda, Jania, Pocockiella*).

4. Moat, 30 cm lower than the main level of the flat, 20–70 m wide, and containing water even at low spring tides. It is floored with mobile rubble, with epilithic algae (*Pocockiella, Schizothrix, Cladophoropsis;* with *Jania, Halimeda,* and *Turbinaria*). Corals cover less than 5% of the floor, and include *Porites, Pocillopora, Acropora,* and *Millepora.*

5. Boulder zone, a narrow zone with boulders up to 1 m in diameter but generally much smaller, prolonged landward by thin gravel tongues of different ages, usually 50 m long but exceptionally up to 250 m long.

6. Marine grass beds on the main reef platform, extending for more than 1000 m. Poorly sorted sands and gravels overlie a rock surface, and these are colonized by a turf of *Thalassia hemprichii* and *Cymodocea ciliata.* Much of this turf dries at low-water spring tides. In deeper areas, especially in gaps between islands, there are large coral colonies, especially of *Porites* and *Heliopora,* forming microatolls, the tops of which are colonized by *Turbinaria ornata.*

7. Beach.

2. Lagoon Reef

By contrast the lagoon reef is only about 100 m wide; it is submerged at all stages of the tide; and it has luxuriant growth of coral. The main zones are:

1. Lagoon slope, which extends to the lagoon floor at 30–35 m. The upper 10 m are dominated by foliaceous corals, mainly *Echinopora lamellosa;* between 10 and 25 m *Acropora formosa* is dominant, though much of it is dead, and other corals include *Fungia, Herpolitha limax,* and *Halomitra philippinensis;* and below 25 m there is a mixed coral zone, with large sheets of *Diploastrea heliopora,* branching *Acropora syringodes,* more massive *Favites flexuosa,* and foliaceous *Leptoseris incrustans.*

2. Reef edge, extending to a depth of 3 m. This is a zone of vigorous coral growth, mainly dominated by foliaceous and branching corals. The main species are *Echinopora lamellosa, Acropora convexa, A. hyacinthus,* and *A. corymbosa.*

3. Outer zone. A zone 15–20 m wide with a mixed coral association covering up to 60% of the surface. Branching and foliaceous corals are important (*Acropora formosa, A. convexa, A. digitifera, Pocillopora meandrina, Echinopora lamellosa*) but massive corals such as *Leptoria phrygia* and *Goniastrea retiformis* are also present.

4. *Acropora formosa* zone, about 20 m wide, with a coral cover of 60–70%. *A. formosa* itself forms about 70% of the coral cover. Other corals present include *Galaxea, Psammocora, Favites abdita,* and *Favia speciosa.*

5. Mixed coral zone, 30 m wide, with branching, foliaceous, and massive corals increasing in cover toward the lagoon. Species present are *Acropora digitifera, A. convexa, A. formosa, Pocillopora damicornis, Leptoria phrygia, Goniastrea retiformis,* and *Heliopora.*

6. Inner zone, a zone 35 m wide with no living corals but with dead colonies in very turbid water. The bottom is covered with mobile sand, with patches of *Thalassia hemprichii* and, near the beach, *Padina commersonii* and other benthic algae.

VIII. Reef Morphology and Sea-Level Change

A. ANOMALOUS REEF FEATURES

Three sets of features may be considered anomalous in terms of present reef growth and processes: elevated reefs, submerged reefs, and certain features of reef surfaces now close to present sea level.

Elevated reefs are well developed in the western Indian Ocean and along the margins of the Red Sea. They are absent from the central Indian Ocean (except for a very recent raised reef in south India), but raised reefs and reef-associated sediments are found in the eastern

Indian Ocean, in the Andamans (Sewell, 1925), Nicobars (Sewell, 1922), western Australia, Houtman's Abrolhos, and Christmas Island. Though widely developed, their absence from the central Indian Ocean throws doubt on any purely eustatic explanation of their origin. Elevated inter-tidal notches in reef limestones have also been described from the Red Sea by Guilcher (1955) and correlated with similar features described by Fairbridge (1948a,b) from western Australia: raised notches are, however, absent from other limestone coasts, for example, at Aldabra and in East Africa, which makes eustatic explanations suspect.

The main submerged reefs are the extensive submarine banks of the Mascarene Plateau and the Chagos Archipelago, which commonly show marked bevelings at levels of 33–90 and 8–20 m depth. Platforms beneath modern reefs are also very evident in the northern Maldives at depths of 50–75 m, and minor bevels have been described elsewhere, e.g., at 8–15 and 18–22 m in the Seychelles (Lewis, 1968).

Anomalous features of reef flats include drying intertidal surfaces too high for modern coral growth and of probably erosional origin; emerged and lithified coral conglomerates, some of which have been described as emerged reef rocks, and reef islands and raised beaches. In spite of their ambiguity, all of these have been used as evidence of former slightly higher sea levels, and workers such as Fairbridge (1948a) in Houtman's Abrolhos and Weydert (1969) at Tuléar, Madagascar, have interpreted them in terms of eustatic sea-level changes. Gardiner (1931) and Sewell (1935) both assumed that raised reefs and reef islands indi-cate emergence following a postglacial high stand of the sea, but in general little attention has been paid to the widespread submerged features.

B. Sea-Level History

The history of the most recent rise in sea level is now well established with the aid of radiocarbon dating. From 15,000 to 8000 years B.P. sea level rose throughout the world at approximately 1.0–1.25 cm yr^{-1}. There is growing evidence for a continued but slower rise of sea level, at from 0.03 to 0.08 cm yr^{-1}, over the last 4000 years. There is little confirmation in recent work for the proposed series of Holocene high sea levels and stillstands, at up to 5 m above the present level, originally defined from the southeast Indian Ocean and later applied throughout the world by Fairbridge (1961). Fairbridge's scheme is, however, often used to explain anomalous features of reefs (e.g., Weydert, 1969; Lewis, 1968), generally without consideration of the process rates involved.

The last glacial low sea level reached —120 m at 15,000 years B.P., and the sea then rose fairly rapidly to —20 m at about 7000 years B.P. Modern reef growth related to present sea level must have begun, therefore, about 5000 years ago. There are nine radiocarbon dates for Indian Ocean samples which suggest slightly higher sea levels than the present during this period, but it is still not possible to confirm the existence of a Holocene transgression.

There is some radiocarbon evidence for a stillstand up to +5 m about 30,000 years ago, and uranium dating of raised reef limestones indicates extensive reef-building near present sea level between 70 and 160×10^3 years. These data are reviewed by Stoddart (1971). Thus it is unlikely that reef features formed during the last 5000 years could compare in magnitude with those formed during these late Pleistocene sea-level stands, or with those formed during earlier Pleistocene times. The complexity of the Pleistocene in the reef seas is only now being appreciated, and paleotemperature data for the equatorial Indian Ocean indicate that here, at least, sea temperatures did not fall sufficiently during glacial maxima to inhibit reef growth (Oba, 1969). From work on modern reefs it is apparent that much present-day reef growth only patchily veneers older and more extensive structures. Thus the outstanding problem in Indian Ocean reef studies is to determine the relative importance of past and present processes in the development of reef features, both in areas where modern reefs are flourishing, and in those where they are not.

References

Agassiz, A. (1903). *Mem. Mus. Comp. Zool. Harvard* **29**, 1.
Alagarswami, K., Lal Mohan, R. S., James, D. B., and Appukuttan, K. K. (1968). *Bull. Cent. Mar. Fish. Res. Inst.* **4**, 1.
Aleem, A. A. (1966). *Abstr. Pap., Int. Oceanogr. Congr., 2nd, 1966* p. 6.
Alexander, C. S. (1968). *Z. Geomorphol. Suppl.* [N.S.] **7**, 133.
Andrews, C. W. (1900). "A Monograph of Christmas Island (Indian Ocean)." Trustees of the British Museum (Natural History), London.
Baissac, J. de B., Lubet, P. E., and Michel, C. M. (1962). *Rec. Trav. Sta. Mar. Endoume* **39**, 239.
Baker, B. H. (1963). *Mem. Geol. Surv. Kenya* **3**, 1.
Baker, B. H., and Miller, J. A. (1963). *Nature (London)* **199**, 346.
Barnes, J., Bellamy, D. J., Drew, E. A., Jones, D. J., Kenyon, L., Lythgoe, J., Rosen, B., and Whitton, B. A. (1971). *Symp. Zool. Soc. London* **28**, 87.
Battistini, R. (1964). "Etude géomorphologique de l'extrême-sud de Madagascar." Editions Cujas, Paris.
Battistini, R. (1965). *Madagascar Rev. Geogr.* **7**, 1.

Battistini, R. (1966a). *Mem. Mus. Hist. Natur., Paris, Ser. A* [N.S.] **41**, 7.

Battistini, R. (1966b). *Bull. Ass. Fr. Et Quat.* **3**, 191.

Battistini, R. (1969). *Bull. Ass. Fr. Etude Quat.* **6**, 229.

Berthois, L., and Battistini, R. (1969). *Madagascar Rev. Geogr.* **15**, 7.

Biewald, D. (1964). *Z. Geomorphol.* [N.S.] **8**, 351.

Blanc, J. J., Chamley, H., and Froget, C. (1966). *Rev. Trav. Sta. Mar. Endoume, Fasc. Hors-Ser., Suppl.* **5**, 24.

Braithwaite, C. J. R. (1971). *Symp. Zool. Soc. London* **28**, 39.

Brander, K. M., Humphreys, W. F., and McLeod, A.A.Q.R. (1971). *Symp. Zool. Soc. London* **28**, 397.

Caye, G., and Thomassin, B. (1967). *Rec. Trav. Sta. Mar. Endoume, Fasc. Hors-Ser., Suppl.* **6**, 25.

Chassé, C. (1962). *Rec. Trav. Sta. Mar. Endoume, Fasc. Hors-Ser., Suppl.* **1**, 237.

Coppinger, R. W. (1884). "Report on the Zoological Collections Made in the Indo-Pacific Ocean During the Voyage of H.M.S. 'Alert' 1881–2." Trustees of the British Museum (Natural History), London.

Coutière, H. (1898). *Bull. Mus. Hist. Natur., Paris* **4**, 87 and 155.

Crossland, C. (1902). *Proc. Cambridge Phil. Soc.* **11**, 493.

Crossland, C. (1904). *Proc. Cambridge Phil. Soc.* **12**, 36.

Crossland, C. (1907). *J. Linn. Soc. London, Zool.* **31**, 14.

Darwin, C. R. (1842). "The Structure and Distribution of Coral Reefs." Smith, Elder & Co., London.

Davies, D., and Francis, T. J. G. (1964). *Deep-Sea Res.* **11**, 921.

Davies, P. S., Stoddart, D. R., and Sigee, D. (1971). *Symp. Zool. Soc. London* **28**, 217.

Davis, W. M. (1918). *J. Geol.* **26**, 198, 289, and 385.

Day, J. H. (1969). "A Guide to Marine Life on South African Shores." A. A. Balkema, Cape Town.

den Hartog, C. (1970). *Verh. Kon. Ned. Akad. Wetenschap., Afd. Natuurk.* **59**, 1.

Dérijard, R. (1965). *Rec. Trav. Sta. Mar. Endoume, Fasc. Hors-Ser., Suppl.* **3**, 1.

Ehrenberg, C. R. (1834). *Abh. Kgl. Akad. Wiss. Berlin, Phys. Kl.* p. 381.

Einsele, G., Genser, H., and Werner, F. (1967). *Senckenbergiana Lethaea* **48**, 359.

Emery, K. O. (1956). *Bull. Amer. Ass. Petrol. Geol.* **40**, 2354.

Evans, G. (1966). *Phil. Trans. Roy. Soc. London, Ser. A* **259**, 291.

Fairbridge, R. W. (1948a). *J. Roy. Soc. West. Aust.* **33**, 1.

Fairbridge, R. W. (1948b). *J. Roy. Soc. West. Aust.* **34**, 35.

Fairbridge, R. W. (1961). *Phys. Chem. Earth* **4**, 99.

Faure, G., and Montaggioni, L. (1970). *Rec. Trav. Sta. Mar. Endoume, Fasc. Hors-Ser., Suppl.* **10**, 271.

Faure, G., and Montaggioni, L. (1971). *In* "Symposium on the Indian Ocean and Adjacent Seas (Cochin)," Mar. Biol. Ass., India.

Fisher, R. L., Johnson, G. L., and Heezen, B. C. (1967). *Geol. Soc. Amer., Bull.* **78**, 1247.

Forbes, H. O. (1879). *Proc. R. Geogr. Soc.* **1**, 777.

Francis, T. J. G., and Shor, G. G., Jr. (1966). *J. Geophys. Res.* **71**, 427.

Friedman, G. M. (1968). *J. Sediment. Petrol.* **38**, 895.

Fryer, J. C. F. (1911). *Trans. Linn. Soc. London* (2) **14**, 397.

Gardiner, J. S. (1901). *Proc. Cambridge Phil. Soc. Math. Phys. Sci.* **11**, 22.

Gardiner, J. S. (1903–1906). "The Fauna and Geography of the Maldive and Laccadive Archipelagoes, Being the Account of the Work Carried on and of Collections Made by an Expedition During the Years 1899 and 1900." Cambridge Univ. Press.

Gardiner, J. S. (1906). *Geogr. J.* **28**, 313 and 454.

Gardiner, J. S. (1907). *Trans. Linn. Soc. London* (2) **12**, No. 1, 111.

Gardiner, J. S. (1907–1936). *Trans. Linn. Soc. London* (2) **12-19**.

Gardiner, J. S. (1931). "Coral Reefs and Atolls." Macmillan, New York.

Gardiner, J. S. (1936). *Trans. Linn. Soc. London* (2) **19**, 393.

Garth, J. S. (1971). *In* "Symposium in the Indian Ocean and Adjacent Seas (Cochin)," Mar. Biol. Ass., India.

Gibson-Hill, C. A. (1948). *J. Malaya Branch Roy. Asiatic Soc.* **21**, 68.

Gibson-Hill, C. A. (1950). *Bull. Raffles Mus.* **22**, 1.

Glennie, E. A. (1936). *Sci. Rep. John Murray Exped. 1933–1934* **1**, 95.

Goreau, T. F. (1904). *Bull. Sea Fish. Res. Sta. Isr.* **35**, 23.

Gravier, C. (1911). *Ann. Inst. Oceanogr. (Paris)* **2**, 1.

Guilcher, A. (1955). *Ann. Inst. Oceanogr. (Paris)* [N.S.] **30**, 55.

Guilcher, A. (1956). *Ann. Inst. Oceanogr. (Paris)* [N.S.] **33**, 64.

Guilcher, A. (1958). *Mem. Inst. Sci. Madagascar, Ser. F* **2**, 89.

Guilcher, A. (1960). *Bull. Soc. Geol. Fr.* [7] **1**, 337.

Guilcher, A. (1971). *Symp. Zool. Soc. London* **28**, 65.

Guilcher, A., Berthois, L., Battistini, R., and Fourmanoir, P. (1958). *Mem. Inst. Sci. Madagascar, Ser. F* **2**, 117.

Guilcher, A., Berthois, L., Le Calvez, Y., Battistini, R., and Crosnier, A. (1965). *Mem. ORSTOM (Off. Rech. Sci. Tech. Outre-Mer)* **11**, 1.

Guppy, H. B. (1889). *Scot. Geogr. Mag.* **5**, 281, 457, and 569.

Hackett, H. E. (1969). *Proc. Inst. Seaweed Symp., 6th, 1968* p. 187.

Harrison, R. (1911). *Proc. Zool. Soc. London* p. 1018.

Hass, H. (1965). "Expedition into the Unknown: A Report on the Expedition of the Research Ship *Xarifa* to the Maldive and the Nicobar Islands and on a Series of 26 Television Films" (Transl. by G. Edwards). Hutchinson, London.

Heezen, B. C., and Tharp, M. (1966). *Phil. Trans. Roy. Soc. London, Ser. A* **259**, 137.

Hodgkin, E. P., and Michel, C. (1963). *Proc. Roy. Soc. Arts Sci. Mauritius* **2**, 121.

Hodgkin, E. P., Marsh, L., and Smith, G. G. (1959). *J. Roy. Soc. West. Aust.* **42**, 85.

Kassas, M., and Zahran, M. A. (1967). *Ecol. Monogr.* **37**, 297.

Kendall, C. G. St. C., and Skipwith, P. A. D'E. (1969). *Geol. Soc. Amer., Bull.* **80**, 865.

Kinsman, D. J. J. (1964). *Nature (London)* **202**, 1280.

Klausewitz, W. (1967). *"Meteor" Forschungsergeb. Reihe D* **2**, 44.

Klunzinger, C. B. (1877–1879), "Die Korallthiere des rothers Meeres." Verlag der Gutmann'schen Buchhandlung, Berlin.

Kohn, A. J. (1964). *Atoll Res. Bull.* No. 101, p. 1.

Kohn, A. J. (1967). *Amer. Natur.* **101**, 251.

Kohn, A. J. (1968). *Ecology* **49**, 1046.

Laughton, A. S. (1966). *Phil. Trans. Roy. Soc. London, Ser. A* **259**, 150.

Lawson, G. W. (1969). *Trans. Roy. Soc. S. Afr.* **38**, 329.

Le Pichon, X., and Heirtzler, J. R. (1968). *J. Geophys. Res.* **73**, 2101.

Lewis, M. S. (1968). *J. Geol.* **76**, 140.

Lewis, M. S. (1969). *Mar. Geol.* **7**, 95.

Lewis, M. S., and Taylor, J. D. (1966). *Phil. Trans. Roy. Soc. London, Ser. A* **259**, 279.

Loya, Y., and Slobodkin, L. B. (1971). *Symp. Zool. Soc. London* **28**, 117.

McElhinny, M. W. (1970). *Nature (London)* **228**, 977.

Macfadyen, W. A. (1930). *Geogr. J.* **75**, 27.

McIntire, W. G., and Walker, H. J. (1964). *Ann. Ass. Amer. Geogr.* **54**, 582.

Macnae, W. (1968). *Advan. Mar. Biol.* **6**, 73.

Macnae, W. (1971). *Phil. Trans. Roy. Soc. London, Ser. B* **260**, 237.

Macnae, W., and Kalk, M. (1958). "A Natural History of Inhaca Island, Mocambique." Witwatersrand Univ. Press, Johannesburg.

Matthews, D. H., and Davies, D. (1966). *Phil. Trans. Roy. Soc. London, Ser. A* **259**, 227.

Maxwell, W. G. H. (1968). "Atlas of the Great Barrier Reef." Elsevier, Amsterdam.

Mergner, H. (1967). *Z. Morphol. Oekol. Tiere* **60**, 35.

Mergner, H. (1971). *Symp. Zool. Soc. London* **28**, 141.

Montaggioni, L. (1970). *C. R. Acad. Sci.* **270**, 663.

Nair, R. V. R., and Pillai, C. S. G. (1969). *In* "Symposium in Corals and Coral Reefs (*Mandapam Camp*)," Mar. Biol. Ass., India.

Nesteroff, W. (1955). *Ann. Inst. Oceanogr. (Paris)* **30**, 1.

Niino, H., and Oshite, K. (1966). *Rec. Oceanogr. Works Jap.* [N.S.] **8**, 27.

Oba, T. (1969). *Sci. Rep. Tohoku Univ., Ser. 2* **41**, 129.

Oldham, C. F. (1895). *J. Asiatic Soc. Bengal* **64**, 1.

Ormond, R. F. Q., and Campbell, A. C. (1971). *Symp. Zool. Soc. London* **28**, 433.

Ortmann, A. E. (1889). *Zool. Jahrb.* **4**, 493.

Ortmann, A. E. (1892). *Zool. Jahrb.* **6**, 631.

Pichon, M. (1964). *Rec. Trav. Sta. Mar. Endoume, Fasc. Hors-Ser., Suppl.* **2**, 79.

Pichon, M. (1967). *Cah. ORSTOM (Off. Rech. Sci. Tech. Outre-Mer), Ser. Oceanogr.* **5**, 31.

Pichon, M. (1971). *Symp. Zool. Soc. London* **28**, 185.

Pichon, M. (1967). *Rec. Trav. Sta. Mar. Endoume, Fasc. Hors-Ser., Suppl.* **7**, 57.

Pillai, C. S. G. (1967). Thesis, University of Kerala.

Pillai, C. S. G. (1971). *Symp. Zool. Soc. London* **28**, 301.

Plante, R. (1964). *Rec. Trav. Sta. Mar. Endoume, Fasc. Hors-Ser., Suppl.* **2**, 205.

Price, J. H. (1971). *Phil. Trans. Roy. Soc. London, Ser. B* **260**, 123.

Rice, M. E. (1971). *In* "Symposium on the Indian Ocean and Adjacent Seas (Cochin)," Mar. Biol. Ass., India.

Rodolfo, K. S. (1969). *Geol. Soc. Amer., Bull.* **80**, 1203.

Rosen, B. R. (1971a). *Symp. Zool. Soc. London* **28**, 163.

Rosen, B. R. (1971b). *Symp. Zool. Soc. London* **28**, 263.

Rosen, B. R. (1971c). *Atoll. Res. Bull.* No. 149, p. 1.

Rossi, L. (1954). *Riv. Biol. Colon.* **14**, 23.

Sauer, J. D. (1961). "Coastal Plant Geography of Mauritius." Louisiana State Univ. Press, Baton Rouge.

Sauer, J. D. (1967). "Plants and Man on the Seychelles Coast." Univ. of Wisconsin Press, Madison.

Schäfer, W. (1967). *Senckenbergiana Lethaea* 48, 107.

Scheer, G. (1964). *Zool. Jahrb. Syst.* 91, 451.

Scheer, G. (1967). *Senckenbergiana Biol.* 48, 421.

Scheer, G. (1969). *In* "Symposium on Corals and Coral Reefs (Mandapam Camp)," Mar. Biol. Ass., India.

Scheer, G. (1971). *Symp. Zool. Soc. London* 28, 329.

Setchell, W. A. (1930). *Proc. Pac. Sci. Congr., 4th, 1929* vol. 3, p. 265.

Sewell, R. B. S. (1922). *J. Bombay Natur. Hist. Soc.* 28, 970.

Sewell, R. B. S. (1925). *Mem. Asiatic Soc. Bengal* 9, 1.

Sewell, R. B. S. (1932). *Geogr. J.* 79, 449.

Sewell, R. B. S. (1935). *Mem. Asiatic Soc. Bengal* 9, 461.

Sewell, R. B. S. (1936a). *Sci. Rep. John Murray Exped. 1933–1934* 1, 63.

Sewell, R. B. S. (1936b). *Sci. Rep. John Murray Exped. 1933–1934* 1, 109.

Shor, G. G., Jr., and Pollard, D. D. (1963). *Science* 142, 48.

Snead, R. E. (1967). *Ann. Ass. Amer. Geogr.* 57, 550.

Sriramachandra Murty, V., Easterson, D. C. V., and Bastian Fernando, A. (1969). *Bull. Cent. Mar. Fish. Res. Inst.* 5, 1.

Stoddart, D. R. (1966). *Atoll Res. Bull.* No. 116, p. 1.

Stoddart, D. R. (1967). *Atoll. Res. Bull.* No. 118, p. 1.

Stoddart, D. R. (1969a). *In* "Symposium on Corals and Coral Reefs (Mandapam Camp)," Mar. Biol. Ass., India.

Stoddart, D. R. (1969b). *Biol. Rev.* 44, 433.

Stoddart, D. R. (1970). *Atoll Res. Bull.* No. 136, p. 1.

Stoddart, D. R. (1971). *Symp. Zool. Soc. London* 28, 3.

Stoddart, D. R., and Taylor, J. D. (1971). *Atoll Res. Bull.* No. 149, p. 1.

Stoddart, D. R., and Yonge, C. M. (1971). *Symp. Zool. Soc. London* 28.

Stoddart, D. R., Taylor, J. D., Fosberg, F. R., and Farrow, G. E. (1971). *Phil. Trans. Roy. Soc. London, Ser. B* 260, 31.

Sugden, W. (1963). *J. Sediment. Petrol.* 33, 355.

Talbot, F. H. (1965). *Proc. Zool. Soc. London* 145, 431.

Taylor, J. D. (1968). *Phil. Trans. Roy. Soc. London, Ser. B* 254, 129.

Taylor, J. D. (1971a). *Phil. Trans. Roy. Soc. London, Ser. B* 260, 173.

Taylor, J. D. (1971b). *Symp. Zool. Soc. London* 28, 501.

Taylor, J. D., and Lewis, M. S. (1970). *J. Natur. Hist.* 4, 199.

Teichert, C. (1947). *Proc. Linn. Soc. N.S.W.* 145.

Teichert, C., and Fairbridge, R. W. (1948). *Geogr. Rev.* 38, 222.

Temple, P. H. (1970). *Tanzania Notes Rec.* 71, 21.

Thomassin, B. (1969). *Rec. Trav. Sta. Mar. Endoume, Fasc. Hors-Ser., Suppl.* 9, 59.

Thorpe, L. (1928). *J. Linn. Soc. London, Zool.* 36, 479.

Trueman, N. A. (1965). *J. Geol. Soc. Aust.* 12, 261.

Vacelet, J., and Vasseur, P. (1965). *Rec. Trav. Sta. Mar. Endoume, Fasc. Hors-Ser., Suppl.* 4, 71.

Vasseur, P. (1964). *Rec. Trav. Sta. Mar. Endoume, Fasc. Hors-Ser., Suppl.* 2, 1.

Voeltzkow, A. (1897–1905). *Abh. Senckenb. Naturf. Ges.* 21, 26, 27.

Voeltzkow, A. (1903). *Z. Gesell. Erdk. Berlin* p. 560.

Voeltzkow, A. (1904). *Z. Gesell. Erdk. Berlin* pp. 274 and 360.
Voeltzkow, A. (1905). *Z. Gesell. Erdk. Berlin* pp. 89, 184, and 285.
Voeltzkow, A. (1906). *Z. Gesell. Erdk. Berlin* pp. 102 and 179.
Voeltzkow, A. (1917). *Reisen in Ostafrika* 3.
von Marenzeller, E. (1907). *Denkschr. Akad. Wiss. Wien* **80**, 27.
Wainwright, S. A. (1965). *Bull. Sea Fish Res. Sta. Isr.* **38**, 40.
Walther, J. (1888). *Abh Math.-Phys. Kl. Kgl. Säechs. Ges. Wiss.* **14**, 437.
Walther, J. (1891). *Petermanns Mitt. Erganzungshaft* **102**, 1.
Weeks, L. A., Herbison, R. N., and Peter, G. (1967). *Amer. Ass. Petrol. Geol. Bull.* **51**, 1803.
Weiss, H. (1966). *Rec. Trav. Sta. Mar. Endoume, Fasc. Hors-Ser., Suppl.* **5**, 165.
Wellman, P., and McElhinny, M. W. (1970). *Nature (London)* **227**, 595.
Wells, J. W. (1954). *U.S. Geol. Surv., Prof. Pap.* **260–1**, 385.
Wells, J. W., and Davies, P. S. (1966). *Atoll Res. Bull.* No. 116, p. 43.
Werth, E. (1901). *Z. Gesell. Erdk. Berlin* **36**, 115.
Westoll, T. S., and Stoddart, D. R. (1971). *Phil. Trans. Roy. Soc. London, Ser. B* **260**, 1.
Weydert, P. (1969). *C. R. Acad. Sci., Ser. D* **268**, 482.
Wood-Jones, F. (1910). "Coral and Atolls." Reeve, London.
Yentsch, A. E. (1962). "A Partial Bibliography of the Indian Ocean." Woods Hole Oceanogr. Inst., Woods Hole, Massachusetts.

3

BIKINI AND ENIWETOK ATOLLS, MARSHALL ISLANDS

Harry S. Ladd

I. Introduction

A. LOCATION

Bikini and Eniwetok, lying some 200 mi apart in the Northern Marshall Islands (Fig. 1), may be regarded as typical of a fairly large group of deep-sea atolls. Though larger than average in size, they are typically annular in shape and lie well within the main atoll area in the Pacific, the postulated former Darwin Rise, a large triangular area extending from Hawaii southeastward to the Tuamotus, then northwestward to Palau. Because both atolls were used repeatedly as testing grounds for atomic weapons, they have been intensively studied by biologists, geologists, oceanographers, and geophysicists.

The two atolls are similar in many of their surface features and this

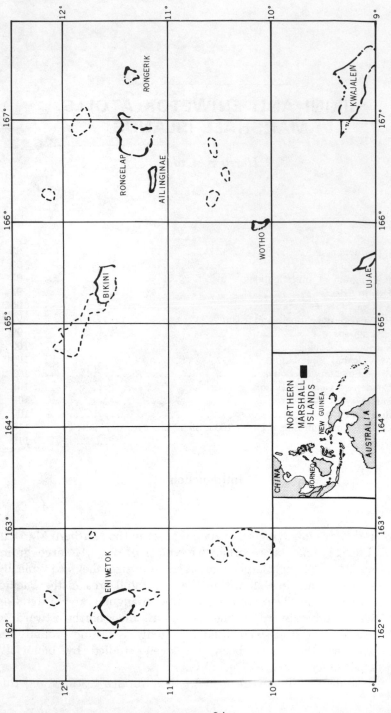

Fig. 1. Location of Bikini, Eniwetok, and nearby atolls and guyots (dashed lines) in the Northern Marshall Islands (after Emery et al., 1954).

fact will be recognized in the brief descriptions presented. Both are known to have had lengthy and somewhat complicated geological histories, as revealed by drilling and by geophysical surveys of several sorts. Major steps in their history are summarized.

B. PROGRAM OF INVESTIGATIONS

In 1945 the United States government decided to carry out a series of tests of its newly developed atomic weapons and Bikini Atoll was selected as a site. This atoll was chosen because it was remote from population centers, major fishing areas, and steamer lanes, yet conveniently located relative to Kwajalein, then a major naval base. Little was known about the atoll, but it was one that had not been damaged by military activities during World War II. The first test program, significantly labeled "Operation Crossroads," was carried out in July 1946; a total of 42,000 men, civilians and servicemen, took part.

Before the first bomb was detonated, comprehensive surveys—geological, biological, and oceanographic—were carried out by teams of scientists. (Geophysical surveys of several sorts were also made, mostly after the initiation of bombing.) Many of the surveys were repeated immediately after each of the two bomb tests and, during the following year, 1947, a resurvey of surface features was made plus deep drilling and additional geophysical work. For control purposes, similar but less detailed scientific surveys were made during Operation Crossroads at three nearby atolls, Rongelap, Rongerick, and Ailinginae. Eniwetok, an atoll lying 200 miles downwind and downcurrent from Bikini, was also investigated. After the conclusion of the first Bikini tests, the nature and thickness of Eniwetok's reef cap were examined by drilling.

Eniwetok was the only atoll of the group studied that had been the scene of major military activities during the war. Direct battle damage there was augmented by the loss of a large fully loaded oil tanker on the windward reef.

After the two Bikini tests that comprised Operation Crossroads were completed, Eniwetok became a base for testing more powerful atomic weapons. Additional nuclear tests were also made later at Bikini. From 1954 to 1958 a total of 21 explosions were made on Bikini, including powerful thermonuclear devices. Many sorts of tests were made at Eniwetok, one islet being completely blown away in the process. Surface vegetation was damaged or obliterated from all but 2 of Eniwetok's 30 islets. A number of the windward islets were connected to each other by causeways to facilitate transportation. These structures barred·

cross-reef circulation of ocean waters in certain areas and radically changed ecological conditions on the reef and in parts of the lagoon. Damage caused by the bomb tests was augmented by pollution from ships occupying the lagoons. At Bikini, during the peak of Operation Crossroads, more than 200 ships were anchored in the 20-mile lagoon. Extensive spraying with pesticides, such as DDT, were carried out on both atolls. The pretest scientific surveys made it possible to assess bomb damage. Though extensive, much of it proved to be temporary.

Testing was suspended in 1958 under international agreement and the natives of Bikini can now be returned to their homes. Atolls and their builders are well-night indestructible.

The varied investigations of Operation Crossroads were supplemented by other surveys in the years that immediately followed. In 1947 an aerial magnetic survey of the Bikini area was conducted cooperatively by the U.S. Geological Survey and the Naval Ordnance Laboratory under the sponsorship of the Office of Naval Research. In 1950 the Scripps Institution of Oceanography and the U.S. Navy Electronics Laboratory cooperated in the Mid-Pacific Expedition. Using two ships, additional seismic investigations in the Bikini area were carried out along with bottom sampling (sediment cores and rock dredging) and bottom photography.

II. Existing Reefs

A. OUTER SLOPES

The seaward rim of the Bikini reef, except for short stretches to leeward, is bordered by a terrace at a depth of about 50 ft. This structure shows clearly on air photographs and was mapped in some detail by soundings. The windward reef of Eniwetok is bordered by a similar terrace, but the extent of its development around the remainder of the atoll was not accurately determined.

Outer slopes of the atolls between the terrace edge and 200 fm average about 38°. At greater depths the slope becomes more gentle until between 2000 and 2500 fm it merges with the floor of the deep sea.

Sediments on the outer slopes consist mostly of coarse coral fragments and segments of *Halimeda* near the surface; with increasing depth the material changes to fine sand and silt. Blocks of reef rock are scattered among the sediments of the outer slope, some of them approaching a ton in weight.

Globigerina sand occurs low on the flanks of the atolls and at depths below 2000 fm grades into red clay.

B. NEAR-SURFACE REEFS AND ISLANDS

Bikini and Eniwetok are typical "rough-water reefs" as that term was defined by P. Marshall [*Rep. Gt. Barrier Reef Comm.* 3, 64–72 (1931)]. Both atolls lie in the belt of the northeast trade winds where their windward sides must withstand a steady wave attack for 9 months of the year. It has been estimated that on Bikini's windward reef the waves dissipate 500,000 hp (Munk and Sargent, 1954). On the south side the reefs are protected from the trades, but periodically they are damaged by powerful long period swells that come up from the southern hemisphere. On the west and northwest the reef edges face relatively calm water, and this is reflected in the configuration and low elevation of the reef margin.

The marginal zone of Bikini's windward reef has been described as the most vital part of the reef because to it the waves bring a constant supply of moving water with oxygen, nutrients, and some food. In this area a definite zonation is developed. Below the level of strong wave action on the seaward side, where light is still strong, there is a zone of rich coral growth. As the reef edge is approached a series of grooves and buttresses is developed. These lead upward to the colorful marginal algal ridge that lies in the zone of surf. Some of the grooves penetrate the ridge as surge channels and continue landward beneath the reef flat. The several parts of this system comprise a most efficient baffle that robs the waves of much of their destructive power and spreads the water to all parts of the marginal zone of the reef. At Bikini a part of the power of the breaking waves is utilized to maintain a water level immediately inside the surf zone that under average trade wind conditions is about 1.5 ft above general sea level. This head permits water to flow downhill over the reef flat to the lagoon, regardless of the stage of the tide. According to Munk and Sargent, the average speed is 0.5–1.0 knots, but occasionally the speed reaches 4 knots. This flow sustains the varied organic communities that live on the broad reef areas. Among these are the Foraminifera that live in countless numbers on the reef flat.

On the windward reefs, the marginal algal ridge is usually bordered to landward by a coral–algal zone and, in many instances, by other zones of organic growth. The main reef flat is a combination of sandy areas and smooth rock that, in many places, is veneered by a mat of living Foraminifera held in place by fibrous algae. When the Foraminifera die their tests are carried lagoonward and become the main constituent of the beach sands that encircle the island.

An examination of the windward reefs of Eniwetok reveals the same types of structures as seen on Bikini, but long sections of the reef in Eniwetok's surf zone are dead, due possibly, I believe, to wartime oil pollution.

The reefs on the south and southwest side of Bikini become adjusted to the small waves and swells that normally strike that coast. They may develop overhanging edges of living organisms. The severe storm swells from the south and, possibly, typhoons that strike at long intervals not only throw large blocks of reef rock upon the reef flat, but also loosen sections of the edge; these latter are broken up or sink to the shallow terrace that is developed along the coast. The irregular outlines of some of these blocks match those of the reentrants in the reef edge.

The reefs of the northwest side of Bikini are protected from waves engendered by the trades and from severe storm swells from the south. These reefs are straight or smoothly scalloped. The marginal zone supports a fairly rich growth of corals and algae, but it rises only a few inches above the main reef flat.

Small islands are found at intervals on the reefs of both Bikini and Eniwetok. The islands are larger and more numerous on the windward reefs than on the reef to leeward. All of the islands are built of reef debris, largely sand, piled up by currents, waves, and wind. Most of them rise only 10–15 ft above the reef flat. Bikini has a total of 26 named islands, Eniwetok has 33 (counting the one that was completely removed by an atomic blast). The shores of most islands are fringed with sand beaches and, in many places, the sands of the present beach partly mask layers of hard beach rock.

Where a zone of beach rock is exposed, it lies between the reef flat and the beach. At Bikini such rock is found along 15% of the beaches. The hard layers rise 1–3 ft above low tide and dip toward the water (sea or lagoon) at an angle of about 9° and the layers may extend as a rock floor beneath the unconsolidated sands. Beach rock is apparently forming today within beach sands. Where exposed, the part of the belt immediately above low tide is scoured by solution and the higher part of the belt is eroded by wave action.

Along straight stretches of coast the layers of beach rock, like the layering to be seen in the existing beach, is regular and generally conforms to the coastline, but near the ends of islands the beds of beach rock and the layers of beach sand become irregular in attitude and texture. These changes are probably due to shifting currents.

Air photographs of the reef flat in areas near existing islands frequently show the strike of now truncated beds of beach rock and this evidence is easily checked on the ground. It indicates clearly that some islands were once larger than they are now and that others have moved an appreciable distance from the seaside toward the lagoon. Other changes are indicated in some areas where the dip and strike of the beach rock layers depart widely from those of the existing beach.

Boulder ramparts are well developed on most of the islands along the south side of Bikini. These ridges of sand and rubble are probably built, moved, or destroyed only by the waves of major storms. On some islands the ramparts are double or even triple, each, probably, the result of a single storm.

The larger islands of Bikini appear to be stable under existing conditions, some indeed may be slowly growing. Smaller islands on the windward side apparently are being eroded on their seaward sides, but much of the sand eroded from that side is worked around to leeward by waves and currents and is added to the lagoon beach. By this means the island may slowly migrate across the reef toward the lagoon.

C. Lagoons

Bikini's lagoon, in which the second atomic explosion of the Marshall Island series was detonated, was intensively studied before and after the test. Plotting of nearly 40,000 soundings revealed a fairly flat lagoon floor, except for numerous coral knolls, and the maximum depth was determined as 32 fm. In places around the margin of the basin there is a 10-fm terrace up to 2000 yd in width. Soundings were not sufficiently dense to permit complete mapping of the terrace, but in well-sounded areas closed depressions extending to a depth of 8 fm below the terrace surface were found. The irregular shapes of these depressions suggested that they may have been formed by solution during periods of lowered sea level. Coral knolls, variable in width and height, were found in all parts of the lagoon, their number totaling hundreds.

The foraminiferal sands of the beaches and the reef flats are carried into the lagoon to form a ring several miles in width inside the annular reef. At somewhat greater depths the alga *Halimeda* grows in abundance, and its platy debris, mixed with finer sands, covers much of the lagoon floor. Coral debris from the knolls and from the coral thickets that grow sporadically at lower levels in the lagoon is also a constituent of the sand blanket that smooths the lagoon floor between knolls. In the deepest parts of the lagoon there is insufficient light for luxuriant *Halimeda* growth and Foraminifera tests again form an important constituent in the bottom sediment. These Foraminifera are of a different type from those found on the reef and in the beach sands.

Bikini's lagoon was found to have a primary circulation, an overturning circulation driven by the prevailing trade winds, and a secondary circulation, a rotary circulation made up of two counterrotating compartments (von Arx, 1954). The circulation patterns showed more strength and stability during the winter months than during the summer season when the trades were weakened.

Studies of Eniwetok's lagoon were aided by work done prior to Operation Crossroads. In 1944 the U.S. Navy made about 180,000 soundings in the 20-mile body of water. When plotted these soundings revealed detailed topography, similar in major features to those determined at Bikini. The chart (Emery *et al.*, 1954, chart 5) showed a wide terrace at about 10 fm with shallow irregular depressions and one depression $2\frac{1}{2}$ miles long with its bottom 8 fm below the terrace surface. Eniwetok's lagoon is slightly deeper (maximum 35 fm) than that of Bikini and there is a fairly large area deeper than 32 fm. Coral knolls were so numerous that all could not be charted. More than 2000 were measured. The lagoonal sediments were similar to those of Bikini in composition and distribution.

Each of the main reefs—the one at Bikini and the one at Eniwetok—is cut by a single pass that is approximately as deep as the deeper parts of the lagoon behind it. On Bikini the deep pass is near the middle of the south side; on Eniwetok it is on the southeast side. The more numerous shallow passes that cut the reefs of both atolls are only as deep as the terrace that is present in the lagoon and on the seaward side of the main reef.

III. Subsurface Geology

A. Volcanic Foundations

Bikini and Eniwetok are limestone caps on the tops of volcanoes that rise 2 miles above the floor of the deep sea. At Bikini the presence of a basaltic foundation was indicated by samples of volcanic rock dredged from the lower slopes of the atoll and from Sylvania Guyot that adjoins the atoll on the west. At Eniwetok the basaltic foundation was penetrated by the drill.

The abundance of pyroclastic material and the highly vesicular character of some of the volcanic rocks of the Bikini area indicate that the eruptions did not take place at great depth. This interpretation is supported by the recovery of what appeared to be a rounded basalt pebble from Sylvania Guyot. The exact age of the volcanic rocks is not known, but the presence in them of fresh undevitrified glass makes it unlikely that they are older than late Paleozoic and they may be considerably younger (Macdonald, *in* Emery *et al.*, 1954, pp. 120–124). The olivine basalt that was cored below Eniwetok's reef cap apparently formed a part of a flow. No pyroclastic material was found at Eniwetok, although such materials may exist under some parts of the atoll.

The flatness of Sylvania Guyot adjacent to Bikini and the occurrence of the above-mentioned basalt pebble suggest that the surface of Sylvania Guyot and probably the guyots that underlie Bikini and Eniwetok were truncated by erosion. The date of this action is suggested by the occurrence of phosphatized *Globigerina* ooze obtained from cracks in a tuff-breccia from the surface of Sylvania Guyot which proved to be early Eocene (Tertiary *a*) in age. These are the oldest fossiliferous rocks from the Marshall Islands (Hamilton and Rex, 1959).

B. SEDIMENTARY SECTION

The limestone cap of Eniwetok has been shown to exceed 4000 ft in thickness and its volume is estimated at 250 mi^3. Geophysical data indicate that Bikini's cap is about the same thickness.

The bulk of the limestones that make up the caps is Tertiary in age. The major subdivisions of these and younger beds are shown in Table I.

During the deep drilling on Bikini, and later on Eniwetok, one of the most interesting discoveries was the recognition of thick sequences

TABLE I

MAJOR STRATIGRAPHIC SUBDIVISIONS RECOGNIZED IN
HOLES DRILLED ON BIKINI AND ENIWETOK[a]

Stratigraphic divisions	Bikini, depth (ft)	Eniwetok, depth (ft)
Post-Miocene	0–700	0–615
Upper Miocene		
Tertiary *g*	700–980	615–860
Lower Miocene		
Tertiary *f*	980–1166	860–1080
Tertiary *e*	1166–2556+	1080–2780
Upper Eocene		
Tertiary *b*		2780–4610

[a] (After H. S. Ladd, *U. S., Geol. Surv., Prof. Pap.* **531**; (1960); based on Cole, 1959).

of limestones in which aragonitic fossils, such as corals and many molluscs, were exceptionally well preserved. These fossiliferous beds were sandwiched between other thick sequences from which all aragonite had been dissolved, leaving molds or calcitic replacements with some dolomite. The leached and recrystallized calcitic intervals were believed to record periods when the top of the atoll stood above water for appreciable times and was subjected to solution by meteoric water, the unleached sections never having been above the sea to receive such treatment. The exact boundaries of the zones were difficult to draw because of the low percentage of core recovery obtained in the Marshall Island drill holes. After all drilling was completed, Schlanger made an intensive study of the petrology of all sediment cores and cuttings from Eniwetok and was able to delimit the leached and unleached zones with fair accuracy. He was also able to tie the zones of Eniwetok to those found under Bikini some 200 miles to the east.

Schlanger agreed with the emergence interpretation and he called the top of the partially leached zones separating each pair of contrasting sequences *solution unconformities*. His diagrammatic representations of the major stages in the development of such a feature are shown in Fig. 2 (Schlanger, 1963, pp. 994–998).

The three stages shown in Fig. 2 are explained as follows. In stage 1 the volcanic foundation or basement, truncated by wave action, is covered by aragonite-rich sediments. These sediments increase in thickness as the foundation subsides or sea-level rises. In stage 2 the sediment-covered mass is partly emerged due to uplift of the foundation or a fall in sea level. The sediments of the emerged part are leached and recrystallized under atmospheric conditions. Calcite replaces aragonite and a hardened irregular surface is developed. In stage 3 submergence again occurs and a new cap of aragonite-rich sediment is deposited above the hardened surface of the solution unconformity.

Gross and Tracey studied the oxygen and carbon isotopic composition of limestones and dolomites from Bikini and Eniwetok drill holes. In their preliminary report [Science **151**, 1082–1084 (1966)] they noted that aragonitic sediments were isotopically identical with unaltered skeletal fragments, whereas the recrystallized limestones exhibited isotopic variation that they credit to alteration in meteoric waters during periods of emergence.

Major stratigraphic units of the Eniwetok and Bikini sedimentary sections and the relations of the solution unconformities to stratigraphic boundaries are shown in Fig. 3. It will be noted that one unconformity occurs at the top of the Eocene (Tertiary *b*), another at the top of early Miocene (Tertiary *e*), and a third above the top of late Miocene (Tertiary *g*). The third is probably Pleistocene.

The Miocene sediments in all three of the deep holes drilled on Eniwetok yielded a variety of pollen and spores. Included were high-island forms that indicated to Leopold (1970) the presence of an upland forest community.

From the Miocene beds drilled on Bikini and Eniwetok were recovered shells of land snails of a type (*Ptychodon*) that characteristically inhabit only high limestone islands [H. S. Ladd, *J. Paleontol.* **32**, 183–198 (1958)].

The two deepest holes drilled on Eniwetok, F-1 on Elugelab and E-1 on Parry Island, are 20 miles apart on the windward side of the atoll. In hole F-1 the sediment section measured 4610 ft, in hole E-1 4,170 ft. The lithologic and faunal units recognized in the upper parts of the two holes are similar to a depth of 1400 ft and they can be clearly correlated with units recognized in the deepest hole drilled on Bikini 210 miles to the east. The lower parts of the two Eniwetok holes, however, are quite different. The lithologic differences were recognized in the field when the holes were drilled, but they were not fully understood until laboratory studies by Todd and Low (1960) noted abundant Globigerinids and other deep water Foraminifera in parts of the otherwise nearly barren lower F-1 sediments. Todd and Low concluded that the fossiliferous beds in the lower part of F-1 were deposited in moderately deep water, probably on the outer slope of the atoll. The limestones in E-1 contained shallow water fossils from top to bottom. The Todd-Low findings are portrayed graphically by Schlanger in his paleoecology logs, here reproduced as Fig. 4.

A number of Quaternary limestones from drill holes on Eniwetok have been dated by radiocarbon methods and by methods involving the ionium–uranium ratios and studies of this sort are continuing [E. A. Olson and W. S. Broecker, *Amer. J. Sci.* **257**, 1–28 (1959); M. Rubin and C. Alexander, *Amer. J. Sci., Radiocarbon Suppl.* **2**, 129–185 (1960); J. W. Barnes, E. J. Lang, and H. A. Potratz, *Science* **124**, 175–176 (1956); D. L. Thurber, W. S. Broecker, R. L. Blanchard, and H. A. Potratz, *Science* **149**, 55–58 (1965)]

In addition to the deep holes on Bikini and Eniwetok, many shallow holes were drilled to depths of 100–200 ft. Most of these were on Eniwetok and were in unconsolidated sediments save for beach rock layers at intertidal levels. Two holes, however, showed more hard rock than the others. One was Bikini No. 3, located close to the reef edge at the south end of Bikini Island, the other, RU-2, was on Runit Island,

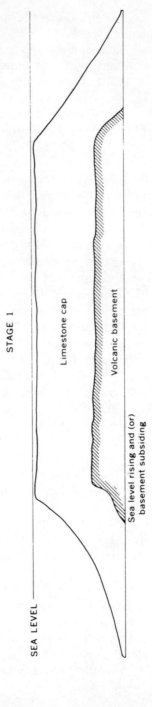

STAGE 1

Limestone cap

Volcanic basement

SEA LEVEL

Sea level rising and (or)
basement subsiding

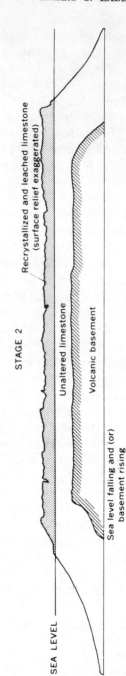

STAGE 2

Recrystallized and leached limestone
(surface relief exaggerated)

Unaltered limestone

Volcanic basement

SEA LEVEL

Sea level falling and (or)
basement rising

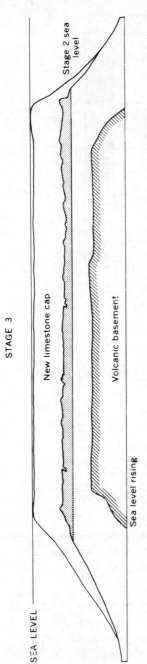

Fig. 2. Diagrammatic sections showing stages in the development of a solution unconformity (after Schlanger, 1963).

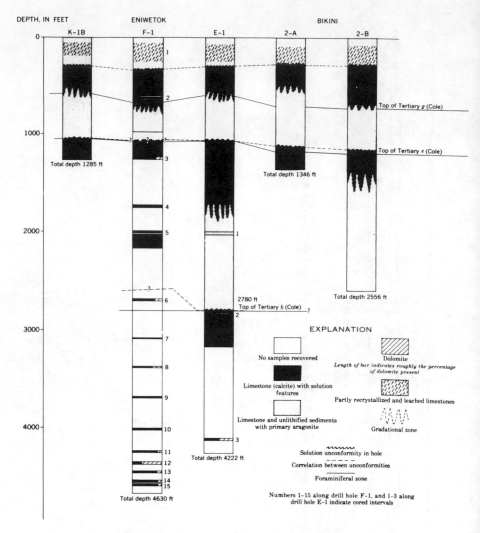

Fig. 3. Drill hole sections below Eniwetok and Bikini atolls showing lithology and relations of solution unconformities to stratigraphic boundaries (after Schlanger, 1963).

Eniwetok, where the reef is exceedingly narrow. The Bikini Island hole was carried to 178 ft, the Runit hole to 135 ft. The two holes mentioned may have penetrated the hard rim that is supposed to extend downward from the wave-resisting reef edge of an atoll. The postulated wall contains the unconsolidated lagoonal sediments as the rim and walls of a pail contain water.

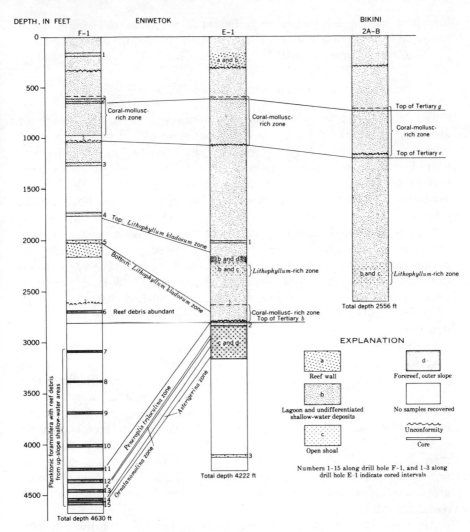

Fig. 4. Correlation of depositional and faunal zones in drill holes on Eniwetok and Bikini (after Schlanger, 1963).

In addition to the numerous holes drilled on the islets of Eniwetok, a series of four holes were cored on the seaward reef. Two were drilled on a rocky groin extending out on the reef flat, two others on the reef flat itself. All four holes penetrated hard limestone for 6–15 ft. This plate is thicker than any consolidated beds found at shallow depths under the islets (Ladd and Schlanger, 1960, pp. 884–890). It is not known if such a solid plate underlies atoll reef flats elsewhere, but a comparably thick crust was drilled beneath a fringing reef off Samoa [L. R. Carey, *Carnegia Inst. Wash. Publ.* 413, Vol. 27, 53–98 (1931)].

IV. Condensed Geologic History

The basaltic volcanoes that underlie Bikini and Eniwetok were built above the surface of the sea in pre-Tertiary time. When volcanic activity ceased, the mounds were partly truncated by wave action. The effectiveness of wave erosion was probably aided by slow subsidence, a process that probably began at the cessation of volcanic activity. Reefs were established on parts of the platforms in late Eocene time. Reef growth continued throughout the remainder of the Tertiary, but the sites of growth were shifted laterally on at least three occasions when the reef caps were elevated above the sea and subjected to erosion under atmospheric conditions. During two of these intervals of emergence in the Miocene the atolls stood as high islands and the land flora and fauna that characterize such islands were established. After each such emergence, subsidence resumed and the eroded caps were covered by new limestones.

During the Pleistocene other changes of elevation relative to the sea occurred. The deeper parts of the lagoon floor and the deep passes may have been developed when the sea stood several hundred feet lower than it does now. During the warmer interglacial stages of the Pleistocene, reefs developed on the prepared surface, growing upward more rapidly around the margins than elsewhere. The reef is thought to have flourished over the wide area now covered by the shallow terraces inside and outside the lagoon. The existing reef has grown up during the postglacial rise of sea level, the shallow passes representing areas where, for various reasons, possibly largely ecological, the new reef did not flourish.

After the last glacial epoch, sea level in the Bikini–Eniwetok area seems to have risen to a height several feet above its present level and reef growth was stimulated. When the sea sank to its present level waves attacked the low emerged reefs and it is probable that many of the existing islands were formed at that time.

With so much known about the origin and history of Bikini and Eniwetok and considering their location well within the main atoll area (the postulated Darwin Rise), there is a temptation to look upon them as typical of all Pacific atolls, but this temptation should be resisted. The histories of nearby atolls in the Marshall Islands may prove to be similar to those established for Bikini and Eniwetok, but elsewhere in the Pacific the histories as regards age, rate of reef building, periods of emergence, and erosion may be quite different. Such differences have already been revealed by drilling at Funafuti in the Ellice Islands, Mururoa in the Tuamotu Islands, and Midway in the Hawaiian group.

V. Summary of Results

The results of the investigations of Bikini and Eniwetok may be summarized under four heads:

1. Detailed charts of the atoll slopes, the surface reefs, and the lagoons, including types of sediments and their distribution.

2. Fairly detailed descriptions of many of the animals and plants living on and around the atolls and the nature of the physical environments in which the life exists.

3. From dredging on the lower slopes of Bikini and from deep drilling on Eniwetok, a confirmation, for the first time, of the major tenet of Darwin's theory, that open sea reefs develop on slowly subsiding volcanoes, the subsidence totaling thousands of feet.

4. Descriptions of Cenozoic fossil faunas and floras and the establishment, in the open Pacific, of a standard geological section extending into Eocene time. This section has proved useful for reference purposes.

Acknowledgments

Appreciation is expressed to Dr. J. I. Tracey, Jr. of the U.S. Geological Survey and Dr. S. O. Schlanger of the University of California at Riverside for their critical reviews of the manuscript.

References

A voluminous literature dealing with the reefs of Bikini and Eniwetok has appeared during the twenty-five years that have elapsed since the first bomb tests made the atolls the subject of intensive scientific investigatons. Many reports in scientific journals and several complete volumes have been published, some by the government, others by private agencies. Many specialized reports deal primarily with the effects of bombing on the environment. These have had a limited circulation and some remain classified. Final reports on various aspects of reef studies have been brought together and published by the United States Geological Survey as Professional Paper 260. The first chapter, with its box of plates and charts, deals with the regional setting and the initial geological and geophysical surveys connected with Operation Crossroads. Later chapters are more specialized. More than 40 authors have contributed to one or more units. The series, now essentially complete, consists of 35 chapters totaling more than 1100 pages, some 600 plates and figures, and about a dozen charts. The chapters are listed below in chronological order.

Chapters of Professional Paper 260—Bikini and Nearby Atolls

Chapter	Date	Author(s)	Title	Pages
A	1954	Emery, K. O., Tracey, J. I., Jr., Ladd, H. S.	Geology of Bikini and nearby atolls	1–265
B	1954	von Arx, William S.	Circulation systems of Bikini and Rongelap lagoons	265–273
C	1954	Munk, Walter, H., Sargent, Marston C.	Adjustment of Bikini Atoll to ocean waves	275–280
D	1954	Robinson, Margaret K.	Sea temperature in the Marshall Islands area	281–291
E	1954	Sargent, Marston C., Austin, Thomas	Biologic economy of coral reefs	293–300
F	1954	Johnson, Martin W.	Plankton of Northern Marshall Islands	301–314
G	1954	Cooper, G. A.	Recent brachiopods	315–318
H	1954	Cushman, Joseph A., Todd, Ruth, Post, Rita J.	Recent Foraminifera of the Marshall Islands	319–384
I	1954	Wells, John W.	Recent corals of the Marshall Islands	385–486
J	1954	Dobrin, M. B., Perkins, Beauregard, Jr.	Seismic studies of Bikini Atoll	487–505
K	1954	Raitt, Russell W.	Seismic-refraction studies of Bikini Kwajalein Atolls	507–527
L	1954	Alldredge, L. R., Keller, Fred, Jr., Dichtel, W. J.	Magnetic structure of Bikini Atoll	529–535
M	1954	Johnson, J. Harlan	Fossil calcareous algae from Bikini Atoll	537–545
N	1954	Todd, Ruth, Post, Rita J.	Smaller Foraminifera from Bikini drill holes	547–568
O	1954	Cole, W. Storrs	Larger Foraminifera and smaller diagnostic Foraminifera from Bikini drill holes	569–608
P	1954	Wells, John W.	Fossil corals from Bikini Atoll	609–617
Q	1954	Hartman, Olga	Marine annelids from the Northern Marshall Islands	619–644
R	1955	Mao, Han-Lee, Yoshida, Kozo	Physical oceanography in the Marshall Islands area	645–684

Chapters of Professional Paper 260—Bikini and Nearby Atolls
(*Continued*)

Chapter	Date	Author(s)	Title	Pages
S	1957	Raitt, Russell W.	Seismic-refraction studies of Eniwetok Atoll	685–698
T	1957	Revelle, Roger, Emery, K. O.	Chemical erosion of beach rock and exposed reef rock	699–709
U	1958	Swartz, J. H.	Geothermal measurements Eniwetok and Bikini Atolls	711–741
V	1957 (1959)	Cole, W. Storrs	Larger Foraminifera from Eniwetok Atoll drill holes	743–784
W	1959	Hamilton, Edwin L., Rex, Robert W.	Lower Eocene phosphatized *Globigerina* ooze from Sylvania Guyot	785–798
X	1960	Todd, Ruth, Low, Doris	Smaller Foraminifera from Eniwetok drill holes	799–861
Y	1960	Ladd, Harry S., Schlanger, Seymour O.	Drilling operations on Eniwetok Atoll	863–905
Z	1961 (1962)	Johnson, J. Harlan	Fossil algae from Eniwetok, Funafuti, and Kita-Daitō-Jima	907–950
AA	1962	Swartz, J. H.	Some physical constants for the Marshall Islands area	953–989
BB	1963	Schlanger, Seymour O.	Subsurface geology of Eniwetok Atoll	991–1066
CC	1964	Todd, Ruth	Planktonic Foraminifera from deep-sea cores off Eniwetok Atoll	1067–1100
DD	1964	Wells, John W.	Fossil corals from Eniwetok Atoll	1101–1111
EE	1964	Brown, D. A.	Fossil Bryozoa from drill holes on Eniwetok Atoll	1113–1116
FF	1964	Cooper, G. A.	Brachiopods from Eniwetok and Bikini drill holes	1117–1120
GG	1964	Kier, Porter M.	Fossil echinoids from the Marshall Islands	1121–1126
HH	1964	Roberts, Henry B.	Fossil decapod crustaceans from the Marshall Islands	1127–1131
II	1969 (1970)	Leopold, Estella B.	Miocene pollen and spore flora of Eniwetok Atoll, Marshall Islands	1133–1185

Other References

The chapters of Professional Paper 260 cover a number of organic groups that make up the living fauna of the reefs and lagoons, but the living flora is not included. The marine and island floras were fully reported upon by William Randolph Taylor in a volume entitled "Plants of Bikini and other Northern Marshall Islands" [*Univ. Mich. Stud., Sci. Ser.* 18, 1–227 (1950)].

A brief report on the fishes of Bikini was published by Leonard P. Schultz in 1948 [*Smithson. Inst., Annu. Rep.* pp. 301–316 (1948)] and from 1953 to 1966 three volumes by Schultz and collaborators appeared as *U.S., Nat. Mus., Bull.* 202; these monographs covered most of the fishes of the Marshall and Mariana Islands.

In 1954 the U.S. Atomic Energy Commission established a marine biological laboratory at Eniwetok. It was designed as a base of operations for studies related to atomic tests but was opened to biologists interested in conducting fundamental biological studies. A number of publications have resulted. A well-known example: "Trophic structure and productivity of a windward coral reef community on Eniwetok Atoll" by Howard T. Odum and Eugene P. Odum [*Ecol. Monogr.* 25, 291–320 (1955)]. Results of the six-weeks study by the Messrs. Odum are still being discussed and tested.

4

GEOMORPHOLOGY AND GEOLOGY OF CORAL REEFS IN FRENCH POLYNESIA

J. P. Chevalier

I. General Survey

French Polynesia includes 110 islands of oceanic type, situated inside the andesite line and spread over a surface between lat. 7°51′S and 27°40′S between long. 134°W and 155°W. They form four groups aligned in a SE–NW orientation: Marquesas Islands, Tuamotu Archipelago, Society Islands, and Austral Archipelago (Fig. 1). These lines are connected eastward to the great ridge of the eastern Pacific which is a large submarine north–south oriented chain (Menard, 1964). The

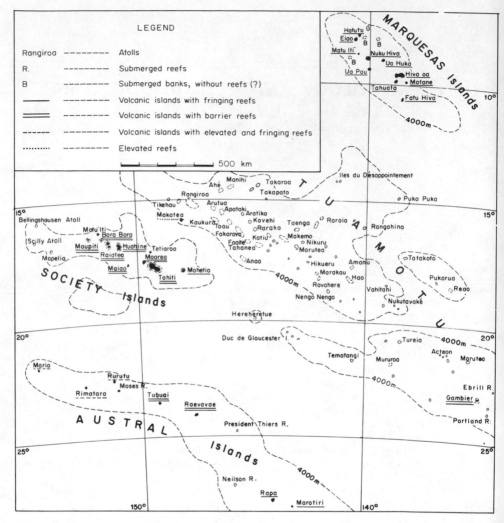

Fig. 1. General map of French Polynesia.

topography is rough and presents fractured areas through which lava poured out erecting volcanic structures that formed "high" islands or served as supports to reefs.

A. CLIMATOLOGICAL DATA

French Polynesia lies in a wet tropical zone. A hot season with heavy rains (generally from November to the end of April) and drier cool season (from May to the end of October) can be distinguished. The temperature decreases from north to south. The annual average temperature is 26.9°C at Nuku-Hiva (Marquesas) and 20.6°C at Rapa (Austral Islands).

The trade wind is the prevailing one; it blows from northeast to southeast, with an easterly dominance. In southeastern Tuamotu and in the Austral Islands these wind conditions are disturbed by southerly and westerly squalls. Anywhere else squalls from north and northwest are also observed.

Cyclones do not occur frequently, the fiercest ones having taken place in 1878, 1903 (517 persons killed), 1906, and 1933. They generally have a W–E or a NW–SE orientation. The cyclones often bring about the formation of boulder ramparts that can be seen on the atolls, or the accumulation of bioclastic materials. At Rimatara, after the 1903 cyclone, the reef was covered by a coral sand hill 3 m high.

B. Hydrological Data

The temperature of the seawater gradually decreases southward and eastward. At the southernmost of the Austral Islands (Rapa), the limit of temperature suitable for coral growth is attained. The following tabulation presents a few examples of mean seawater temperature in the vicinity of different islands.

	Marquesas	Tahiti	Gambier	Rapa
Mean temperature: (°C) (February)	26.5	27	25.5	23
Mean temperature: (°C) (August)	26	25.5	21.5	20

The islands are often battered on the south and southwest by a heavy swell, stronger than on the leeward side and due to the numerous gales of the high southern latitudes.

Tides are of small amplitude (maximum 1 m) and are semidiurnal.

II. Coral Reefs of the Marquesas Islands

The Marquesas are composed of volcanic, partly collapsed, high islands established on a system of perpendicular cracks, the mean direction being N 83°E and S 7°W (Chubb, 1930). These volcanic structures date from Miocene to Pleistocene times. The islands are edged by high cliffs which extend below sea level and are due to breaks.

Some coasts are almost entirely bare of reefs. These are situated at
the entrance of bays especially when they open northward or westward.
The reefs are built up of lithothamnion and corals which are numerous,
but poor in species, sometimes only one (Crossland, 1927). Clear of
the bays along the cliffs isolated coral colonies grow, forming the beginning of a fringing reef (Davis, 1928).

The small development of reefs in the Marquesas Islands is due, according to Dana (1849), to a too-rapid subsidence. In Agassiz's opinion
(1903), the presence of loose materials hindered the settling of coral
larvae. Besides the causes indicated by Agassiz, Crossland (1927)
thought that the following hypothesis has to be added: at regular intervals, cold waters upwell along the islands and inhibit the growth of
reefs, except in the more sheltered bays. This author thought he found
evidence of a periodical destruction of corals. Ranson (1952) agreed
with him on the subject. Davis (1928) believed that during glacial times
the Marquesas Islands were situated in a fringe zone outside of the
region of coral growth. The reef-building organisms were killed by cold
currents which are an extension of Humbolt current. During the warming
up of postglacial times, the reef organisms could not settle actively on
these islands because of their geographical isolation.

III. Coral Reefs of Tuamotu Archipelago

A. GENERAL ASPECT

The Tuamotu Islands are arranged in several linear series (less precisely definite in the eastern part of the archipelago) and a few NE–SW
lines. Four groups of islands and shallows may be recognized: (1) A
small archipelago composed of volcanic high islands, the Gambier
Islands, made up of two calderas (definitely collapsed), around which
a barrier reef has developed; the lagoon is wide opened toward the
sea (Fig. 6). (2) Eighty atolls, the largest being Rangiroa which is
the largest but one in the world (80 km long). Tuamotu atolls are
characterized by a narrow reef rim which indicates a centripetal growth
(according to Newell, 1956). They are composed of an ancient cemented
coral conglomerate or an ancient emerged reef, and they are broken
by incomplete channels open toward the lagoon and called hoa, and
by scanty and narrow passes. (3) An emerged reef island, Makatea,
which has been raised to 113 m high. (4) A few reef banks reaching
or almost reaching the sea surface, in the southeast of the archipelago:
Ebrill Reef, Portland Reef.

These islands were built undoubtedly at the beginning of the Tertiary, as the supposed Recent nummulitic age of limestones in Makatea and the discovery of Eocene Foraminifera typical of shallow waters, on the walls of a submarine mountain now 980 m deep, at the westernmost of the archipelago, bear witness (Cole, 1959). It seems that in the Tuamotu Islands the subsidence was more pronounced in the northwest than in the southeast, since volcanic rocks emerge in the Gambier Archipelago and are, respectively, 400 and 800 m deep at Mururoa and at Takaroa.

Several scientific expeditions, the more important being those of the "Challenger" (1885) and the "Albatross" (Agassiz, 1903), visited these islands. However, the reefs of only the three following islands have been the subject of detailed investigations: Raroia (Newell, 1956), Rangiroa (Stoddart, 1969), and Mururoa (Chevalier *et al.*, 1968).

B. STUDY OF AN ATOLL: MURUROA (Fig. 2)

Mururoa was the site of French nuclear testing. So at the instigation of the Service Mixte de Contrôle Biologique (laboratory depending on the Direction des Centres d'Expérimentations Nucléaires, Paris) several

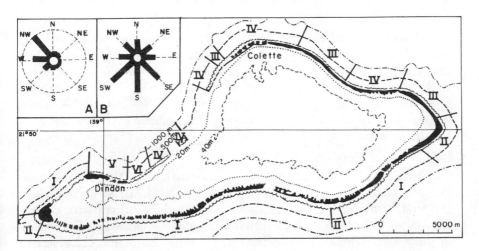

Fig. 2. Mururoa Atoll. Islets are in black. The roman numerals indicate the various types of outer reefs in the different zones of the reef rim. Dindon and Colette points indicate the sites of deep drill-holes. In the lagoon, patch reefs are not shown. (A) Distribution along the reef rim of passages between the lagoon and the ocean (passes, hoa, reefs bearing no islets). (B) Mean width of the outer reef flat. (According to Chevalier *et al.*, 1968.)

studies were undertaken on this atoll—geomorphological (Chevalier *et al.*, 1968), sedimentological (Trichet, 1969), and geological (Deneuf-bourg, 1969; Labeyrie *et al.*, 1969; Fontes *et al.*, 1969). Other Tuamotu atolls, besides Mururoa, have been investigated in the same program, (still unpublished works).

1. The Reef Rim

Uninterrupted but for a large pass cutting the northwest side, it is 150–500 m wide and composed of the following structures.

a. Ancient Coral Conglomerate. This is a continuous formation which does not exceed 1.3 m above high tide level (south rim of the atoll). It is formed by a 1–3-m thick accumulation of cemented bioclastic, very coarse or fine-grained, material (calcarenites) derived from the outer slope of the atoll. The elaborate cementation of this detrital formation definitely took place in the intertidal zone and is an indication of a slight lowering of sea level. Carbon-14 determination gives a date of 3000–4000 years. This conglomerate was submitted to erosion which formed a small escarpment 30–40 cm high, that can be seen principally on the southern rim of the atoll. The hollowing of channels (toward the lagoon) or hoa and the mushroom-shaped limestone outliers (*feo*) are also ascribed to erosion.

b. Unconsolidated Structures. They are younger, made of bioclastic materials and form islets called *motu* in the Paumotu language. They are in a greater number on the windward side and facing the southern swell, where they constitute unbroken formations several kilometers long. Most often these islets present a concave seaward side, where the roughest materials are to be found. Whereas the inner part of the *motu* is stabilized, a gradual shifting of reef debris can be observed from the sea toward the lagoon along the walls of the *motu*.

c. Outer Reef. It is isolated from the lagoon by the ancient conglomerate and the islets, except in the west and in the northwest of the atoll. It has its maximum development in the southern part, clearly because of the strong swell which promotes the growth of building organisms (in particular the algae) and destruction of the old emerged conglomerate. From the inner edge it comprises.

1. The outer flat is not deeper than 50 cm at low tide. In its inner part, it is composed of the ancient eroded conglomeratic limestone, outliers of which still remain (*feo*) nearly on a line facing the swell. In its outer part, on the other hand, it is an autochthonous limestone

which bears calcareous algae and generally few corals. In this area parts of the old reef still show and on them organisms (lithothamnion especially) are growing (Fig. 3).

2. The algal ridge, which is situated at the outer edge and built in particular by *Porolithon, Chevaliericrusta,* and *Jania;* the swell always breaks upon it.

3. The outer slope. Following a small escarpment 2–4 m high, the outer slope averages about 15°–30° and is broken by two terraces lying approximately 8–10 and 20 m deep. The building process by algae and corals is important. Below 20 m the slope is 45° down to 300–500 m, then the gradient decreases gradually.

The morphology of the algal ridge and of the upper part of the outer slope is largely influenced by the swell. At Mururoa six different reef types may be recognized (Fig. 4).

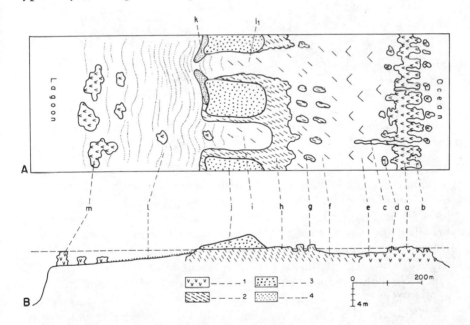

Fig. 3. The reef rim in the southern part of Mururoa. (A) Topography. (B) Section through the reef rim. (a) Algal ridge; (b) grooves; (c) surge channels; (d) dead emerged reef; (e) outer part of the reef flat; (f) inner part of the reef flat; (g) *feo;* (h) emerged reef conglomerate; (i) closed hoa; (i_1) functional hoa; (j) islet (*motu*); (k) off-shore bars; (l) gently sloped platform of the lagoon margin, covered with coarse sand; (m) coral formations; (1) built roof; (2) conglomerate; (3) bioclastic materials of the motu; (4) loose materials (off-shore bar, sands of the lagoon coast). (According to Chevalier *et al.,* 1968.)

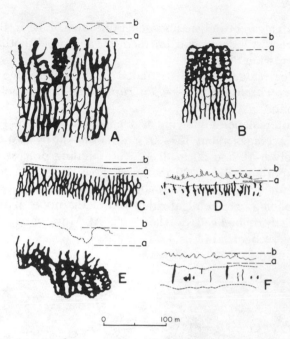

Fig. 4. Different types of outer reefs at Mururoa. (A) Type I; (B) type II; (C) type III; (D) type IV; (E) type V; (F) type VI; (a) algal ridge; (b) reef flat. On the reef front, the grooves are in black and the spurs in white. (According to Chevalier *et al.*, 1968.)

Type I: This is typical of the reefs facing the swell of the southern rim. The algal ridge, 30–50 m wide, emerges 80 cm above low tide level. It is furrowed on its edge by a system of deep, narrow, and branched grooves and spurs. Grooves sometimes entirely cross the algal ridge and extend through the outer reef flat, forming surge channels. They extend beneath the sea to a depth of approximately 20 m. This type is probably similar to the type IB1 recognized at Bikini by Emery *et al.* (1954).

Type II: This type is very similar to type I, differing only by the presence of secondary grooves at right angles to the preceding ones and creating a "pillar structure." A heavy swell is the controlling factor in its formation.

Type III: The algal ridge is less developed than in the two other types. It is only 10–20 m wide and it does not exceed 20–30 cm above low tide level. The more irregular, narrow, and unramified grooves cut the algal ridge only in its outer part and outline low spurs. Type III is very similar to type IA at Bikini. It is influenced by the swell of the eastern part, which is weaker than in the southern part.

Type IV: The algal ridge, 5–12 m wide is low, just above sea level. Besides calcareous algae, corals have a great part in the reef-building. The grooves are not branched and are well separated. This type can be seen in areas sheltered from the strong swell northwest of the atoll.

Type V: In this type the grooves form wide passages, the spurs are very irregular, forming scattered or grouped knoll reefs covered with reef-builders.

Type VI: This type is located west of the atoll, in a sheltered zone. The algal ridge is submerged. The distance between the grooves is considerable and they are disrupted and shallow. It corresponds with type IIA at Bikini, described by Emery *et al.* (1954).

The grooves and spurs system is clearly influenced by the swell and seems to be the result of both selective building of reef organisms and destruction by storms and cyclones. At present, construction seems to prevail over destruction since here and there (principally on the leeward side) the grooves become covered with reef-builders.

d. Hoa. This term is given to shallow channels (50 cm to 2 m deep) that open toward the lagoon and are cut out of the reef conglomerate. There is a distinction between nonfunctional hoa, which are incomplete and end in a small escarpment and which do not connect lagoon water and seawater except during stormy weather, and functional hoa which gradually come to the level of the reef flat. Lagoonward offshore bars may obstruct the hoa partly or entirely. There are also dry hoa, either filled by an accumulation of detrital materials, or rendered dry by a slight and recent lowering of sea level. Hoa obviously originate in erosion as Newell (1956) suggested took place at Raroia. The erosion would have started from the numerous notches crossing the conglomerate due to slightly greater density in the reef.

e. Lagoon Edge. Lagoonward the reefs lie on a more or less wide platform formed by the ancient reef conglomerate, or by older reef limestone. On the windward side (N and W of the atoll), they constitute a continuous flat 5–20 m wide, which is dead on the surface, but alive on its edge.

Leeward reefs (E and S of the atoll) are disrupted and composed of a great number of patch reefs reaching the sea surface and of submarine knoll reefs scattered or aligned in the wind direction SE–NW, or even along the shore 100–200 m from it. Their distribution is under the control of winds and the dynamics of ocean waters coming into the lagoon during storms and causing a heavy bioclastic sedimentation on the southern rim.

2. The Lagoon

The bottom is formed by a small platform, 40–50 m deep in the central part, which slopes up gently eastward, southwestward, and southward. Except close to the lagoon fringe, patch reefs are distant from one another. In contrast, knoll reefs which do not reach sea surface are very numerous. A lot of them are in a process of destruction. On the lagoon shore two submarine terraces, respectively, 6–8 and 20–25 m deep, were discovered.

In the lagoon, sedimentation is very important. Close to the reef rim and to the patch reefs the sediments are nearly rough sands made of the debris of corals (*Acropora*), lithothamnion, Foraminifera (*Amphistegina lessoni, Marginopora vertebralis*) and mollusc shells. In the center there are fine sediments, but no calcareous mud can be observed. At Mururoa, *Halimeda* is very scanty.

3. Communications between Lagoon and Ocean

Such are assured through functional hoa over windward reef plates or through the shallow (8–9 m deep) and large (4500 m wide) pass, which is formed by a small platform covered with numerous knoll reefs. Reef-builders flourish.

4. Subreef Geology

About a hundred shallow drillings, between 15 and 30 m deep, and three deep ones were carried out at Mururoa. Two of the latter penetrated to the volcanic substrata at depths of 415 and 438 m, (Fig. 5), respectively. The drills encountered only coral debris before they reached the andesitic rocks. The simplified cross section is as follows.

1. At the top is a not very coherent conglomeratic layer about 100 m thick with great amounts of coral debris, lithothamnion, and Foraminifera (*Homotrema rubrum*). But included are three harder layers. One is superficial (at the level of the ancient conglomerate referred to above). The other two are situated between levels −6 and −20 m and are correlated with eustatic changes.

2. A middle layer made of conglomeratic finely crystallized coral limestone which is generally massive and coherent (250–300 m thick).

3. A lower detrital structure made of coral sands and gravels (40 m thick).

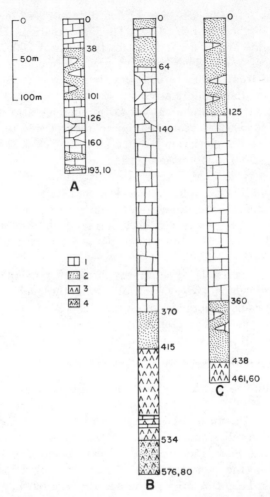

Fig. 5. Schematic sections of the deep drillings at Mururoa atoll. (A) E. Dindon; (B) W. Dindon; (C) Colette (cf. Fig. 2). (1) Subcompact coral limestone; (2) sands, gravels, and coral debris; (3) unaltered volcanic formations; (4) volcanic breccia formations.

4. Andesitic formation of Hawaiian type with the upper part made of breccia crowded with coral reef fragments.

Downward, the aragonite disappears at approximately 30 m depth and is replaced by calcite. Dolomitic limestones appear suddenly at about 100 m depth; they are magnesian limestones, not true dolomite.

Dating of the andesite sampled at 438 m depth, by K/Ag, indicates

7 million years ±1 (Pliocene time). The different reef levels have been dated by ^{14}C method or by $_{234}Ur/_{230}Th$ determination. The part of the reef conglomerate above sea level dates from about 3000 years (sample from a depth of 7 m). Then without any intermediate stage, there is a jump in age from 8000 to 100,000 years, clearly corresponding with the elevation of the atoll relative to sea level during the last glacial age (Würm.). At a depth of about 20 m there is also a jump in age from 200,000 to 500,000 years related to the relative uplift which occurred during the last glaciation but one (Riss).

From these determinations and from as yet unpublished results of seismic tests, it is concluded that Mururoa originated in Pliocene time by the formation of two volcanos of Hawaiian type, around which reefs grew, even before the complete cessation of volcanic action. A slow subsidence interrupted by emergences during glacial times occurred until the present day. Exposed limestones do not go back to the Tertiary as Agassiz thought (1903).

At present Mururoa is affected by three phenomena: (1) Erosion and dissolution which are evident all along the reef rim, tending to destroy the ancient emerged conglomerate; (2) heavy sedimentation in the lagoon and on the reef rim; (3) reef-building, which prevails particularly on the outer edge and on the outer slope.

5. Comparisons with Other Atolls of Tuamotu Archipelago

From one atoll to another, important morphological modifications can be observed.

The height of the ancient reef shows variation, 3.5 m above sea level at Matahiva, 5 m at Niau, and 6 m at Anaa. Furthermore at Rangiroa, Guilcher et al. (1969) and Stoddart (1969) found that the feo on the southern rim of the atoll represented remnants of an ancient in situ reef raised to a height of 2.5 m, whereas at Mururoa they are composed of reef conglomerate. The ancient reef is at least 15,000–20,000 years old and probably older (Guilcher et al., 1969). Veeh (1966) dated Anaa feo and found they were 80,000–100,000 years old. In these atolls the ancient reef, where it was in a position to grow, would have been erected during the interglacial period (between Riss and Würm).

Among the 80 atolls of the Tuamotu Archipelago, only 10 have two passes and 21 have only one pass. In the others the turnover of lagoon waters is carried out through the functional hoa and over the reef rim. The waters go out over the leeward side of the atoll (Guilcher et al., 1969). Thus these atolls are submitted to an heavy sedimentation which tends to fill them. Among the closed atolls can be distinguished deep atolls: Marutea Sud, etc.; shallow atolls: Reao, Matahiva, and Anaa where the lagoon is divided into several basins by a few exposed sand banks; Puka-Puka with a 4-m deep lagoon filled in great measure by a fine calcareous mud; Niau with a lagoon separated from the sea and communicating with it only by cracks, etc.; and atolls with filled-up lagoon: Nukutavake, Tepote, Tikei, and Akiaki.

This process of evolution is controlled by the initial size of the atoll, the depth of the lagoon, communications existing between it and the ocean, and lastly, by hydrodynamical factors.

C. MAKATEA ISLAND

On this entirely calcareous island with its highest point reaching 113 m can be seen (Fig. 11B) (Agassiz, 1903; Ranson, 1956; Obellianne, 1962; Doumenge, 1963) a peripheral rim made of strongly recrystallized, magnesian coral limestones which have been doubtfully determined as of late Tertiary age; an intermediate zone composed of shelly limestones, a detrital formation on the inner edges of the ancient lagoons; and, in the center, chalky limestones occupying the area of the ancient lagoons and containing calcium phosphate; the limestones contain *Borelia* cf. *pygmaeus* from Eocene time.

The high cliffs which show several marine notches at heights of 10.35 and 55 m, are bounded by a very narrow fringing reef. A submarine terrace 15 m deep was also recognized.

In Agassiz's opinion, Makatea is a raised Tertiary reef with no lagoon. According to his theory the lagoon-like basin observed at the present time was formed by an erosion process. Observers who visited the island afterward admitted, however, that Makatea is an old faro built during the Tertiary on submarine volcanos, subsequent to a subsidence estimated to have amounted to 300 m from the slope aspect. This uplift took place in Miocene times and was accompanied by fracturing and collapse. As soon as emersion began, the reaction of phosphatic ions and of birds feces (carried by rainwater) on the limestone resulted in the formation of tricalcic phosphate. The almost exhausted phosphate layers are not worked at present.

IV. Coral Reefs of the Society Islands

A. GENERAL ASPECT

In the Society Islands we can observe, from SE to NW, a gradual increase in depth of the different ridges which form the archipelago, an increase in the age of volcanic structures, and a greater physiographical maturity. Thus we can distinguish (1) the volcanic high islands in the southwestern part. Mahetia is a volcanic cone, practically unchanged. The cliffs are low and the fringing reefs not very important. At Tahiti, which is surrounded by a narrow barrier reef, the two volcanic cones are easily recognizable, but the center of each caldera is partly eroded. Erosion was even more active at Moorea, Huahine, Raiatea, Tahaa, and particularly at Bora-Bora, all of which are surrounded by barrier reefs. Maupiti shows only a volcanic peak fringed by a large reef bounding a narrow lagoon (Fig. 6). (2) The atolls, all of which are situated northwest of the archipelago, except Tetiaroa: they comprise Motu Iti, Mopelia, Scilly, and Bellingshausen atolls.

A great number of authors studied the reefs of the Society Islands, in particular, Davis (1918, 1928), Setchell (1922, 1926), Crossland (1928, 1939), Stark and Howland (1941), Guilcher (1968), and Guilcher *et al.* (1966, 1969).

B. TYPE OF HIGH ISLAND: TAHITI

Tahiti is composed of two ancient volcanoes in a NW–SE alignment and joined by an isthmus (Fig. 6). Although subject to considerable erosion, one can recognize the two calderas and the slopes of the volcanoes averaging 6°–10° which end in cliffs that are higher in the southeastern part. A narrow coastal plain (lacking at the southeast end) separates the cliffs from the sea. Three volcanic stages have been distinguished; during these basalt lava and tephritic flows were formed which have been dated from 2 to 20 million years. In the center of the caldera a nepheline gabbro which has been sampled, may be as old as 150 million years (middle Jurassic time). The island is surrounded by fringing reefs and a chain of barrier reefs limiting a narrow lagoon. *Porolithon onkodes* and, in some places, *Chevaliericrusta polynesiae* are the principal builders of Tahitian reefs (Setchell, 1926; Denizot, unpublished work) and are more numerous than corals which are abundant only on the outer slope. There are no raised reefs.

1. Fringing Reefs

In general there is not much living material. The reef-building organisms, because of the heavy alluvial deposits, flourish only on the edge (calcareous algae, but also corals). Some reefs are dead, killed by sediment (especially in Phaeton Bay) and others entirely capped with alluvial deposits. Reefs facing the passses and subjected to the

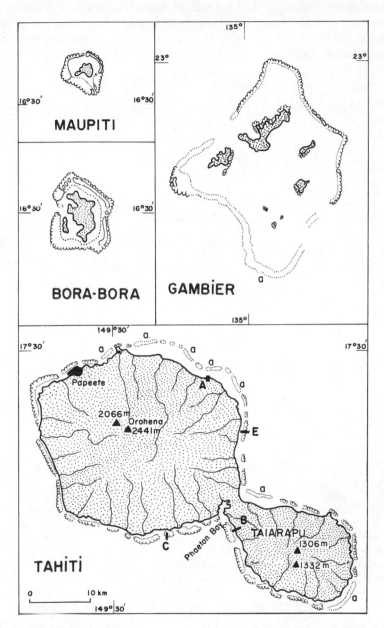

Fig. 6. Different types of high islands surrounded by barrier reefs. Every island is on the same scale. (a) Submerged reefs. (A, B, C, E) Cross sections of various types of Tahiti (see Fig. 7A,B,C,E).

swell, or situated in zones lacking barrier reefs, are wider, and are broken only at the mouths of large rivers. On the outer edge there is a small, low algal ridge, hardly exposed at low tide and made overall of *Porolithon onkodes*. Corals are abundant on the outer slope (Fig. 7A). The

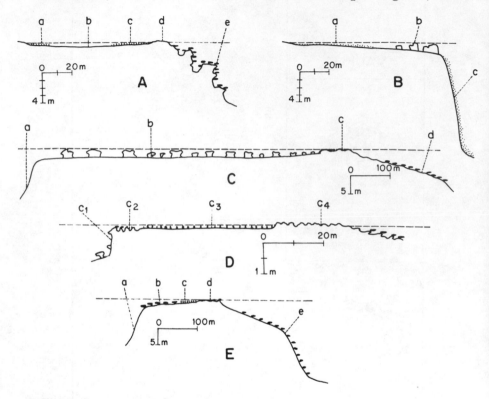

Fig. 7. Cross sections of various types at Tahiti (see situation of the reefs on the Fig. 6). (A) Fringing reef in the Tiarei district (reef facing the swell); (a) basaltic blocks; (b) inner flat partly covered with sargassoes; (c) outer part of the reef flat covered with algae (sargassoes, *Turbinaria*); (d) small algal ridge; (e) outer slope covered with corals. (B) Fringing reef in the Vairao district (sheltered reef); (a) dead reef flat covered with volcanic mud; (b) outer part of the reef flat bearing coral structures; (c) outer slope (sands with corals and algae). (C) Barrier reef in the Mataiea district; (a) lagoonward inner slope; (b) inner reef flat bearing small coral masses; (c) algal ridge and reef flat reaching the sea surface; (d) outer slope covered with corals. (D) Detail of the c zone on the Fig. 7C; (c_1) small inner escarpment; (c_2) zone with soft algae (sargassoes and *Turbinaria*) and scanty corals; (c_3) zone with sargassoes; (c_4) zone with *Porolithon*. (E) Barrier reef in the Hitiaa district; (a) inner lagoonward slope; (b) inner edge with corals; (c) zone with sargassoes; (d) low coral and algal ridge; (e) outer slope covered with corals (according to author's observations). The water level is at low tide.

sheltered fringing reefs, on the contrary, are disrupted, the outer edge being covered with corals which are more numerous than lithothamnion and the vertical outer slope is generally bare of building organisms (Fig. 7B). Besides lithothamnion, fringing reefs bear a great number of algae (*Sargasso, Jania, Turbinaria*).

2. Barrier Reefs

A belt of barrier reefs broken by passes surrounds the island. North, east, and southeast, they are approximately 10 m deep. The reefs are capped by only a few islets made of detrital unconsolidated material. Nowhere do raised reefs emerge.

West and south of the island barrier reefs are very wide, up to 1500 m; in places they grade into fringing reefs. They are composed (Fig. 7C,D) of an almost emerged surface edged outside by a low litho-thamnion ridge (*Porolithon onkodes*) and inside by a small escarpment 1 m high, and of a submerged flat sloping gently down toward the lagoon where small coral formations arise. The inner slope of the reef is covered with sands composed of calcareous algae (lithothamnion, *Peyssonnelia calcea, Halimeda*) and corals. Northward and eastward, on the other hand, the barrier reef is much narrower. Corals are generally more abundant on the outer edge (Fig. 7E).

The outer slope of the barrier reefs is comparatively gentle to a depth of approximately 70 m. Then, according to the borings conducted by the "Challenger" (1885), it becomes steeper (between 30° and 45°) to 300 m and thereafter decreases gradually. Volcanic materials appear at a depth of about 300 m. Below 800 m the bottom is made exclusively of volcanic mud. The rim of the reef is generally furrowed by shallow grooves extending into the water to a depth of not more than 20 m.

3. The Lagoon

This is deeper to the eastward on the windward side, where depths of 40–50 m are observed, than in the western part where the bottom does not exceed 30 m depth. Patch reefs and knoll reefs are numerous in the western and southern part. They are alive on the top, but subject to erosion on the walls.

The sedimentation in the lagoon was investigated by Guilcher *et al.* (1969). It is very complicated because of the rough surface of the island and of the narrowness of the lagoon. Four types of sediments have been distinguished.

1. Sediments with a small part of fine materials and a small percentage of $CaCO_3$ ($<26\%$) at river mouths.

2. Sediments with a small part of fine materials and a great percentage at $CaCO_3$ (>60%) close to the reefs.

3. Sediments with an important part of fine materials and a small percentage of $CaCO_3$ (<25%) facing river mouths, but in the center of the lagoon.

4. Sediments with an important part of fine materials and a great percentage of $CaCO_3$ (>60%) around the small volcano (Taiarapu) where the hydrographical system is less developed than on the chief volcano and where consequently less terrigenous sediments pour into the lagoon.

The study of sedimentation in the Tahitian lagoon led Guilcher et al. (1969) to state the following law: In wet tropical lands, the percentage of detrital and organogene sediments in the lagoons behind barrier reefs depends principally on the ratio of the surface area of the central volcanic noncoralline islands to the area of the lagoon.

4. The Passes

They correspond to three types (Crossland, 1928): (1) deep passes through which the lagoon floor slopes down gradually toward the ocean (e.g., Taunoa Pass). Such are located opposite wide river mouths; (2) shallow passes in which a step separates the floor of the lagoon and that of the ocean (e.g., Papeete Pass); (3) wide gaps only a few meters deep, crowded with reefs and such that only small boats are able to traverse them.

C. COMPARISON WITH OTHER HIGH ISLANDS

The morphology of Moorea reefs is similar to that of Tahiti reefs, but rather different from that of Bora-Bora reefs. Moorea reefs were investigated by Stark and Howland (1941), and especially by Guilcher et al. (1969).

At Bora-Bora, the reef rim reaches 1–2 km in width and is formed, like the Tuamotu atolls, by an ancient reef (elevated to 0.50–0.70 m), conglomerates, and old beachrock bearing islets (motu), which are located almost entirely in the eastern part (Fig. 6). The raised reef may date from 2250 years (^{14}C determination). The rim is incised by a great number of temporary and permanent functional hoa and by only one pass (20 m deep) to the west. The radially striated morphology and the sand tongues stretching into the lagoon are affected by the swell coming into the lagoon through the hoa and out through the pass.

The lagoon, because of rocky spurs extending under water, is divided into several shallow basins in the east (due to the gradual invasion of sands brought from outside through hoa), but to the west the depth exceeds 40 m in places. The lagoon contains few patch reefs.

The hydrology and sedimentology were thoroughly investigated at Bora-Bora (Guilcher *et al.*, 1969). The sediments of the lagoon, in spite of the existing volcanic island, are essentially calcareous (percentage of CaCO$_3$: 98–100%), coarse close to the reef and finer grained in the deep part of the lagoon (between 10 and 45 m). They are composed mainly of coral and lamellibranch debris.

D. Mopelia Atoll

Mopelia is the one atoll in Society Islands to have been studied in detail (Guilcher, 1968, Guilcher *et al.*, 1969). It exhibits the principal characteristics of the Tuamotu low islands.

The reef rim, about 1 km wide, is made up of an ancient reef not exceeding 1 m high and going back to 3450 years (by ^{14}C determination). That old emerged reef is found again at the top of the patch reefs in the lagoon. The islets, made of bioclastic materials and surmounting the rim, are located on the eastern part and are certainly influenced by the trade wind. There is only one narrow (less than 50 m wide) and shallow (4 m deep) pass (Fig. 8).

Hydrodynamical factors control the reef morphology. The flow of water brought by the south and southeast swell may raise the lagoon level. Discharge of waters is through the pass and over the reef rim on the windward side. To these water movements are ascribed the sand and gravel tongues in the western part of the atoll, the free ends of which stretch northwestward and not toward the lagoon.

The low algal ridge is made overall of *Porolithon onkodes* and *P. craspedium*. It is furrowed by a spurs-and-grooves system; the spurs are formed by the old reef with lithothamnion which tend to fill the grooves and to create the "room-and-pillar structure" that can be observed to a depth of not more than 7 m on the outer slope. The latter slopes gently to a depth of 15 m, more steeply to 25 m below that, and thereafter very abruptly.

The lagoon is divided into several basins and the floor goes down to approximately 40 m depth. It contains a great number of patch reefs and knoll reefs covered with corals, *Halimeda*, and mother-of-pearl oysters. Guilcher *et al.* (1969) show that, in spite of the one pass being narrow, a turnover of lagoon water does occur, even near the bottom when it is affected by the heavy swell coming from the south. This characteristic allows organisms to live at any depth.

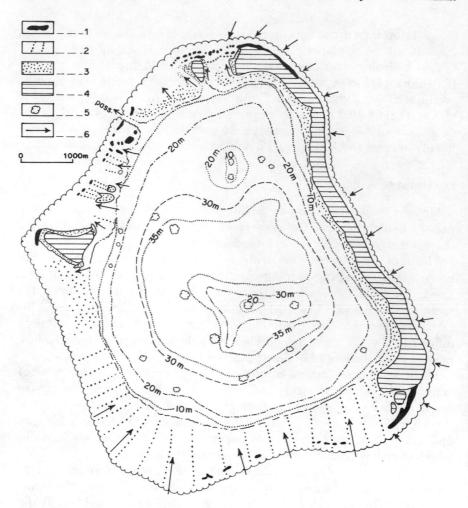

Fig. 8. Morphological map of Mopelia. (1) Old reef; (2) radial rays on the reef; (3) sands, gravels, and pebbles; (4) *motu* bearing vegetation; (5) patch reefs; (6) direction of surge currents. Isobaths are expressed in meters. (According to Guilcher *et al.*, 1969, but simplified.)

Sediments of the lagoon were thoroughly investigated. They contain from 99.33 to 99.99% $CaCO_3$ (insoluble residues include spicules of sponges and diatoms). They are heterometric, rather coarse on the reef and close to it, and finer grained in the deeper parts. They are composed mainly of coral debris, Foraminifera (Miliolidae, Orbitolinidae, Rotalidae), *Halimeda,* and Bryozoa (Fig. 9 and 10).

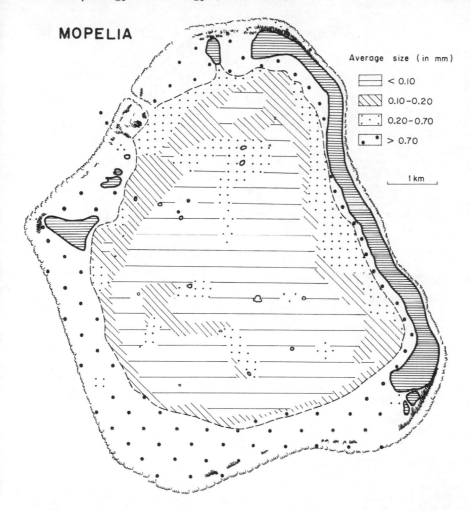

Fig. 9. Mopelia. Granulometric distribution of sediments after the average size in millimeters. (According to Guilcher *et al.*, 1969.)

E. ORIGIN OF THE SOCIETY ISLANDS REEFS

Many scientists have studied the origin of Society Islands reefs and particularly of Tahiti. Darwin (1842) saw here clear proof of his subsidence theory, each stage being represented, from fringing reefs to atolls. Dana (1886) added some precision to Darwin's ideas and was the first to explain the formation of the now filled bays in Tahiti by a sinking of the island. He thought the thickness of the barrier reef to be approximately 250 ft (75 m).

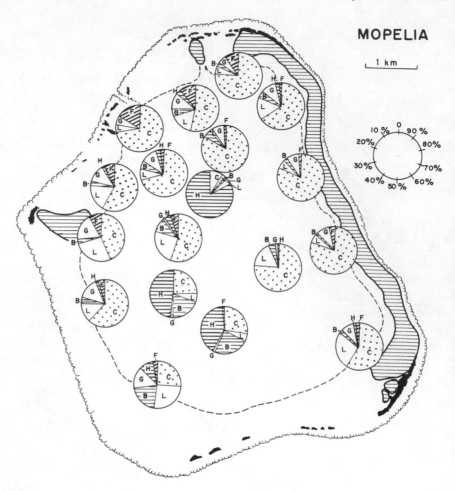

Fig. 10. Mopelia. Sedimentary composition of lagoon deposits (upper part of only 0.36 mm). B, Bryozoa; C, Coral debris; F, Foraminifera; G, Gasteropods; H, Halimeda; L, Lamellibranch. (According to Guilcher *et al.*, 1969.)

Murray (see "Challenger" Reports, 1885) denied the subsidence. According to him, reefs settle on an abrasion platform. The dissolving effect of water contributes to the deepening of the lagoon. Agassiz (1903) expressed this idea again, making it more precise. According to him, in the Society Islands, under the influence of the trade wind, the erosion process is more important in the east and the southeast than elsewhere. This would explain, in Tahiti, the formation of a deep lagoon around Taiarapu and of the submerged barrier reefs.

Daly (1915) did not study the Society Islands in particular; but in his opinion, his well-known theory of glacial control explained completely the formation of the reefs in this lagoon.

Davis (1918, 1928) did not deny the existence of a large marine abrasion platform which was formed at the close of volcanic activity. Nevertheless, following this period, during which the great cliffs of the high islands were formed, subsidence, estimated to have amounted to 120–150 m at Tahiti, and 300–600 m at Moorea and at Bora-Bora, took place. A barrier reef formed on the edge of the abrasion platform and later when alluvial deposition decreased on the lagoon margins, fringing reefs were initiated. The atolls situated northwestward resulted from still greater subsidence.

In contrast, Setchell (1926) in his explanation of the origin of the Tahitian reefs did not agree with the subsidence theory, but accepted the growth of fringing and barrier reefs on an abrasion platform, formed, it seems, during glacial times. This author originally suggested that the first reef-building organisms would have been solely calcareous algae. Referring to the growth of these organisms, Setchell dated the initiation of the first Tahitian reefs at 180,000 years. According to him, the barrier reefs were submerged subsequent to a cessation of coral growth during later volcanic eruptions.

Crossland (1928, 1939), like Setchell, put aside subsidence theory. With Murray and Agassiz, he contended that reefs built on a platform made of detrital materials surrounding high islands were continuous from shore to the outer edge, as was proved by the presence (very rare) of basaltic stones on the outer part of the Tahitian barrier reef. The lagoon and passes were formed, according to Crossland, by erosion of radial and circular cracks. At Moorea a collapse of the southeastern part occurred, drowning the cliffs of the southern coast. The two big fiords which exist in the northern part of the island, were created by the explosive breaching of the crater, according to this author. In Crossland's opinion the barrier reef in Tahiti would not exceed 70 fm in depth and would have been built up after the Würm glacial time.

Williams (1933) was the first in fact, to apply the glacial control theory to Tahitian reefs. He thought the abrasion platform, on which the barrier reef rests, to be 350–400 ft wide and that its initiation goes back 50,000 years. Investigations at Bora-Bora led Stark and Howland (1941) to explain the origin of this island and of its reefs by subsidence along with collapse of the western part of the volcanic structure. Nearer our time, Obellianne (1955) thought subsidence alone to be sufficient to explain the origin of reefs. The absence of barrier reefs northeast and southeast of Tahiti was due, according to this author, to collapse of some parts of this island.

On the barrier reef near Papeete two drill holes sunk in 1965 encountered the volcanic substratum at approximately 90 and 100 m depth (Deneufbourg, 1971). This significant result perhaps supports the glacial control theory. In Tahiti it would not be necessary to apply subsidence theory which, however, seems to be very likely at Bora-Bora and in the northeast atolls.

V. The Austral Islands

A. GENERAL REVIEW

According to Chubb (1927b) the islands are aligned on three SE–NW oriented ridges, including from north to south, a ridge bearing Rurutu Island; a ridge stretching from President Thiers Bank to Maria Island and further; and a ridge stretching from Marotiri to Rapa and Nelson Reef.

Tubuai Archipelago exhibits more varied morphological types than other island groups in Polynesia. There is no difficulty in distinguishing (1) volcanic high islands made of basalts, with trachytes ranging in age from 5 million years (Rapa) to 30 million years (Tubuai), some with only poorly developed fringing reefs (Rapa, Marotiri), others surrounded by barrier reefs (Raevavae, Tubuai); (2) high islands formed both by volcanic structures and raised reefs (Rurutu, Rimatara); and (3) reef banks reaching the surface (Moses Reef) or submerged (Nelson Reef 4 m deep and the President Thiers Reef 30 m deep).

Explaining the origin of the islands, Chubb (1927b) suggested the existence of a monoclinal fold running northeastward. On its crest, only raised reefs are to be seen, bounded by narrow fringing reefs; in contrast, on the southeast side, affected by subsidence, are high islands surrounded by barrier reefs. Furthermore, the island alignments are in tiers so that, as the fold stretches toward the northeast, the locus of maximum folding shifts laterally. No worker later than Chubb agrees with this hypothesis. In addition to Chubb, Aubert de la Rüe (1959) and Obellianne (1955) investigated the raised reefs of the Austral Islands.

B. TYPE OF HIGH ISLAND WITH FRINGING REEFS: RAPA

Made entirely of basalt, Rapa is the most southern island in French Polynesia and is situated at the southern limit of coral reef distribution.

In spite of high cliffs, there is around the island a gently sloping platform reaching 3 km in width. It could have served as a foundation for a barrier reef. But at Rapa the reefs are all fringing and exist only in bays or at their entrance (Chubb, 1927b). They are formed of calcareous algae and also corals which are never exposed. The submarine platform surrounding the island is formed by a reef pavement made of lithothamnion covered with sand and sargassoes; corals are not abundant and do not contribute to reef-building. Builder organisms are obviously hindered in their growth by the relatively low temperature of the water, as is proved also by the presence of some particular seaweeds.

C. TYPE OF HIGH ISLAND SURROUNDED BY BARRIER REEFS: RAEVAVAE

At Raevavae, the island consists of both basalt and trachyte, the products of volcanic eruptions which occurred in Pliocene times (Obellianne, 1955). Fringing reefs are living only on their edges and are essentially built by calcareous algae.

The reef rim is an ancient conglomerate formed by cemented coral colonies, reaching 80 cm above sea level and capped with unconsolidated material (islets or *motu* and ramparts). On the leeward side there is no true algal ridge. On the south and east the reef faces a heavier swell and is more developed. It is however, low and broken by shallow grooves and spurs which extend everywhere onto the outer gently inclined slope. The barrier reef is broken by two passes, north and south of the island.

The lagoon consists of a wide outer shallow zone 2–6 m deep, the bottom being formed by sands containing *Halimeda* and by a few coral colonies, and an inner channel 1 km wide and 10–20 m deep. The lagoon is in the process of being filled.

According to Obellianne (1955) the eruption of trachytes which marked the end of volcanic activity (after the basaltic stage) brought about uplift at the eastern of the island of up to 100 m (as elevated marine limestones at this level bear witness) and the formation of a N–S fracture which would have initiated the two passes. The erosion formed a submarine platform on which fringing and barrier reefs arose, perhaps thanks to a subsidence.

D. RURUTU (Fig. 11A)

Rurutu, with its highest point 400 m above sea level, is made up of basalts, coral sands, and, up to 100 m, of detrital (in part dolomitic) limestones. It is edged by a narrow fringing reef. Terraces raised to a

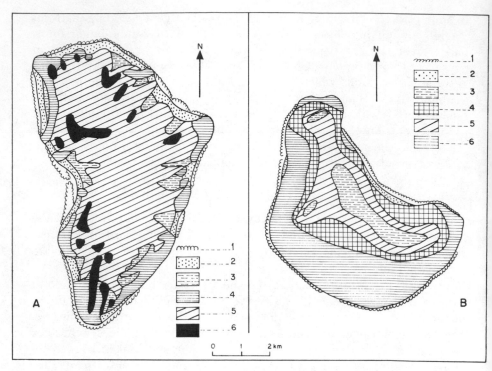

Fig. 11. Geological map of Rurutu (A) and of Makatea (B). (A) (1) fringing reefs; (2) recent marine formations; (3) recent alluvial deposits; (4) raised coral limestone; (5) laterites coming from the weathering of volcanic rocks; (6) basalt. (B) (1) fringing reefs; (2) recent marine formations; (3) chalky limestones (ancient lagoons); (4) coral and shelly limestones alternating with each other; (5) shelly limestones; (6) massive coral limestones. (According to Obellianne, 1955, 1962, but simplified.)

height of 200 and 260 m., situated in the volcanic part of the island, represent, it seems, ancient sea levels (Chubb, 1927a). There is also a terrace at a height of 6–8 ft and another one, not so well defined, at 1 or 2 ft elevation. Chubb explained this morphology simply by the elevation of a volcanic island lying on the crest of a monoclinal fold and surrounded by a fringing reef. Obellianne (1955) added glacial eustatism to the raising process, whereas Davis (1928) thought that the island was subjected to subsidence before the elevation of the marine limestones to a height of 100 m.

Rimatara exhibits a similar formation, the coral limestones being raised to a height of 10 m.

VI. Conclusions

The principal points which may be drawn from this study are as follows.

1. In spite of their geological and geomorphological variety, all the islands in French Polynesia show similar characteristics: they have a volcanic origin and they served as foundations for more or less developed fringing and barrier reefs or for atolls. The reef-building began, it seems, in Eocene times (?) at Makatea, in the Pliocene at Mururoa, and the Pleistocene time in the Marquesas Islands, at Mahetia, and at Tahiti. The upper part of these formations dates from the last interglacial period or the holocene.

Although the islands or the reef banks lie on SE–NW-oriented ridges, a single comprehensive explanation cannot be given for the origin of all Polynesian reefs. Subsidence took place at the time of origin of the Tuamotu atolls, as the drillings on Mururoa reefs strikingly demonstrate. It does not explain, however, the morphology of all the reef islands. Subsidence perhaps did not occur at Tahiti and at some of the islands of the Tubuai Archipelago (Rurutu). In this case the theories of an abrasion platform with no variation of sea level, and of glacioeustatism seem to be essential. The oscillation of sea level during glacial time is easily recognizable: it is apparent, even in areas where subsidence occurred, in the Tuamotu for example, and seems, furthermore, to have played an important part in the building up of barrier reefs.

2. At the present time reef-building by living organisms is essentially restricted to the outer sides of reefs. Crossland (1928) thought that coral growth declined and that reef-building did not increase, at least in the Society Islands. Recent work shows, however, that on the outer side corals and calcareous algae are active. Lithothamnion are indeed the principal builders of the outer edge of the atolls and barrier reefs. On the other hand, it is noted that a slow filling of the lagoon by terrigenous alluvial deposits or by bioclastic materials is taking place.

Selected References

Agassiz, A. (1903). The coral reefs of the tropical Pacific. *Mem. Mus. Comp. Zool. Harvard* 28, I–XXXIII, 1–410.

Aubert de la Rüe, E. (1959). Etude géologique et prospection minière de la Polynésie Française. *In* "Recherche géologique et minéralogique en Polynésie Française," pp. 7–43. Paris.

"Challenger" Expedition. (1885). "Report of the Scientific Results cf the Voyager . . ." (under the superintendance of Sir C. W. Thomson and J. Murray), Narrative, Vols. I–II.

Chevalier, J. P., Denizot, M., Mougin, J. L., Plessis, Y., and Salvat, B. (1968). Etude géomorphologique et bionomique de l'atoll de Mururoa (Tuamotu). *Cah. Pac.* 12, 1–189.

Chubb, L. J. (1927a). Mangaia and Rurutu a comparison between two Pacific Islands. *Geol. Mag.* 64, 518–522.

Chubb, L. J. (1927b). The geology of the Austral or Tubuai Islands. *Quart. Geol. Soc. London* 83, 291–316.

Chubb, L. J. (1930). The geology of the Marquesas Islands. *Bull. Bishop Mus., Honolulu* 68, 1–71.

Cole, W. S. (1959). Asterocyclina from a Pacific Seamount. *Contrib. Cushman Found.* 10, 10–14.

Crossland, C. (1927). Marine ecology and coral formations in the Panama region, Galapagos and Marquesas Islands and the atoll of Napuka. *Trans. Roy. Soc. Edinburgh* 55, 531–554.

Crossland, C. (1928). Coral reefs of Tahiti, Moorea and Rarotonga. *J. Linn. Soc. London* 36, 577–620.

Crossland, C. (1939). Further notes on the Tahitian Barrier Reef and lagoons. *Linn. Soc. London, Zool.* 40, 459–474.

Daly, R. A. (1915). The glacial control theory of coral reefs. *Proc. Amer. Acad. Arts Sci.* [2] 51, 175–251.

Dana, J. D. (1849). "Geology. U.S. Exploring Expedition." Philadelphia, Pennsylvania.

Dana, J. D. (1886). A dissected volcanic mountain (Tahiti), some of its revelations. *Amer. J. Sci.* [3] 32, 247–255.

Darwin, C. (1842). "The Structure and Distribution of Coral Reefs," 2nd ed. London.

Davis, W. M. (1918). Les falaises et les récifs coralliens de Tahiti. *Ann. Geogr.* 27, 241–284.

Davis, W. M. (1928). The coral reef problem. *Amer. Geogr. Soc., Spec. Publ.* 9, 1–596.

Deneufbourg, G. (1969). Les forages de Mururoa. *Cah. Pac.* 13, 47–58.

Deneufbourg, G. (1971). Etude écologique du Port de Papeete, Tahiti (Polynésie Française). *Cah. Pacifique*, 15, 75–82.

Doumenge, F. (1963). L'île de Makatea et ses problèmes (Polynésie Française). *Cah. Pac.* 5, 41–68.

Emery, K. O., Tracey, J. I., and Ladd, H. S., (1954). Geology of Bikini and nearby atolls. Part I. Geology. *U.S., Geol. Surv., Prof. Pap.* 260-A, 1–265.

Fontes, J. C., Kulbicki, G., and Letolle, R. (1969). Les sondages de l'atoll de Mururoa: Aperçu géochimique et isotopique de la série carbonatée. *Cah. Pac.* 13, 69–74.

Guilcher, A. (1968). Transport of sediments over atoll rims and barrier reefs in South Pacific. *Akad. Nauk SSSR, Inst. Okeonol.* pp. 242–250.

Guilcher, A., Denizot, M., and Berthois, L. (1966). Sur la constitution de la crète externe de l'atoll de Mopelia ou Maupihoa (îles de la Société) et de quelques autres récifs voisins. *Cah. Oceanogra.* 18, 851–856.

Guilcher, A., Berthois, I., Doumenge, F., Michel, A., Saint-Reguier, A., and

Arnold, R. (1969). Les récifs coralliens et lagons coralliens de Mopelia et de Bora-Bora (îles de la Société). *Mem. O.R.S.T.O.M.* **38**, 1–103.

Labeyrie, J., Lalou, C., and Delibrias, G. (1969). Etude des transgressions marines sur l'atoll de Mururoa par la datation des différents niveaux de corail. *Cah. Pac.* **13**, 59–68.

Menard, H. W. (1964). Marine geology of the Pacific. *Earth Sci. Int. Ser.* pp. 1–269.

Newell, N. D. (1956). Geological Reconnaissance of Raroia (Kon Tiki) Atoll, Tuamotu Archipelago. *Bull. Amer. Mus. Natur. Hist.* **109**, 312–372.

Obellianne, J. M. (1955). Contribution à l'étude géologique des îles des Etablissements Français de l'Océanie. *Sci. Terre* **3**, Nos. 3–4, 1–146.

Obellianne, J. M. (1962). Le gisement de phosphate tricalcique de Makatea. *Sci. Terre* **9**, No. 1, 1–60.

Ranson, G. (1952). Note sur la cause probable de l'absence de récifs coralliens aux îles Marquises et de l'activité réduite des coraux récifaux à Tahiti, aux Tuamotu, aux Hawaii etc . . . *C. R. Somm. Soc Biogéogr.* No. 248, pp. 3–11.

Ranson, G. (1956). Observations sur les falaises et les phosphates de Makatea (île des Tuamotu). Le problème de la consolidation des sédiments. *Proc. Pac. Sci. Congr. 8th, 1954* Vol. III A, pp. 909–918.

Setchell, W. A. (1922). A reconnaissance of the vegetation of Tahiti with special reference to that of the reefs. *Carnegie Inst. Wash., Yearb.* **21**, 180–187.

Setchell, W. A. (1926). Phytogeographical notes on Tahiti. II. Marine vegetation. *Univ. Calif., Berkeley, Publ. Bot.* **12**, No. 8, 291–324.

Stark, J. T., and Howland, A. L. (1941). Geology of Bora-Bora, Society Islands. *Bull. Bishop Mus., Honolulu* **169**, 1–43.

Stoddart, D. R. (1969). Reconnaissance geomorphology of Rangiroa Atoll, Tuamotu Archipelago. *Atoll Res. Bull.* **125**, 1–44.

Trichet, J. (1969). Quelques aspects de la sédimentation calcaire sur les parties emergées de l'atoll de Mururoa. *Cah. Pac.* **13**, 1–16.

Veeh, H. H. (1966). Th_{230}/U_{238} and U_{234}/U_{238} ages of Pleistocene high sea level stand. *J. Geophys. Res.* **71**, 3379–3386.

Wiens, H. J. (1962). "Atoll Environment and Ecology." Yale Univ. Press, New Haven, Connecticut.

Williams, H. (1933). Geology of Tahiti, Moorea and Maiao. *Bull. Bishop Mus., Honolulu* **105**, 1–89.

5

CORAL REEFS OF NEW CALEDONIA

J. P. Chevalier

I. General Survey

New Caledonia and Loyalty Islands lying east of the Coral Sea are bounded by the parallels of lat. 18°S and 23°S, and the meridians of long. 163°E and 169°E. This group of reefs forms two chains with a SE–NW orientation, a continuation of the Norfolk Ridge (or New Caledonia Ridge) which extends as far as New Zealand. This ridge sinks slowly northeast of New Caledonia. It is limited to the north and to the east by the New Hebrides Trench and South Fijian Basin (depth 4000–7000 m), and to the west by the Norfolk Trough, the deepest part of the Coral Sea (up to 4000 m).

A. PHYSIOGRAPHICAL AND GEOLOGICAL SURVEY

The Loyalty Islands contain raised reef islands or atolls. On the other hand, New Caledonia presents a more complex structure; 400 km long and 40–50 km wide, it is a mountainous island the highest point being Mont Panie (1639 m). The central chain is asymmetrical. The northeast slope plunges sharply into the sea and the coasts offer very steep profiles (50°–60°), broken near the sea by vertical cliffs 3–10 m high (Davis, 1926) and by deep and narrow bays. The southwest side on the contrary, ends toward the sea in gently sloped hills and presents no cliffs, but large bays filled with a great number of islands.

Davis (1926) described an old peneplain sloping down northwestward from 700 m high at the south end of the island to hardly 100 m high in the northern part.

New Caledonia displays various and strongly folded structures where crystalline schists and especially peridotites and serpentines prevail. Routhier (1953) distinguished three orogenic periods, the latest one corresponding probably to the emergence of the island in Oligocene time.

B. CLIMATOLOGICAL DATA

Situated in a wet tropical zone, New Caledonia presents a hot season with heavy rains (from December to March), a cooler but wet season (from June to August) and two drier seasons alternating with each of them. The mean temperature averages 25.5°C. The trade wind, blowing SE and ESE, predominates, is more regular in the north of the island than in the south, and stronger along the east coast. The trade winds are often interrupted by northwesterly and westerly winds. Whereas the rainfall averages 2 or 3 m on the east coast, the west coast is drier (0.90–1.20 m).

Cyclones occur frequently during the hot season. They come from the north and move south or southeast, causing severe damage to the reefs.

C. OCEANOGRAPHICAL DATA

New Caledonia and the Loyalty Islands are surrounded by waters with temperature suitable to coral growth, as Table I shows.

TABLE I

MEAN TEMPERATURE OF THE SEAWATER

	New Caledonia		Loyalty Islands	
	North	South	Maré	Uvéa
February	26.8°	26°	26°	26.5°
August	23.5°	22°	22.5°	23°

The variation of water temperature in the lagoon between winter and summer is a little more pronounced and sometimes in winter the temperature may fall significantly. Thus on the 29th of August 1942 it dropped to 16° at 7 AM at Nouméa.

Lagoon circulations have been little investigated. They are correlated with tides. In the passes currents are often very strong and may be accelerated or slowed down by changes in wind direction.

A heavy swell resulting from storms occurring in high austral latitudes often breaks on the south reefs. The highest amplitude of the semidiurnal tides ranges from 1.50 to 1.70 m.

D. THE CORAL REEFS (Fig. 1)

The Neo-Caledonian reef structures show a great variety in the morphological patterns. They include, especially on the west and east coasts, two barrier reefs which extend northward far beyond emerged lands; these barrier reefs enclose around the island a large lagoon containing "knoll reefs," "patch reefs," "table reefs," "circular reefs," and "fringing reefs." Furthermore in the north we find atoll-shaped reefs: the D'Entrecasteaux Reefs, whereas in the south, Island of Pines is surrounded by a wide range of reef types. To this group of recent reefs may be added raised reefs: the Loyalty Islands, Island of Pines, and the fossil reef of Yaté-Touaurou.

The Neo-Caledonian reefs have rarely been investigated. Chambeyron (1875, 1876) published a topographical and hydrographical study on the subject and Davis (1926) searched for the origin by making a detailed physiographical analysis of the island. Other works include: Risbec (1930), Haeberle (1952b), and Avias (1959), which give a general, but brief survey of the Neo-Caledonian reefs and related problems. The raised reefs have been still less studied (Sarasin, 1925; Davis, 1928; Haeberle, 1952a; Koch, 1958b). Nevertheless, after the French expedition to the coral reefs in New Caledonia 1960–1963 (Singer-Polignac Expedi-

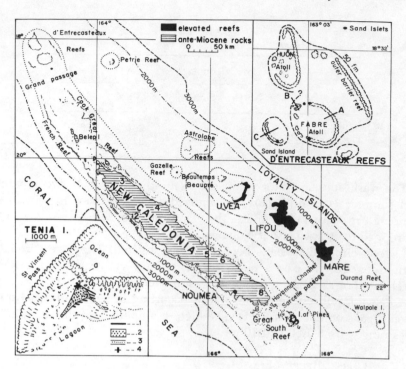

Fig. 1. Coral reefs of New Caledonia and Loyalty Islands. (1) Saint Vincent Bay; (2) Koné; (3) Pouébo; (4) Tuo; (5) Canala; (6) Thio; (7) Kouakoué; (8) Yaté. Details of D'Entrecasteaux Reefs (according to the marine chart, Haeberle, 1952b; Taisne, 1965). Cross sections A, B, C of the reefs (see Fig. 2). Details of Tenia Island (Saint Vincent Bay). (a) Depths 1.5–3 m (beginning of the building of a little faro). (1) Beachrock; (2) sandy islet; (3) sandy offshore bar; (4) place of the deep drilling (according to aerial photographs).

tion) a few articles have been published: Taisne (1965), Guilcher (1965), Chevalier (1968), and others will follow.

II. D'Entrecasteaux Reefs

These form a well separated group distinct from other Neo-Caledonian reefs in their morphology and situation. Chambeyron (1876) and Haeberle (1952b) described them, but very briefly. They were visited during the Singer-Polignac Expedition (as yet unpublished observations).

They lie on an irregular-shaped platform with steep slopes. These are more gentle southeastward toward the "Grand Passage" which separates them from the New Caledonian reefs. The "Grand Passage" goes down to nearly 1300 m in depth.

D'Entrecasteaux Reefs include two circular reefs, Huon and Fabre atolls, two oblong reefs northeast and southwest, and a few table reefs. They border the platform on which they are built (Fig. 1).

A. ATOLLS

Huon and Fabre atolls are 22 and 18 nautical miles long, respectively. The reef rim, from 500 to 1000 m in width, uninterrupted except on the leeward side where large passes occur, bears only a few islets made of coral sand and surrounded by beachrock. At low tide it is submerged to a depth of between 50 cm and 1 m. Except on its inner, lagoonward side, the reef rim contains only a few reef-building organisms, certainly because of the cyclones which frequently reach these northerly reefs. There is no algal ridge. The structure of this rim is practically similar all around the atoll. On Fabre atoll the following zones can be distinguished (Fig. 2A):

1. Reef edge submerged under 50 cm of water; contains few corals and little *Lithothamnion;* is slightly furrowed by shallow grooves extending down the outer slope

2. Reef flat covered with marine phanerogams, bearing a few corals (low microatolls)

3. Slope covered with sands and coral heads

4. Steeper slope covered with sands containing *Lithothamnion,* corals, and Foraminifera

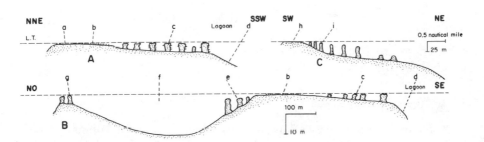

Fig. 2. Cross sections of the reef rim of Fabre atoll (A and B) and of the Southwest Reef (C) (see localization Fig. 1). (a) Outer edge of the reef; (b) zone with marine phanerogams; (c) sandy plate with coral heads; (d) inner slope; (e) knoll reefs of the inner edge toward the lagoon; (f) incomplete lagoon of the faro; (g) patch reefs on the outer edge; (h) reef rim; (i) patch reefs. The A and B figures are on the same scale (author's observations).

Sometimes the reef rim tends to form a little "faro"; northwest of the Fabre atoll it recurves outward and a series of little patch reefs poor in reef-building organisms tend to isolate an approximately 20 m deep pseudolagoon (Fig. 2B).

The floors of the lagoons of these two atolls slope gently westward, certainly because of important sedimentation in the eastern part. Fabre atoll drops to 60 m in depth and contains numerous knoll reefs which do not reach the surface, and comparatively few patch reefs with tops which are close to the water surface. On the other hand, Huon atoll is not so deep (40 m) and is overcrowded with patch reefs. Passes on Fabre atoll do not exceed 10–20 m in depth and form a step between the bottom of the lagoon and the outer slope. Nothing is known about the passes of Huon atoll.

B. Oblong Reefs

Chambeyron (1876) gave only a little information on the "Outer Barrier Reef" which is situated in the eastern part of the d'Entrecasteaux Reefs. It stretches for about 20 miles and emerges above sea level during low tide except in the center where it is submerged to depths of between 2 and 4 m.

The rim of the horseshoe-shaped reef runs along the submarine platform and bears a sandy islet. The lagoon floor is strongly inclined northeastward and contains fewer and fewer reefs as we proceed away from the reef rim. The structure of the latter is similar to that of atolls (Fig. 2C).

III. Reefs Around New Caledonia

These are bounded in the north by the "Grand Passage" and in the south by the Havannah Pass.

A. Barrier Reefs

These form two SW–NE alignments which border the submarine platform and are 1600 km long. Whereas in the central part they are not far from the coast, though near enough in places to form "offshore reefs" (at Koné approximately 15 km long), the distance from the coast becomes much greater at the north and south ends. In the north, the two barrier reefs (the French Reef and the Cook Great Reef) are situated 20–30 km from the Belep Islands, and nearer to the equator, they border a lagoon without emerged land. Close to the "Grand Passage" the two barrier reefs are not so continuous, being broken by a great number of passes which end abruptly. In the south, the barrier reef running along the eastern coast sinks to 10 m depth and ends at the mouth of the Havannah Pass. Southwest, the barrier reef lies far from the coast and recurves. It becomes broken and ends in isolated patch reefs.

The width of the barrier reefs varies greatly: from 200 to 4000 m (average: 300–500 m).

The barrier reefs of the eastern and western coasts are not similar. The former are more irregular, broken by a great number of wide passes and sometimes even (in Tuo area) they form a succession of well-separated steps (Guilcher, 1965). Furthermore they can be submerged under a few meters of water (area of Canala, Kouakoué, Yaté). On the other hand, the western passes are less numerous and narrower. The coral heads bordering the inner part of the barrier reef are more numerous than in the east.

On the barrier reef the following zones can be distinguished: (Fig. 3A,B).

1. The outer slope, which is generally steep and irregular. There is a terrace at about 20 m depth, that can be seen near Tuo (Guilcher, 1965) near Canala, in the Saint Vincent area, and another not so well defined between 6 and 10 m deep; between these two terraces the slope is very steep and the walls of the reef are in places vertical. Downward, the gradient is 6–50% to a depth of about 1000 m, then the slope becomes gradually more gentle. On the contrary the eastern side of the South Great Reef slopes much more gradually. Between this and Island of Pines the bottom does not exceed 500 m in depth (Taisne, 1965; Guilcher, 1965).

Generally the outer slope supports a great number of reef builders (corals and *Lithothamnion*) to about 50 m in depth. It is also crossed by a grooves and spurs system formed by reef-building and erosion processes. Most of the grooves are shallow and well separated and are up to 20 m deep. They incise only slightly the reef edge, but are more developed on the barriers of the eastern coast. They may be lacking on the southwestern coast.

2. The reef plate, which lies under a few decimeters of water at low tide or emerges here and there. There is no *Lithothamnion* ridge at the outer edge. In this part of the reef calcareous algae play a less important role than corals in the reef building. The edge of the eastern barrier reefs is often dead, only covered with marine phanerogams, the

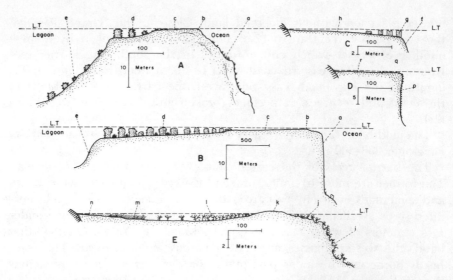

Fig. 3. Cross sections through various types of New Caledonian reefs. (A) Barrier reef near Canala (eastern coast). (B) Barrier reef of Saint Vincent Bay (western coast) (see localization Fig. 4). (C) Fringing reef of the eastern coast at Hugon Island, Saint Vincent Bay (localization Fig. 4). (D) Fringing reef of Bogota near Canala (localization Fig. 6). (E) Fringing reef of Mathiew Island, Saint Vincent Bay (localization Fig. 4): (a) outer slope; (b) outer edge; (c) outer plate; (d) gently sloped plate with coral heads; (e) inner slope; (f) outer slope of the fringing reef; (g) large microatolls made of *Porites;* (h) reef plate capped with sand; (i) outer slope covered with corals; (j) reef edge with *Acropora;* (k) accumulation of dead *Acropora* branches; (l) area rich in branched *Acropora;* (m) zone with marine phanerogams; (n) sandy flat; (p) vertical outer slope; (q) outer reef flat with *Platygyra;* (r) inner part of the reef plate containing few building organisms (author's observations).

organisms having been killed by frequent cyclones. On the other hand, the rim of the southwestern reefs is rich in encrusting corals. Surge channels do not occur frequently.

The depth of the reef plate decreases slightly toward the lagoon and with the deepening colonies of corals and encrusting *Lithothamnion* becomes more numerous. Microatolls are numerous. The inner edge of the reef runs in an uneven line and is furrowed also by spurs and irregular grooves that are much shallower than on the outer slope. On the Great South Reef, however, they may extend up to 10–12 m in depth (Guilcher, 1965)

3. A gently sloped flat, the width of which varies considerably and

which bears coral heads or little patch reefs, often in a great number and rich in corals, and which are aligned in the direction of swell.

4. An inner slope covered with organogene sands and with only a few coral structures.

Because of the throwing back of the swell, barrier reefs often recurve at entrance to passes, generally toward the lagoon but sometimes outward (northeast of Saint Vincent Pass, Fig. 1). Consequently a complete or an unfinished "faro" may arise (Gatope faro; Koniene faro, 10 km long, in the Koné area; end of Doiman reef (see Fig. 7A); faro of the eastern barrier of the Great South Reef). The depth of a faro does not exceed generally a few meters. Exceptionally it may reach 20 m (Guilcher, 1965).

On the barrier reefs, deposits of detrital material are not very important. The sandy islets are sparse, principally situated near passes and occurring more frequently on the southwestern than on the northeastern barrier, the former being more sheltered. The islets are formed with rough bioclastic material but have no gravel ramparts.

B. THE LAGOON

Its width shows considerable variation, up to 60 km in the south part. The lagoon is generally deeper along the eastern than along the western coast of the island except at the south end.

The deepest lagoons exceed 70 m in depth. Depths of 40–50 m occur in many places. The topography of the floor of the lagoon is not the same in different regions. At the Great South Reef, it is a gently westerly sloping flat, on which patch reefs arise. In the Tuo area, it has a more broken appearance and Guilcher (1965) pointed out the presence of an ancient subaerial, but now submerged mountain. At a number of places round the island there are submarine valleys which become deeper with increasing distance from the land and which link up with the passes (for example, Saint Vincent Bay, Taisne, 1965; Fig. 4). The result is that most of the passes through the barrier reefs are very deep and face the great rivers, or at least the mouths of ancient hydrographic systems. The lagoon contains a great number of reefs.

1. The Fringing Reefs (Fig. 3C,D,E)

The development of fringing reefs may be hindered by the lack of a suitable marine platform along certain cliffs of the eastern coast, or again by alluvial deposits near a river mouth. Elsewhere, on the contrary,

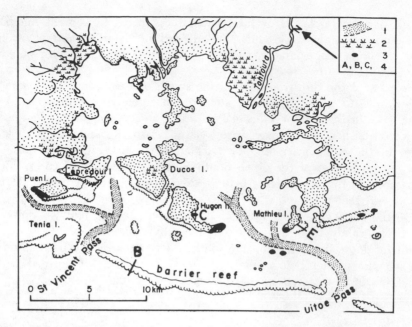

Fig. 4. Saint Vincent Bay (west coast). (1) Submarine valleys; (2) marshes; (3) eolianites; (4) cross sections through different reefs (see Fig. 3). (According to Avias, 1965; Taisne, 1965.)

they may extend several kilometers in width, certainly under the influence of the prevailing winds and swell. Consequently, fringing reefs are well developed on the west coast of the different islands of Saint Vincent Bay where it faces the ocean, whereas on the other side, the coast presents small inactive reefs or no reef at all (Fig. 4).

The morphology and building of reefs vary considerably from one place to another, depending on the coast topography, on sedimentation in the lagoon, and on the direction of winds or swell.

Fringing reefs are exposed or submerged to a depth of only a few decimeters at low tide. They may be separated from the coast by a boat channel a few meters deep, or may hug the coast. In the latter case the inner part is overgrown by filamentous algae and marine phanerogams or invaded by alluvial deposits or even in places by a mangrove swamp (Fig. 5). Some reefs are entirely dead and covered with mud (marsh of Mara on the west coast, Avias, 1949; Baltzer, 1965). Others bear sandy islets.

The edge and the outer slope of fringing reefs are generally rich in corals. In areas with strong lagoon surge, particularly in the eastern

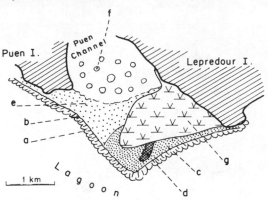

Fig. 5. Fringing reef of Puen Island and of Lepredour Island (Saint Vincent Bay). (a) Reef front with living corals; (b) accumulation of dead corals; (c) sandy flat covered with marine phanerogams, microatolls formed with dead corals 10 cm deep at low tide; (d) series of dead *Acropora;* (e) plate covered with mud under 10–60 cm of water at low tide (a few living algae and corals); (f) coral heads with living corals; (g) mangrove swamp (according to an aerial photograph and to author's observations).

lagoon, the outer ridge presents an irregular spurs and grooves system (for example, Bogota Reef near Canala).

2. Lagoon Reefs

Lagoon reefs show great variation in shape and distribution from one place to another. On the whole they are more numerous in the western lagoon than in the eastern. The following can be observed:

(a) "Knoll reefs" are reefs which do not reach the sea surface. Many of the deepest ones are dead. They are in great numbers near the western barrier reef, where they are often accompanied by "coral heads" (Saint Vincent Bay).

(b) "Patch reefs" which reach the surface. Their morphology and distribution depend on swell and winds, but also on submarine topography. Thus in Tuo lagoon, reefs form an inner barrier behind the outer barrier (Guilcher, 1965). Such double barriers occur at other places on the eastern coast: at Puebo, between Port Bouquet and Thio, near Canala, where the inner barrier reaches 9 km in length, and is bordered along its two slopes by depths of 30–50 m (Fig. 6). We will reexamine the origin of this morphology. Some reefs are very elongated, like Cimenia Reef in the south, which extends 9 miles and certainly rests on an ancient submarine mountainous basement (Guilcher, 1965, Fig. 7D).

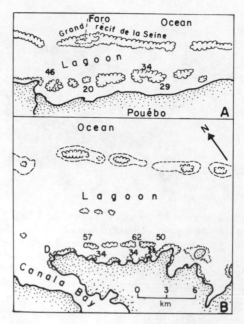

Fig. 6. Double barrier reefs in Pouébo (A) and Canala (B) regions. (D) Shows the place of the cross section through the Bogot fringing reef (see Fig. 3D).

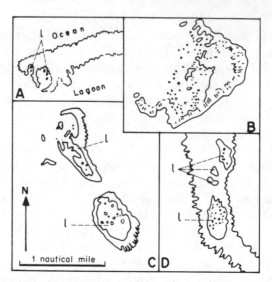

Fig. 7. Different types of reef morphology. (A) Double faro of Doiman (area of Tuo); (B) horseshoe-shaped reef (Great South Reef); (C) faro of Ndito (Great South Reef); (D) Cimenia Reef (Great South Reef). (According to Guilcher, 1965).

As a consequence of swell refraction reefs recurve (Fig. 7B). If this process is prolonged a faro appears with a lagoon having a depth in some cases of over 20 m (Ndito Faro, at the Great South Reef; Fig. 7C). All intermediate structures between patch reefs and faros can be seen, especially in the south of New Caledonia (Guilcher, 1965).

(c) Reefs with sandy islets are in greater numbers on the west coast than on the less sheltered east coast; the morphology of islets depends on trade wind and swell.

3. *Lagoon Sediments*

These were investigated only at some places, in particular in the Tuo area and at the Great South Reef (Guilcher, 1965), and also at a few points of Saint Vincent Bay (Gambini, 1959; Salvat, 1964).

The island affects greatly the sedimentation because of the rough surface and rainfalls. The percentage of $CaCO_3$ increases with distance from the coast. It is only 6–20% close to the coast, 85% in the outer part of the Tuo lagoon, and reaches as much as 98% at some localities on the Great South Reef.

Lagoon sediments are characterized by a strong heterometry and an irregular distribution. They are composed principally of coral debris, calcareous algae (*Lithothamnion, Halimeda*), and Foraminifera (*Marginopora*). Near the coast, brought by the rivers, red ferruginous sediments and cobalt, and nickel-rich sediments settled in the lagoon and sometimes even beyond it.

It seems that in the lagoon, sedimentation generally prevails over construction.

IV. Island of Pines Reefs

Between Island of Pines and New Caledonia the following reef zonations are found (Figs. 1 and 8).

1. A group of reefs in an almost SW–NE direction, which has never been investigated and which is situated between two passes 40 and 70 m deep, respectively, the Havannah channel and the Sarcelle Pass.

2. Around Island of Pines, a group of reefs situated on a gently sloping platform is composed of fringing reefs, barrier reefs, faro, table reefs, which are often oblong, in a SE–NW alinement forming a succession of steps and bounding lagoons with wide openings: Duamoeo Bay (25–45 m deep); Gadji Berth (30–50 m); and Vao Berth (6–10 m).

3. Separated from the former by a channel 10–20 m deep, an atoll-

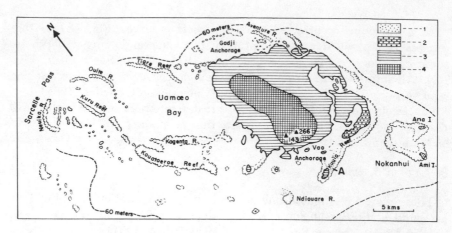

Fig. 8. Reefs at Island of Pines. (1) Reef islands formed by loose detrital organic materials; (2) eolianites; (3) raised reef limestones; (4) peridotite. (A) indicates the place of the cross section through Kuuia Reef (see Fig. 9).

shaped reef: Nokanhui (2–8 km long). It is open to the sea through three small passes which do not exceed 3 m in depth. The lagoon is overcrowded with patch reefs and coral heads and does not reach 12 m in depth.

In comparison with reefs surrounding New Caledonia proper, Island of Pines reefs are somewhat different.

1. *Lithothamnion* take a predominant part in the reef-building to such a point that some of the reefs are almost entirely formed by these encrusting algae. This phenomenon is certainly due to heavy swell since calcareous algae have a great extension principally on the southernmost reefs (Nokanhui, Kuuia, Ndionare). On the reefs facing the trade wind, as Tiare Reef, they play a less important role.

Many reefs around Island of Pines have a low algal ridge, rising not more than 1 m above low-tide level and having spurs built by *Lithothamnion* and separated by narrow grooves. Spurs themselves are sometimes furrowed by little shallow channels, so that the structure of the reef recalls, on a smaller scale, that of "room and pillars structure" of Marshall atolls. Corals are in a great abundance only on the outer slope (Fig. 9).

2. The grooves and spurs system is generally well marked. Some reefs present a great number of surge channels facing the waves.

3. A small number of reefs contain true ramparts. Thus at Nokanhui, Ami Island is formed by a great rampart 3 m high composed of coral boulders.

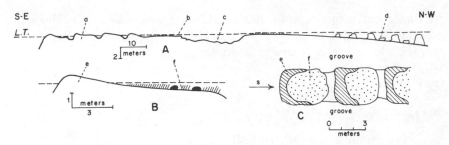

Fig. 9. Cross section through Kuuia Reef, Island of Pines (see localization Fig. 8). (A) General view of the cross section; (B) detail of an "algal head"; (C) detail of a spur of the algal zone. (a) Algal ridge; (b) plate reef 30 cm deep at low-tide level with marine phanerogams; (c) old groove; (d) sandy flat with small formations rich in corals and *Lithothamnion;* (e) living *Lithothamnion;* (f) dead *Lithothamnion* covered with marine phanerogams and Alcyonaria; (g) swell (author's observations).

V. Raised Reefs

Raised reefs are situated in the south of New Caledonia (area of Yaté and Island of Pines) and in the Loyalty Islands which are entirely formed of coral limestones.

A. New Caledonian Raised Reefs

1. Yaté Région

The raised reef covers an area 25 km long and 2 km wide. It does not exceed 3.5 m above highest tide level, but it slopes gently northwestward. It consists of a porous *Lithothamnion* limestone; corals and Foraminifera (*Amphistegina, Marginopora, Alveolinella*) (Arnould and Avias, 1955). We are concerned here with an ancient fringing reef, which presents, at some points at least, a wide boat channel now filled by a marsh. The age of the reef has not been yet determined.

2. Island of Pines (Fig. 8)

In its center, the island consists of peridotite which forms a peneplain about 100 m high with a remnant of an earlier more elevated structure. Around the ancient central part, an important formation of fringing reefs, perhaps of barrier reefs on the east coast and of patch reefs, was built. At the present time these reefs are 10–50 m above sea level and higher eastward than westward. They are composed of corals, *Lithothamnion, Halimeda,* shells, and Foraminifera. Their investigation is in hand.

Younger structures rest on this coral limestone: eolianites in the south of Gu Bay and arising up to 20 m high, old beachrocks, and sands. On this old reef rests, here and there, the recent reefs.

The age has not yet been precisely determined: Pleistocene according to Routhier (1953).

B. LOYALTY ISLANDS

These comprise the following islands (Fig. 1):

1. Walpole Island

Walpole is a little island made of coral limestone which has been elevated to 75 m high and rich enough in phosphate to have been worked in the past (Wright, 1924; Aubert de la Rüe, 1935).

2. Maré Island (Figs. 10 and 11)

Maré is an atoll which was raised, higher southwestward than northeastward, and which is surrounded by a recent fringing reef.

The ancient reef rim is unbroken only eastward and southward (i.e., facing the trade wind). In contrast it presents a great number of passes on the leeward side. Its height varies from 60 to 138 m. It is formed of little table reefs or faros and for the most part rises vertically above the old lagoon. The ancient rim is made of built or detrital limestones (reef organisms being reworked in place).

The ancient lagoon is a northeastward sloping plate situated 45–70 m high. It is made of detrital limestones, dolomitic limestones, and dolomites (J. Levaud, 1965, unpublished work). It contains a few patch reefs. In its center three basaltic peaks rise to a height of a few meters (Sarasin, 1925). The presence of basaltic blocks in the limestones strongly suggests a volcanic origin (Koch, 1958a).

The atoll is broken at different heights by 15 terraces (some of them are not well defined) representing ancient sea levels. The lower terraces (height: 12; 8–9; 3–4; 2–2, 5; 1, 5 m) have horizontal surface and an eustatic origin. The higher terraces slope northeastward and are slightly distorted. Two terraces play an important part; the terrace III is certainly an ancient abraded plate at lagoon level and forms a basement for the reef rim. The upper part of a faro was also built at terrace I level. The old passes are situated level with these two terraces.

The age of the basalt was estimated by the K–Ag method to be 29 ± 4 million years (late Oligocene or early Miocene). The oldest limestones around volcanic buttress would be from late Miocene or Pliocene (Chevalier, 1968).

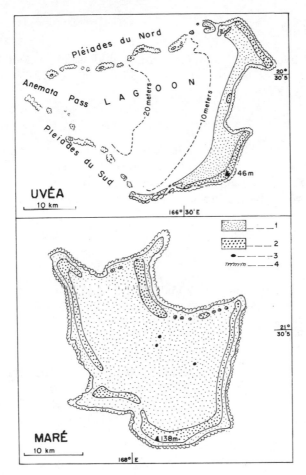

Fig. 10. Uvéa Island and Maré Island (Loyalty Islands). (1) Emerged reef limestones; (2) ancient reef rim; (3) peaks of basalt rocks; (4) recent reefs.

Fig. 11. North–South cross section through Maré. Basalt peak is in black.

3. Tiga Island

Lying between Maré and Lifu, this island is an ancient little raised atoll. On its sides four terraces, 4–5, 28–34, 60–70, 72–78 m high, can be observed (Koch, 1958a).

4. Lifu Island

Lifu is also an elevated atoll, with dolomitic rocks, which does not exceed 60 m in height. The old reef rim, 200–2000 m wide, surrounds the ancient lagoon which slopes gently to the southwest. The oldest limestone would be Miocene (discovery of *Flosculinella* sp., F. Le Jan, 1964, unpublished work). On the walls of this raised island Chambeyron (1875) recognized five sea levels.

5. Uvéa Island

This island is an almost atoll and comprises (a) an ancient atoll which has been raised on the eastern side only; it is formed of an old reef rim rising to a maximum height of 46 m and of detrital limestones, along an ancient lagoon; (b) a series of recent reefs lying on this ancient atoll in the west; and (c) a recent lagoon in the western part, with a gradual and gentle slope. Sarasin (1925) thinks that the age of the coral limestone is Miocene. The discovery of *Meretrix erycina* confirms this opinion (M. Beaufort 1964, unpublished work). Four sea levels can be observed on its slopes.

6. Beautemps-Beaupré

This is a half-atoll perhaps lying on an old submerged atoll (Haeberle, 1952b) and sloping downward to the northwest. Southeast of the atoll an ancient limestone, in part raised to 3.5 m above low tide level, is still upright. Nearer the northwest, the Astrolabe Reefs are half-atolls, the reefs of which do not rise above the surface.

VI. Subsurface Geology of the Reefs

Drilling was carried out in 1965 on the western barrier reef near Tenia Island to obtain useful data on reef-building (Fig. 1).

The reef substratum was found to be 226 m deep and is formed with phtanites loded by dolomites from early Eocene. Above them the following formations were discovered (Avias and Coudray, 1967; Coudray, 1971).

1. Grey detrital organic limestones (from 226 to 130 m deep). These begin with a coral breccia, elements of which are partly dolomitized and surmounted by a phtanitic sand layer. The base of these limestones is well consolidated, but has a cavernous nature. The top layer is finer and less consolidated. They contain corals, *Lithothamnion*, Foraminifera, and rare terrigenous fragments. They are interrupted between 220 and 200 m and between 185 and 181 m by harder breccia (coral banks and beachrock). These limestones are partly dolomitized, the percentage of $MgCO_3$ averaging 30% at some levels.

2. A succession of white bioclastic and biochemical limestones (130–83 m deep). These are made up of little fragments of calcareous algae containing approximately 10% of $MgCO_3$ in the lower part. Toward the upper part they include more corals, Foraminifera, and shells, many of which are mussels. This structure seems to be due to precipitation of calcareous muds in the lagoon owing to a supersaturation of $CaCO_3$. The presence of cracks filled with red silt, and the resolution of organisms, including shells, leads to the conclusion that emersion took place on completion of deposition of this formation (85 m deep).

3. Series of calcareous muds formed by the accumulation of still finer fragments of algae (0.1–1 mm) and Foraminifera; these muds contain up to 15% of $MgCO_3$ (83–70 m deep).

4. Fine sands rich in Foraminifera and which look like the sand of the recent beaches are between and possibly indicative of emersion (70–66 m deep).

5. Reef limestones rich in corals and *Lithothamnion* (66–11 m deep) and broken by bioclastic layers. A reddened zone, 40 m deep, containing recrystallized corals, probably indicates an emersion of the reef.

6. Sand and unconsolidated blocks (11–0 m deep). This structure shows the changes in a plate reef which is hardly submerged at low tide. The sand, 3.7 m in thickness, gave rise to the Tenia sandy islet.

VIII. Origin of the Reefs

A. Davis' Studies (1926)

Davis was the first to try to explain coherently from physiographic data, the origin of the New Caledonian reefs. So he postulated an unequal subsidence, greater at each end of the island, and stated that it averages 200 m based on his analysis of the slopes of the valleys. He pointed out three cycles in the process of evolution of the coral reefs (Fig. 12).

First cycle: New Caledonia, wider than at present, has become a peneplain. Davis does not date this time.

Second cycle: The east coast was raised forming an arch, whereas the west coast sank slightly into the sea. Davis thought that this rising was more developed in the southeast of New Caledonia because of the presence in the actual morphology of a raised peneplain which slopes northwestward. During this period the east coast would be exposed to marine abrasion. Cliffs and an abrasion platform would gradually form, but no barrier reef could be built on it. On the contrary, on the sunken western coast, fringing and barrier reefs would grow. So in Davis' mind, barrier reefs of the western coast are older than those of the eastern coast.

Third cycle: A rapid subsidence occurred breaking the second cycle. Slight in the central part of the island, where the barrier reef is sometimes close to the coast, it was more important at the ends. This subsidence is proved according to Davis by the submergence of the lower part of the valleys (Saint Vincent Bay, Canala Bay) and that of the cliffs of the eastern coast. Consequent on the sinking, barrier reefs and fringing reefs would arise edging this coast. Nevertheless, the subsidence being rapid, the barrier reefs on the eastern coast would be more disrupted or even submerged. The third cycle could have been disturbed by a few local risings which brought to light old reefs at Island of Pines and in the vicinity of Yaté.

Davis recognized the lowering of the sea level during glacial periods

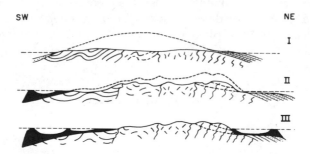

Fig. 12. Origin and evolution process of coral reefs (in black) considering the hypothesis of the three cycles (I, II, III). Cross sections through New Caledonia. (According to Davis, 1926, but simplified.)

Fig. 13. A faro in the Great South Reef, New Caledonia.

when he explained the origin and building of New Caledonian reefs, but he assigned to it no part in the reef morphology.

B. Later Works

Davis' work served as a base for later investigations. Routhier (1953) agreed with it unreservedly and added new morphological and geological data in favor of this hypothesis. He related the first cycle to late Oligocene or to recent Miocene. Furthermore Routhier pointed out the existence of three old sea levels at 15 m, 3–6 m, and 1.5–2 m high, which certainly bear witness to recent and slight movements of the basement, except perhaps the latter which would be of eustatic origin.

Avias (1953a,b, 1959) changed considerably Davis' conception and assigned an important part to glacioeustatism. According to him, after the building of the peneplain that he related to Miocene times (first cycle), a weak erosion occurred during late Neogene. During the glacial period, the deepening of the valleys was ascribed to the lowering of sea level and the ocean, rising again at the end of this time, drowned the seaward ends of the valleys and allowed the conversion of fringing reefs into barrier reefs. In addition to sea-level variations, epirogenic movements raised certain regions (Island of Pines, Yaté region) and lowered others (between Island of Pines and New Caledonia in the region of the Great Pass). Avias discovered also old marine terraces, the most recent being situated 0.30 m above high tide. Correlating with the Great Barrier Reef in Australia, Avias thought that the whole of the New Caledonian reefs date from Quaternary time.

The observations made during the Singer-Polignac Expedition from 1960 to 1963 demonstrate the importance in the present morphology of movements of the basement and of eustatic movements. Geological studies showed the existence of a tectonic Plioquaternary structure (Coudray, 1969). Hydrographical researches at different places in New Caledonia (Tuo Bay, Island of Pines, Saint Vincent Bay) indicated the existence of a former relief now sunken (Guilcher, 1965; Taisne, 1965). In Saint Vincent Bay in particular, old river valleys were found now under 70–80 m of water at their mouth (situated at present time at similar depths to those of the passes) and which bear witness to the influence of glacial eustatism on the building of barrier reefs (Avias, 1965). In contrast, the presence of double barrier reefs found on the eastern coast is related, according to Guilcher (1965), to a two-step subsidence. Haeberle (1952b) had already agreed with this hypothesis for the D'Entrecasteaux Reefs, in explaining the origin of reefs lying outside Huon and Fabre atoll.

C. INFORMATION YIELDED BY THE DRILLING

We have seen before that the drilling through the barrier reef inter-sected a 225-m thick layer of reef sediments, but that the true reef began at a depth of 70 m and was certainly related to a rise of sea level, after the glacial period. These formations have not yet been dated. They seem, however, to be of Quaternary time, the base being perhaps Pliocene (Coudray, 1971); as they were formed in shallow waters, the idea of a subsidence must be accepted. Evidence of emersion at depths of 85 m, 66 m, and 40 m are ascribed to glacial ages. During one of these glacial stages (perhaps the Riss glaciation, which would correspond to the 66 m deep level), eolianites, which were recognized at several points of Saint Vincent Bay, could have been formed (Avias and Coudray, 1965). Coudray (1971) agreed with Davis' hypothesis, but made it more precise. According to him, the subsidence of the third cycle, during which sediment layers were formed, was disturbed by eustatic movements. In middle Pleistocene time, subsidence ceased and the recent barrier reef was formed based on a platform which was formed during a lowering of sea level.

Fig. 14. Uvéa (Loyalty Islands). Lesquine Cliff with old sea levels.

D. ORIGIN OF THE LOYALTY ISLANDS

Although the history of the Loyalty Islands differs from that of New Caledonia, movements of the basement and glacioeustatism controlled the development of the ancient and recent reefs in the same way.

The age of the older raised limestones is Miocene or Pliocene. At that time a group of volcanoes on a fold of the New Caledonian ridge served as a base for the atolls. The islands were probably emerging; they must have been eroded during the different glacial stages. At the end of this period, periodic sinking occurred northwestward of the Loyalty Islands, taking place with interruptions and local warpings (Davis, 1928; Haeberle, 1952b; Chevalier, 1968).

VIII. Conclusions

With its great variety of reefs surrounding emerged lands and with its double chain of barrier reefs that border it, New Caledonia appears to be one of the richest areas in reef structures in the world. Neglected for a long time, they are now a field of extensive geological, morphological, and biological studies. The results of these researches are already significant. In the future, they will enable us to describe precisely the respective stages of subsidence, explained by Davis in a brilliant manner, and also the role of glacioeustatism in their origin and the development of the present morphology.

Selected References

Arnould, A. R. and Avias, J. (1955). Notice explicative de la carte géologique de Nouvelle-Calédonie.

Aubert de la Rüe, E. (1935). Sur la nature et l'âge probable de l'île Walpole (Océan Pacifique Austral). *C. R. Somm. Soc. Geol. Fr.* No. 4, pp. 48–49.

Avias, J. (1949). Note préliminaire sur quelques phénomènes actuels ou subactuels de pétrogénèse et autres dans les marais côtiers de Moindou et de Canala (Nouvelle-Calédonie). *C. R. Somm. Soc. Geol. Fr.* No. 13, pp. 277–279.

Avias, J. (1953a). Variations des lignes de rivage en Nouvelle-Calédonie. *Actes Congr. Int. Quat. 4th* p. 1.

Avias, J. (1953b). Contribution à l'étude stratigraphique et paléontologique des formations antécrétacées de la Nouvelle-Calédonie Centrale. *Sci. Terre* 1, No. 1–2, I–XVI and 1–276.

Avias, J. (1959). Les récifs coralliens de la Nouvelle-Calédonie et quelques-uns de leurs problèmes. *Bull. Soc. Geol. Fr.* [7] 1, No. 4, 424–430.

Avias, J. (1965). Sur l'origine du gîte du chrome de l'île Hugon, baie de Saint-Vincent (Nouvelle-Calédonie). *C. R. Somm. Soc. Geol. Fr.* No. 9, pp. 301–303.

Avias, J., and Coudray, J. (1965). Sur la présence d'éolianites en Nouvelle-Calédonie. *C. R. Somm. Soc. Geol. Fr.* No. 10, pp. 327–329.

Avias, J., and Coudray, J. (1967). Premiers enseignements apportés par un forage réalisé dans le récif barrière de la Nouvelle-Calédonie. *C. R. Acad. Sci.* 265, 1867–1869.

Baltzer, F. (1965). Le marais de Mara. *Cah. Pac.* 7, 69–91.

Chambeyron, L. (1875). Note relative à la Nouvelle-Calédonie. *Bull. Soc. Geogr.* [6] 9, 566–586.

Chambeyron, L. (1876). Le Grand Récif au Nord de la Nouvelle-Calédonie. *Geographie* [6] 12, 634–644. Paris.

Chevalier, J. P. (1968). Géomorphologie de l'île Maré. Les récifs de l'île Maré. Expédition française sur les récifs coralliens de la Nouvelle-Calédonie. *Edit. Fond. Singer Polignac* III, 1–82.

Coudray, J. (1969). Observations nouvelles sur les formations miocènes et postmiocènes de la région de Népoui (Nouvelle-Calédonie): Précisions lithologiques et preuves d'une tectonique "récente" sur la côte Sud-Ouest de ce territoire. *C. R. Acad. Sci.* 269, 1599–1602.

Coudray, J. (1971). Nouvelles données sur la nature et l'origine du complexe récifal côtier de la Nouvelle-Calédonie. *Quatern. Research,* Vol. 1, n. 2, p. 236–246.

Davis, W. M. (1926). Les côtes et les récifs coralliens de la Nouvelle-Calédonie. *Ann. Geogr.* 34, 1–118.

Davis, W. M. (1928). The coral reef problem. *Amer. Geog. Soc., Spec. Publ.* No. 9, pp. 1–596.

Gambini, A. (1959). Sur la composition de quelques sables coquilliers à Foraminifères des lagons de Nouvelle-Calédonic. *Bull. Soc. Geol. Fr.* [7] 1, 431–434.

Guilcher, A. (1965). Grand Récif Sud. Récifs et lagon de Tuo. Expédition française sur les récifs coralliens de la Nouvelle-Calédonie. *Edit. Fond. Singer Polignac* I, 137–225.

Haeberle, F. R. (1952a). Coral reefs of the Loyalty Islands. *Amer. J. Sci.* 250, No. 9, 656–666.

Haeberle, F. R. (1952b). The d'Entrecasteaux reef group. *Amer. J. Sci.* 250, 28–34.

Koch, P. (1958a). Hydrogéologie des îles Loyauté. *Bull. Geol. Nouvelle-Caledonie* No. 1, pp. 135–185.

Koch, P. (1958b). Introduction à la géologie de la Nouvelle-Calédonie et dépendances. *Bull. Geol. Nouvelle-Caledonie* No. 1, pp. 9–22.

Risbec, M. J. (1930). Quelques remarques sur l'allure des récifs frangeants en Nouvelle-Calédonie. *Proc. Pac. Sci. Congr., 4th, 1929* Vol. 2B, pp. 787–795.

Routhier, P. (1953). Etude géologique du versant occidental de la Nouvelle-Calédonie entre le col de Boghen et la Pointe d'Arama. *Mem. Soc. Geol. Fr.* [N.S.] 67, No. 32, 1–271.

Salvat, B. (1964). Etude préliminaire de quelques fonds meubles du lagon. *Cah. Pac.* 6, 94–119; 7, 101–106 (1965).

Sarasin, F. (1925). Forschungen in Neue Caledonien und auf den Loyalty Inseln. *In* "Nova Caledonia" (F. Sarasin and J. Roux, eds.), Vol. A-4, pp. 1–178, Wiesbaden.

Taisne, B. (1965). Expédition française sur les récifs coralliens de la Nouvelle-Calédonie. I. Organisation et hydrographie. *Edit. Fond. Singer Polignac* I, 1–132.

Wright, A. M. (1924). Phosphate from Walpole Island. *N.Z. J. Sci. Technol.* 7, 91–94.

6

CORAL REEFS OF THE NEW GUINEA REGION

F. W. Whitehouse

I. Preliminaries

In any consideration of the coral reefs of the world, New Guinea and the islands close to it have a special place. This vast island, the biggest island by far occurring entirely within the torrid zone, with admirably warm waters on every side of it, might be expected to have particularly favorable conditions for corals all around it. Yet of its long coastline less than one-sixth is fronted by coral reefs of any moment, and the same applies to the islands adjoining it. The Louisiade Archipelago is everywhere abundantly coral-rich; but the groups north of it and warmer still, the Trobriand, Solomon, and Bismarck Archipelagoes, are deficient in reefs. On the other side the Aru Islands close to New Guinea in the Arafura Sea have none at all. Southern New Guinea and its extension into the Louisiades are fronted by a particularly rich barrier reef 500 miles long; and this is separated from the 1200-mile-long Great Barrier Reef by 300 miles of low New Guinea coast in front of which there is not a coral reef of any kind.

New Guinea and its neighbors are thus particularly important in that, while they offer some excellent opportunities for studying coral reefs in tropical waters, they are even more important for finding out why such reefs also may not occur in seas that seem ideally suited to them. So let us consider summarily at first the relation of New Guinea to its coralline neighbors.

Small-size Australia is merely the remnant of a once larger continent. Bryan (1944) has given evidence that it once extended as far to the east as the Marshall Line, which he named, on the far side of the Solomon Islands. New Guinea is an obvious part of the continent, arising from the same continental shelf as Australia.

Australia, as we know it, is the smallest of the continents and it is also the lowest and by far the driest, with its highest peak only 7316 ft above sea level, with no Tertiary fold mountains, with an undue expanse of hot deserts, and without any glaciers or active volcanoes. New Guinea, in contrast, is the complement of all this, with high fold mountains (up to 16,500 ft high), glaciers within 4° of the equator (muddy-looking glaciers it is true, but glaciers nonetheless), abundant water everywhere in many great rivers, with some active volcanoes and no arid lands.

The evidence here for continental drift is overwhelmingly strong, that the combined Australia–New Guinea massif, drifting north, has collided lately with, and pushed aside, island arcs on either side (the Indonesian and Bismarck Archipelagoes), creating within these arcs lines of explosive volcanic caldera of the utmost intensity, and with fold mountains, apparently the latest alpine folds of the world and still not completely quiescent (Zwierzycki, 1926; Behrmann, 1928; St. John, 1970), crumpling the forefront into mountainous folds, broken into separate blocks by long, deep, structural valleys within which flow great rivers.

Those mountains of New Guinea are one of the three great east–west chains of Tertiary fold mountains of the world, the other two being the Alps and the Himalayas, each of which, however, is remote from the sea. The New Guinea fold mountains are mainly Upper Miocene to Upper Pliocene in age (St. John, 1970) and so were the last to form. On the southern side of all three are alluvial lowlands where the land has been depressed between the towering fold mountains of the north and the firm but older country to the south. Southward into all these depressions flow the tributaries of great rivers which then swing axially into the depressions: the Po on the Plains of Lombardy in front of the Alps, the Indus and the Ganges in India, and here the whole concourse of rivers from the Digoel to the Fly, which will be discussed later.

Those great swampy plains of New Guinea are now very little above sea level. But when, eustatically, sea level was 100–200 ft lower, not so long ago, they had generally a more elevated aspect, but probably they were just as swampy. That eustatic lowering was due to great accumulations of land ice over the world. At its peak, for one instance, almost all of Canada was under an ice sheet. Today the snow line in New Guinea is about 14,500 ft above sea level and the highest peaks on the Carsten Toppen, above 16,000 ft, have glaciers. At those eustatically lowered times snow and ice must have been widespread on those ranges and glaciers more abundant, with increased runoff even though, the earth movements being not yet complete, the mountains may still have been rising and not yet have reached their present peak. The alluvia of the Fly–Digoel plains are not very thick. The corresponding alluvial plains in front of the Alps and Himalayas are older and more mature, with gravels 6000–10,000 ft deep in the Indo-Gangetic Plains, and very deep also in the Plains of Lombardy. In ages to come the New Guinea Plains may be much longer, and deep like these.

In New Guinea the rivers rushing south down the slopes to the alluvial plains diverge symmetrically about a recent sigmoidal divide (see Fig. 1 on foldout) which links an almost imperceptible rise between these rivers, a low upland through the Oriomo region, the narrow, shallow ridge of Torres Straits and the insignificant Great Divide of Cape York Peninsula, which lately has suffered a slight uplift.

The high mountain slopes, descending steeply, are everywhere curiously stepped (Detzner, 1922). Road construction work across the highlands, benching into them, has convinced me that these ever-present steps are mainly old, vegetated landslides. In times of marked earthquake tremors, as, for example, last year (1971), similar slides recur and begin freely again. Indeed, after the pronounced tremors last year the long, steep, northern coastal scarp of New Britain had the curious appearance of being one continuous landslide.

Locally, of especial importance to us, are the ocean currents. The warm equatorial currents in each ocean flow by inertia from east to west; and when they meet the western margins they are deflected north and south to begin the cycles of currents in each hemisphere. But the western equatorial margin of the Pacific Ocean is not a complete barrier, as it is in the Atlantic and Indian Oceans, but is a "sieve" of islands. Accordingly the warmest of waters stream through these many gaps into the Indian Ocean, which thus is the only ocean to have abundant coral reefs on its eastern side. Wartime experience on a corvette in the Lesser Sunda Islands has convinced me of the reality of these streamings.

So the great island of New Guinea is peculiar in having warm, coral-loving seas all around it. But where are all the corals?

II. The Coral Reefs

A. REEFS ON THE SOUTHERN SIDE OF NEW GUINEA

Consider now coral reef conditions around New Guinea, beginning on the southern side with Torres Straits. The word "straits," in the plural, is used advisedly. The gap from Cape York to New Guinea is 100 miles wide and is so cluttered with islands and reefs that several passage-ways from the Coral to the Arafura Sea exist (at least nine of them), with only a few of them suitable for navigation.

The rocky islands of the straits are essentially granitic, similar to the basement of the northern part of the Peninsula and of the nearer part of New Guinea, the Oriomo Platform. The only hill on the southern plains of the island, Mabaduan Hill, is of granite porphyry. These islands (Moa, Badu, Horn, Prince of Wales, Hammond, Thursday, etc.) have a pronounced lattice form induced by two sets of fractures at right angles (northeast and northwest), a structure continued to Cape York Peninsula (Richards and Hedley, 1925; Jones and Jones, 1956). Very recently, when sea level was lower, there was a complete barrier across the straits; and it would take a drop in sea level of only 40 ft to re-establish it, which satisfactorily would account for the coexistence in New Guinea and North Queensland of many large animals which are specifically identical, but which can neither fly nor swim—the death adder, the taipan, and the startling green python among the snakes, and two species of wallaby and the cassowary, for example. It may have been the route also of low-stature man ("pygmies") from New Guinea to the eastern scrublands of Australia (Whitehouse, 1972).

The gaps lately were alternately open and closed on several occasions. Other genera are common to the two lands, but not specifically the same—the monotreme *Echidna* (a separate genus in New Guinea), *Varanus*, other reptiles and marsupials, for example—suggesting migra-tions of earlier times with opportunities to evolve into other species.

The straits were last opened, then, at a very recent date, and this seems to be reflected in local tidal peculiarities. In the third volume of the Admiralty Tide Tables (1963) which covers the Pacific Ocean and adjacent seas, the tides in the Torres Strait (particularly the main shipping lane, 15 miles long) are described in these words:

> The restricted channels of Torres Strait connect two areas in which the tides differ very remarkably. The difference does not occur in the diurnal component of the tide, for the range and occurrence of high water of this component are more or less uniform over the whole area and there are no marked diurnal differences between the levels in either entrance to the strait. The contrast occurs between the semidiurnal components of the tide in either entrance. It is spring tide in one when it is neaps in the other and at some phases of the moon it is high water in one when it is low water in the other. In consequence marked differences in the levels in the two entrances occur and the streams through the channels flow from the high level to the low level. The tidal streams are, in contrast to the tides, which may have a relatively large diurnal component, predominantly semidiurnal.

The British Navy is not given to effusive statements and the adjective "remarkable" does not seem to appear elsewhere in this thick volume.

Such tidal maps as we have of the region show that the diurnal cotidal lines proceed regularly through the straits into the Arafura Sea, but that the semidiurnal tides are related to two amphidromic points with discordant cotidal lines, the one in the east near Guadalcanal in the Solomon Islands and the other in the west near the New Guinea coast northeast of Arnhem Land. Conceivably this anomaly is related, in some way, to recent fluctuations in the openings of the straits.

1. The Torres Straits Axis

The 100-mile stretch of Torres Straits is divided into two equal parts. The southern region is known in considerable detail, up to the southern edge of the Orman Reef, a large platform reef. But the 50 miles north of this are quite unknown. Even the shape and the extent of the Orman Reef have not been determined. Apart from a few low islands off the reef-free edge of New Guinea (reef-free, be it known), of which the largest is Boigu, there is nothing on any map or chart; there are no aerial photographs in the comprehensive Queensland Government collection (it is the only part of the State without them) nor any, so far as I know, in any private collection; and I know of nobody who has been within that region. I know nothing of it personally. Since all passages used through the straits are in the south, and all of these are hazardous, I assume that the reefs and islands in the north are so congested and dangerous that no boat has traversed the region.

In the southern part of the province the large rocky islands of Torres Straits lie within a belt 50 miles long and only 20 miles wide; and about them the waters are not more than 40 feet deep. Each island has a narrow zone of poor fringing reefs; and between them, between Prince of Wales Island in the south and Moa and Badu in the north, a distance of about 25 miles, there are four large, narrow, cigar-shaped platform reefs, trending east–west within the region through which the currents stream through the straits; and with very little doubt they are determined by them.

On either side of this axial zone the waters deepen to 60 ft and more, over slopes without coral, with bottoms of sand, mud, and shells, dissected outward possibly by turbidity currents.

2. The Warrior Reefs

Nearly midway between the Torres Straits Axis and the edge of the Great Barrier Reef, where the latter is beginning to lose continuity close to its northern end, where currents are streaming through the Flinders and Bligh Entrances and the Great North-East Channel, there is a group of very large platform reefs aligned almost normal to the New Guinea coast near Daru and extending from it in a very slight arc convex to the east. These are the Warrior Reefs. There are eight of them, generally with narrow gaps between, extending over a length of 80 miles. These great platforms, awash at low water, extend as far south as to be opposite the Orman Reef. In other words, they are a splendid frontier to that unknown northern part of the Torres Straits which, as noted, is *possibly* thickly congested with reefs.

They are wide reefs, up to 8 miles wide, and they arise from water about 10 fm (60 ft) deep. How far the rocky basement is below is not known. But this depth, being greater than the 40 ft from which the closely rockbound, streamlined reefs of the Torres Straits arise, they probably began to form earlier.

On either side, between the Torres Straits on the west and the Great Barrier Reef on the east, there are smaller platforms, many with cays and one or two with basaltic lava flows.

B. The Inhibited Riverine Region to the North

Reference has been made above to the plexus of the Fly, Digoel, and other rivers immediately to the north. On the eastern side these streams have aggregated into the one great combine of deltas and estuaries of the Fly, Kikori, Purari, and other rivers which has a frontage to the Gulf of Papua of over 100 statute miles, through the whole of which pours an incredible surge of very muddy fresh water, bearing great trees with it out to sea. It is a most impressive sight from the air. No less favorable environment for coral reefs could be imagined in tropical seas, and not one reef occurs. If there were any reefs there at any time, the depositing muds have obliterated them. The Fly, whose tributary, the Strickland, is much longer and probably more copious, was first traversed and well described by Sir William Macgregor (1890), in after years to be, nostalgically, Governor of Queensland and the first Chancellor of the University of Queensland. Clearly this land frontage is growing outward, but at what rate we have no means of knowing. The much more mature and far longer known Po river frontage in historically annotated Europe is advancing at the rate of 1 mile in 120 years.

In time to come, as this active deposition extends, these many streams, almost now at coalescing point, will join to be the one Stream of Po, Indus or Ganges type.

On the western side of the sigmoidal divide the Digoel, Eilanden, and other streams have not yet combined into the one delta frontage. Descending from the very highest region of the mountain core they are probably even more voluminous. Their outgoing sediment spreads over the nearer part of the Arafura Sea. As with the eastern streams there are also no coral reefs here.

Far out in the Gulf of Papua, though, over 120 miles from land and presumably away from the influence of the river muds, reefs do occur—the Portlock and Boot reefs, and the Eastern Fields. Portlock and Boot together are aligned, suggesting an origin in keeping with barrier reefs. Still further to the southeast the Osprey Reef is a typical, oval atoll, elongated 16 miles to the southeast. A long way further still, Willis Island, a long reef with a cay, repeats some of these features; and there are others more remote still about the axis of the Coral Sea.

C. THE SOUTHEASTERN BARRIER REEF

Coral reefs begin again at Yule Island, fringing the coast 110 miles southeast of the delta of the Purari River. Reefs then continue past Port Moresby to the eastern tip of the island near Samarai and beyond, into the Louisiade Archipelago, a distance of 500 miles, most of these being barrier-reef types. This barrier has a narrow crest and keeps a steady distance of 5–15 miles from the land, as compared with 20–100 miles for the Great Barrier Reef which is 1200 miles long, north to south. Such conditions continue off the mainland around the southeast tip as far as East Cape, on the northern side of Milne Bay. All in all it suggests a reef of fairly late origin.

The long lagoon between the reef and the shore is shallowed by sediment and is usually less than 10 fm (60 ft) deep. Immediately beyond the barrier the water deepens suddenly to some hundreds of fathoms. The reef thus has grown up along the edge of the continental shelf, as it does in North Queensland. Like the Great Barrier it is broken into sections by narrow gaps, most of them too small for navigation, but sufficient to accommodate ingoing and outgoing tides.

There are, however, three extensive gaps in the structure. The most westerly is from beyond Yule Island easterly to Caution Bay, close to Port Moresby. The ocean part of this has a steep continental slope, from shallow to deep water. Northerly the shallow shelf top averages above 25 fm but deepens on the way. It is deeper than the shelf behind the conspicuous reefs. On the naval charts this line of sudden change is marked as "probable trend of the sunken Barrier Reef," and this may very well indeed be so. It could be suggested that this presumed obliterated extension of the barrier reef was smothered and killed and then worn down by powerful turbidity currents as the greatly muddy, surging sediments from the Fly region grew forward.

The next gap, of 23 miles, is from Hood Point to Marshall Lagoon (localities not specifically shown on the present map). It is of similar kind though shallower, and behind it, lining the shore, are fringing reefs.

After that comes the largest of these gaps, 140 miles, from Table Bay to the first of the pronounced Louisiade reefs. Actually, it seems to be a double gap. Over the eastern part of this distance there are contiguous elongate shoals where the barrier reef should be; but the western part is suddenly without such marked shoaling and is merely an abrupt change of slope.

Regarding these, as Navy does, as sunken parts of the reef, and assuming, following Darwin, that this barrier reef is in a region of recent subsidence, such gaps are, feasibly, sections of greater and maybe later subsidence as a whole. The convenient harbor of Port Moresby, lined with fringing reefs, may be a fourth and very narrow one of these gaps.

Only a very few of the individual reeflets forming this barrier have coral debris piled high enough to be cays. Really they are barely wide enough.

Small regions of fringing reefs occur in places behind the main barrier; and tiny patch reefs and shoals are scattered behind it. In the larger and more mature Great Barrier Reef these lagoon reefs are often large, platform reefs, lagoon atolls, and the like. But here the reeflets have not had space and maybe time enough to grow into such features.

D. THE EASTERN SIDE OF THE ISLAND
(From East Cape to the Huon Peninsula)

The profuse coral development of the southeast coast of New Guinea and its extension throughout the Louisiade Archipelago stops about the narrow peninsula leading to East Cap; and thereafter, along the eastern side of the island, coral reefs are found only as occasional small fringing reefs and patch reefs. Throughout this distance though, shoals, shoal patches, and shoal grounds are more than usually common, in places extending extremely far from the shore. The shoal ground off Hall Point and Dyke Acland Bay extends out for 60 miles; and beyond Goodenough Bay (5000 ft deep) and the D'Entrecasteaux Islands (8400 ft high) another reaches out 75 miles again to the Trobriand Islands. These extensive shoal grounds are most readily explained as composed of reef areas now depressed below a sedimentary cover.

The greatest bay in this sinuous coastline is the Huon Gulf, the terminus of the long, structural Markham Valley. Rather similar in general outline to the Gulf of Papua, it is entirely different in origin. From the Gulf of Papua issues that wide, greatly voluminous spate of exceptionally muddy water described above which has smothered and killed all reefs. The Markham comes forth as one clear stream, rippling over gravel, and probably would have a limited lateral depreciation upon reefs, did any exist.

E. THE NORTHERN SIDE OF THE ISLAND
(From Finschhafen to Memberambo)

This simple-looking coastline, 800 miles long and reaching to within 2° of the equator, is without any active reefs. Behind considerable portions of it the great mountains of New Guinea rise steeply. But landslides, and their debris, which are common on these slopes, are not the prime cause of this deficiency in reefs. Nor is the fact that, in one extensive region along it, the combined structural valleys of the Sepik and the Ramu Rivers break widely through the mountain mass and the great rivers pour a considerable volume of fresh water into the sea. For the western and warmer part of the frontage, beyond the Cyclops Mountains, is mainly a long coastal plain, without landslide slopes, and with smaller rivers.

But coral reefs have existed there in the past. Isolated around the lowest reaches of the Tami River, just across the border into West Irian, there are compact, raised, sub-Recent coral reefs, heavily vegetated, rising to 1200 ft above the sea and pretty level (Schultze, 1914; Zwierzycki, 1926). It is a curiously isolated occurrence. The oscillatory history of this long frontage has yet to be worked out.

F. The Western End of the Island
("The Bird's Head")

Toward the western end, the high mountain axis of New Guinea swings around in a sinuous curve and both coastlines follow suit to give the curious "Bird's Head" (*Vogelkopf*) of the island. The coast has the great, wide indentation of Geelvink Bay behind the neck of the bird, and the gashlike McCluer Gulf to be the beak.

The long, almost reef-free northern coast continues as such into most of Geelvink Bay with only very occasional areas of fringing reefs about some of the small high islands. Then, as the bay changes structurally to more northern alignments on its western side, reefs, drying out at low tides, and shoals become promiscuous and about equally abundant around the peninsula. This continues until the mountain axis is rounded and the coastline becomes part of that great muddy-water, reef-free frontage of the Digoel and other rivers which we have discussed.

What we know of bottom samples in this near-shore zone (J.A.N.I.S., 1944) points to very little coral indeed in the sediments.

Contrasting with this open-sea disposition of coral reefs around the *Vogelkopf*, which extends even to such narrow passages as Sele Strait which is cluttered abundantly with reefs, there are those two very long indentations, McCluer Gulf and Kamrau Bay, which are reef-free. This may be a tectonic reflex, they having not completed the fracturing into new islands, assisted by streams coming down from considerable heights into these constricted gulfs, decreasing their salinity.

G. The Eastern Island Groups

1. The Partitioning of New Guinea

The island groups between New Guinea and the Marshall Line are related and suggestively, they are parts of the one continental mass which, over an interval, has been fractured and dismembered. The earth movements causing all this, infinitesimally slow by human standards, but cosmically leisurely, were effective over a geological age in producing a swirllike pattern and, quite probably, they have not yet completed all their destruction. The great archipelagoes are, as it were, broken parts of the siallic crust bobbing cosmically slowly about upon the sima. Wandering around those vast, deep, gashlike river valleys of New Guinea, map in hand, the Sepik, Markham, Ramu, Bulolo, Waria and so on, noting their simple lineaments and their continuities in some identical planes even though flowing in opposite directions, one is left with the impression that other blocks like the Adelbert Range, the Finisterre Range plus the Huon Peninsula, and the Mumeng and Bowutu Mountains, to mention three large masses, were, and maybe still are, in the very process of breaking away to join other island remnants like New Britain, New Ireland, Bougainville, the Solomons, the Louisiades, the D'Entrecasteaux Group and others; and perhaps in ages to come they may yet do so. Also remote from these, the long, gashlike indentations of McCleur Gulf (150 miles long) and Kamrau Bay (100 miles) which almost have sliced through the Bird's Head, show that, almost, they have added two more islands to the archipelagoes near by. Even some of the larger islands that have already broken away (New Ireland, for instance) seem physiographically ready to break further.

One cause of this fuss seems to be that what was the southern continent, gliding steadily north upon the sima, has encountered Asia, producing within its outlying island chains catastrophic, simatic upsurgings as the great, volcanic caldera of historic Tamboro and Krakatoa, and others of prehistory, as a later and perhaps a milder series than those that affected the leading edge of New Britain, blowing out the great harbor of Rabaul, Lake Dakatua on the Williaumez Peninsula, Lake Wilson on Long Island, Johann Albrecht Harbour on Garove Island, and a few more.

The point at issue is this: That in this current, cosmic bobbing and buckling in the sima, some of these faintly moving blocks are rising, others sinking, some at last are steady, that is, some are in favorable attitudes to attract corals, others not to have them.

It is not just a matter of the blocks pulling away, for there were vertical movements as well, simic chasms like Milne Bay, Goodenough Bay, the region between Tagula and Samarai, and a number more where depths of water between high island masses frequently reach 5000 feet. Milne Bay, within a thickly strewn coral reef region, is a harbor because it is at the end of one of these submerged chasms too deep for coral.

2. *The Louisiade Archipelago*

The strong barrier reef fronting the southeastern coast of New Guinea continues in its steady direction for another 230 miles bounding the continental edge of the Louisiade Islands. Before reaching the fractured end of the main promontory it had sunk. But just past that end it makes again. It weaves profusely around the many islands of the group avoiding, naturally, those watery chasms some thousands of feet deep. The barrier, as now restored, begins with a double face enclosing two curious "atolls," Long Reef and Bramble Haven, with water up to 100 ft deep in each lagoon. Then continues the largest chain, as a complex barrier reef around Tagula Island, and then the final and discrete barrier around Rossel Island.

The Rossel Island reef, 50 miles in longest diameter and 8 miles wide, is the ideal modest barrier reef, a narrow, continuous rim with only two or three extremely narrow tidal gaps leading into a patch-littered lagoon, up to 150 ft deep.

The much larger circum-Tagula barrier is likewise complete in its eastern half; but in the west the reef breaks into bits surrounding elongate, deep lagoons.

And so reefs of sundry modes extend through this region, avoiding the 150-mile-long deep waters toward Misima, the Bonvoulir Islands, and Normanby Island, but occurring in splendid profusion in the shallows.

3. The D'Entrecasteaux Group and Others

When this archipelago, mainly of three large, high islands, Goodenough, Ferguson, and Normanby, isolated from New Guinea by extremely deep water, forms land again, it has passed into a region in which rich coral conditions have virtually ceased. Immediately around these lands there are few reefs. But very shortly after there is a wide area of small islands, reefs, and shoals extending 75 miles north to the Trobriand Islands, so very rich in species. A few of these shoals are very narrow and very long, some of them north–south, others east–west.

Reappearing a long way to the east after an interval with very deep water, Woodlark Island has a barrier reef and Egum is an atoll. They are the major reefs in a coral-enriched belt that stretches for 200 miles. Well north of this are the abyssal depths (to at least 28,670 ft, say, 30,000 ft) of the Solomon Sea, and then the Bismarck Archipelago.

4. The Bismarck Archipelago (with New Britain Greatly Dominating)

This great, arcuate island has a few scattered reefs on both sides, though not a great many, and they are not continuous. Its northern side is the steeper and subject greatly to landslides, as mentioned earlier. But that northern side is, if anything, richer in reefs than the southern, so that it seems unlikely that landslide debris was primarily the cause of the paucity. The sea around and north of the concave face of this island, and its continuation west to New Guinea, is marked by a row of prehistorically extinct, giant, explosive, volcanic caldera, maybe the largest that we know anywhere. Although no doubt at those times effusive outpourings of volcanic ashes killed all reefs that were there, reefs have grown again though not extensively.

5. *New Ireland and the Solomon Islands*

These long continental islands, which extend behind the Marshall Line just beyond the edge of the primal Pacific Ocean, also have close relationships with Australia and New Guinea. Individually and in groups they extend in the general direction of northwest to southeast, parallel to the Marshall Line in this vicinity; and the Solomon Islands are split cleftwise into two nearly parallel groups with this trend by a deep waterway (up to 6000 ft deep), appropriately called The Slot (or New Georgia Sound), with offshoots running featherlike off it—Indispensable Strait, Bougainville Strait, and Iron Bottom Sound, for instance. Within this general assembly the islands are quite irregularly coralliferous.

Those long island members, Choiseul, Santa Ysabel, and Malaita, closely adjacent to the Marshall Line, have flourishing fringing and incipient barrier reefs on all sides. New Georgia, on the other (western) side of The Slot, has reefs almost entirely on the eastern (Slot) side. Koolombangara Island, in the New Georgia Group, has a tight fringing reef all around it, as has Rennell Island, far away and isolated to the southwest. New Ireland, Gaudalcanal, and San Cristoval, great elongate islands west of The Slot or its continuation, are without reefs. And around Bougainville, as around New Britain, reefs are most capriciously developed. All of this variety is in warm, wide oceanic waters, within 12° (800 statute miles) of the equator.

As already noted for New Britain, landslide debris and volcanic ash fallout can hardly be claimed for the variety; and no great river, like the Fly, disgorges killing mud about these archipelagoes. But all show signs of tectonic fracturing; and the variety of reefs is in accord with a variety of tectonic habit. At least three major tectonic episodes have affected this region, the last alpine folds of the world, as represented by the mountains of New Guinea; the fracturing of what was, until recently, the great eastern extension of the Australian continent; and the pressure outward from the primary Pacific Ocean which will be considered presently.

Those islands with tight fringing reefs all around them (Rennell, for instance) may be stable. Others with barrier reefs on all sides, like

Santa Ysabel, may be generally subsiding. Still others (like New Georgia) with barrier reefs on one long side only, are possibly tilting and so allowing that reef side only to subside. While those large islands with no reefs at all may be either rising or else sinking too fast. All this, of course, remains to be proved.

III. Looking Widely

In the warmer, western half of the Pacific, coral reefs lie scattered between the limits of lat. 30°N and 30°S. In the east there are no rocky islands or archipelagoes other than more or less recent volcanic clusters. This is the primary Pacific, or Ur-Pacific, and it is widely deep with large numbers of seamounts and guyots such as the Emperor, Marcus Necker, and Austral Chains.

The Marshall Line, as emphasized by Bryan, is no less significant than the boundary between the simatic, primary Pacific in the east and that portion which has been added to it by the foundering of much of the once large southern sialic continent, giving a generally shallowed region within which lie the continental remnants of Australia, Tasmania, New Guinea, New Caledonia, Fiji, the Louisiades, the Solomons, the Bismarck Archipelago, and so on. The structure and bathymetry of this region have recently been discussed by Krause (1967).

Correspondingly there is a marked contrast in coral reefs. In the far east, with no large islands of old rocks for them to form around, there are no large barrier reefs. There occur the great atolls, Kwajalein (75 miles across), Truk, Eniwetok, Bikini, Funafuti, Jaluit, and very many more arising, as there seems now to be little doubt, from seamounts and guyots as they steadily sank. The underwater slope on these may be, as Darwin noted, steeper than any volcanic cone. The size of an atoll is thus a function of its age, the rate of settling and the size of its foundation. When from such submerged peaks volcanoes arise and reach the surface they are encompassed normally by fringing reefs as at Tahiti, Nauru, and Raratonga, but rarely by barrier reefs. Toward the western limits, near the Marshall Line, these atolls get smaller and then cease. Beyond that the Solomons have other reef types, mainly right against the probably moving Marshall Front.

When next we come upon reefs of sinking types they are the very long barrier reefs, the Great Barrier Reef itself and those others of New Caledonia, New Guinea, and Fiji.

There is evidence that in some regions the reefs, after having started to settle, began to sink too fast and became obliterated by sediment. Noticeably this is so in the large gaps with shoals of reef shapes that interrupt the barrier reef of southeastern New Guinea and in the alternation, around the *Vogelkopf*, of patch reefs, mixed reefs and shoals, and shoals only, a gradation from normal reefs that dry out, to banks of sediment anchored upon some firm foundations and permanently submerged. The largest of those gaps in the New Guinea barrier reef is matched across the Owen Stanley Range on to the other side of the island by a deepened region of similar width across Goodenough Bay and the Cape Vogel Peninsula and a further 100 miles out to sea, as though the one trough structure had continued across the island axis.

In such a tectonically not yet stable island as this there are other areas of small patch reefs, of other obliterated reefs and even, as about Humboldt Bay, of raised reefs, where vertical movements seem to have been at variance with the slow, specific rate of optimum subsidence peculiar to the proper formation of atolls and barrier reefs. And so I make bold to suggest that there is a circumscribed but almost imperceptible rate of subsidence (only) associated with extensive reef formation. Indeed, if there were not some such delicate control, other than adventitious deterrents like freshness of water, turbidity and so on, it would be difficult to see why rigid corals should not be almost as abundant as flexible seaweeds and navigation in warm coastal waters almost impossible. After all, reefs are finely adjusted to water heights, allowing for all tidal vagaries. Most species live permanently wetted; and it is only a few hardy types that can stand a drying-out for only brief periods, and even these usually have dead crests to the colonies.

Tectonically, the following seems to have happened: The Ur-Pacific, settling down and pressing outward in all directions in the process, has reacted against its continental margins, creating near them the great, deep, circumoceanic trenches. On the western side these deeps reach their greatest profundity and are within and close to the edges of the foundered continental margins of Asia and Australia. Here, in that tropical region which nearly concerns us, the greatest gash is the Solomons Trench, 30,000 ft deep, in the New Britain–Solomon Islands enclave. All eleven soundings narrowly along the axis of this trench, as shown in the folding map herewith, are over 25,000 ft deep. Between this and the Marshall Front lie the Solomon Islands, streamlined northwest to southeast and divided into two chains by The Slot, or New Georgia Sound, 6000 ft deep, and in that same direction. This blocky structure has a more than usually varied reef assembly, some big islands (e.g., New Ireland and Guadalcanal) without reefs; some (e.g., Rennell and

Kolombangara Islands) with tight fringing reefs; some (Santa Ysabel and Choiseul Islands) with incipient barrier reefs; some very big islands (like New Britain and Bougainville) with extensive reefs only in a few regions.

West of this irregular province there is a vast tract of ocean without guyots, but in which continental islands continue to occur and which all have barrier reefs, the greatest barrier reefs in all the oceans, fronting Queensland (The Great Barrier Reef), southeastern New Guinea, New Caledonia, and Fiji. The first two of these form the margin of that wide, deep, ocean basin, most aptly termed the Coral Sea and descending to 15,500 ft, which not only is so rimmed, but it has within it some shallower regions all of which have coral reefs. There can be little doubt that its coral rim would have been complete around the warm north, around the Gulf of Papua, but for the immense debouchment into its apex of that great volume of muddy fresh water from the Fly-Purari delta complex.

The slowly subsiding basin of the Ur-Pacific, strewn with atolls since there are no old island masses around which to form long barrier reefs, ceases at the Marshall Line toward which its atolls get smaller. Then at this buckling frontage occurs the block-structure assembly of the Solomon Islands, some of which are sinking, some apparently rising, and some no doubt temporarily or permanently stable. The western part of this province has few reefs, and possibly is stable, except for one probably sinking belt in which occurs the lonely Egum Atoll and the barrier reef of Woodlark Island. Then follows the greatest sinking region of all, the Coral Sea, and its continuation to New Caledonia and Fiji.

All of these long barriers, except the Great Barrier Reef, continue to the ends of their warm land masses. But the Great Barrier Reef ceases at Lady Elliott Island (lat. 24°S), even though some small coral areas go further, into Moreton Bay (27°) and to Lord Howe Island (31°). It is perhaps that the ideal rate of sinking that is necessary for a flourishing reef ceases at Lady Elliott, and that some geographical variation in the rate of general sinking may accord with the known variations in the Great Barrier?

The general theme of this chapter, then, is that not only, as Darwin postulated, are the great coral reefs of the world formed upon subsiding foundations, but also that there is a favorable rate of subsidence, and that such reefs are rarely formed if the rate of vertical movement is above or below that favorable value. The question thus arises: If this is so, what is the rate and the range of this subsidence?

That we cannot answer at present; but the Queensland coast offers an admirable opportunity of testing this conjecture. Along it is the greatest and most varied coral chain in the world, mysteriously ceasing at 24°S when it might have gone to 30°S; with two patches of coral reefs south of this, at 27° and 31°. And along this coastline are many important ports where determinations of mean sea level have been made from time to time over several decades. An analysis of the data from marine surveyors' notebooks might possibly bring to light subsidence rates at several points, within and beyond the coral reef region.

The vast Arafura Sea, beyond this region of pronounced instability, is completely puzzling. Its waters are very warm throughout [Schott (1935) records them as practically uniform at any time]. It has a great variety of coastlines: ria coasts in the Kimberley District of Western Australia; hardrock frontages in Arnhem Land and the Aru Islands; soft plains reaching the sea in the south of the Gulf of Carpentaria with a fall of less than 1 ft/mile and continuing that same slope as the sea floor; swampland coasts in the Oriomo region of New Guinea; and so on. It is continuous across the Torres Straits with the Coral Sea where this is particularly coralliferous. Yet it has no active coral reefs except a few insignificant fringing reefs about its southern margin. Fairbridge (1950) records raised coral reefs in places, of presumably Pleistocene age. Here, particularly, there seems to be a case for active coral reef formation having some limiting epeirogenic control. The raised beaches and reefs had probably no more than eustatic (glacial) status; and the last opening to the Pacific across Torres Straits was very recent, probably according with those poor fringing reefs. The curiously unreal uniformity of its whole sea region suggests that it may have had a similar vertical status throughout, on the firm continental shelf of Australia; though whether it is slowly rising, slowly falling, or static remains to be determined.

Finally, attention may be drawn to one other reef environment. The perfectly streamlined form of the large reefs of Torres Straits, with near-maximum restriction of the waterways, is without doubt due to the powerful, recently established Pacific to Indian Ocean currents, and shows admirably the effects of glacial control. The abnormally wide platforms of the Warrior Reefs which, as the folding map prominently shows, are more like those of Torres Straits than they are to those of the adjacent parts of the Great Barrier Reef, may also be in harmony with these ever-sweeping currents. Yet in the many current-powered straits between the Lesser Sunda Islands only an occasional small reef occurs. These are things that need investigation in time to come.

References

Admiralty Tide Tables. (1963). Pacific Ocean and Adjacent Seas (620 pp.). "Tide Tables," Vol. III. (Hydrogr. Dep.), The Admiralty, London.

Behrmann, W. (1928). Die Insel Neuguinea: Grundzüge ihrer Oberflachengestaltung nach dem gegenwärtigen Stande der Forschung. Z. Gesamte Erdk. Berlin pp. 191–207.

Bryan, W. H. (1944). The Relationship of the Australian Continent to the Pacific Ocean—Now and in the Past. J. Proc. Roy. Soc. N.S.W. 78, 42–62.

Detzner, H. (1922). Ergebnisse von Reisen in Neuguinea 1914–1918. Verh. Geographentages Leipzig 20.

Fairbridge, R. W. (1950). Recent and Pleistocene Coral Reefs of Australia. J. Geol. 58, 330–401.

J.A.N.I.S. (1944). Confidential Reports. Australian Joint Army-Navy Intelligence Studies No. 157, Vol. 3.

Jones, O. A., and Jones, J. B. (1956). Notes on the Geology of some North Queensland Islands. Part 1. The Islands of Torres Strait. Rep. Gt. Barrier Reef Cte. 6, 31–34.

Krause, D. C. (1967). Bathymetric and Geologic Structures of the North-Western Tasman Sea–Coral Sea–South Solomon Sea Area of the South-Western Pacific Ocean. N.Z. Dep. Sci. Ind. Res., Bull. 183.

Macgregor, W. (1890). The Fly River. Georgr. J. 12, 352–374.

Richards, H. C., and Hedley, C. (1925). A Geological Reconnaissance in North Queensland. Rep. Gt. Barrier Reef Cte. 1, 1–28.

Schott, G. (1935). "Georgraphie des indischen und stillen Ozeans." Boysen, Hamburg.

Schultze, L. (1914). Forschungen im Innern der Insel Neuguinea: Bericht des Fuhrers uber die wissentschaftlichen Ergebnisse der deutschen Grenz-expedition in das westliche Kaiser-Wilhelmsland 1910. Mitt. Schutzgeb. Berlin, Suppl. No. 11.

St. John, V. P. (1970). The Gravity Field and Structure of Papua and New Guinea. Aust. Petrol. Exp. Ass. J. 10, 41–55.

van Bemmelen, R. W. (1939). The Geotectonic Structure of New Guinea. Ing. Ned.-Indie 6, 17–27.

Wells, J. W. (1955). A Survey of the Distribution of Reef Coral Genera in the Great Barrier Reef Region. Rep. Gt. Barrier Reef Cte. 6, 1–9.

Whitehouse, F. W. (1972). The Coming of Aboriginal Man to Australia. Mankind (in press).

Zwierzycki, J. (1926). Notes on the morphology and Tectonics of the North Coast of New Guinea. Philipp. J. Sci. 29, 505–513.

7

WATERS OF THE GREAT BARRIER REEF PROVINCE

Dale E. Brandon

I. Introduction

Very little oceanographic study has been done on the Queensland Shelf proper (Fig. 1), which contains the Great Barrier Reef. Here I consider oceanography to mean physicochemical measurements of the water column on the shelf. I am excluding biological observations and those dealing with diurnal changes, or those limited to reef flats and lagoons. The only major physical oceanographic work of which I am aware, was conducted by the Great Barrier Reef Expedition of 1928–1929 under the joint sponsorship of the Great Barrier Reef Committee and a Committee of the British Association for the Advancement of Science;

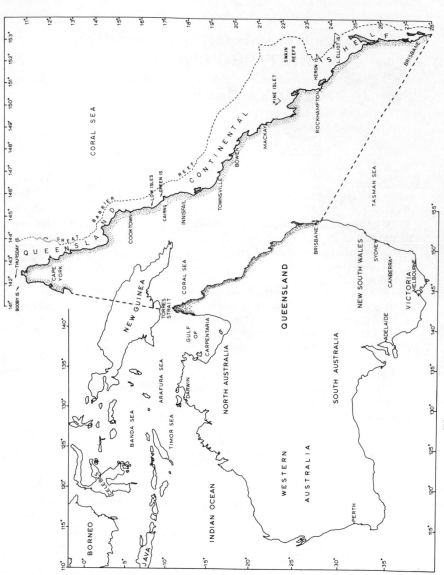

Fig. 1. Australia and the Queensland continental shelf.

the results were published by the British Museum (Natural History). Members of this expedition spent up to a year on Low Isles off the Queensland Coast about 35 miles north of Cairns. Two oceanographic papers resulted from this expedition, one by Orr (1933) and the other by Orr and Moorhouse (1933). Thereafter, either no major oceanographic measurements of the shelf have been made or, if so, they have not found their way into the literature. Some limited data on temperature, salinity, etc., has been published in conjunction with biological papers but these have been confined usually to one location and no interpretation of the oceanographic data was attempted. Some shipboard data have been collected by the Australian Commonwealth Scientific Industrial Research Organization (CSIRO), Division of Fisheries and Oceanography and have been made available to me. These data have not as yet been published, and for the most part, were collected in conjunction with biological studies so that oceanographic interpretation of them are not expected. A marine biological research station exists on Heron Island at the extreme southern end of the Great Barrier Reef, but thus far it is primarily oriented to biological endeavors. It can be seen that only spotty data over this huge area exist and, with the exception of Orr (1933), no attempt has been made to construct a picture of the waters on the Queensland shelf.

From September 1967 through August 1968, I collected water samples over the entire Queensland shelf. Three cruises were made on the M.V. *Cape Moreton*, which is operated by the Commonwealth Lighthouse Service, and one cruise was made on the Yacht *C-Gem* in order to work along the outer barrier and in unsurveyed areas. Cruises were planned to cover as large an area as possible and to obtain as near as practicable a representative sampling over the entire year. Combining these data with those in published and unpublished form it is my desire to establish the basic characteristics of the water on the shelf and to determine what normal changes, modifications, and movements this water undergoes.

MEASUREMENTS AND METHODS

Temperature and salinity were the prime measurements carried out on all the samples. Temperature was measured either by reversing thermometers when Nansen cases were made, or by special immersion-type thermometers for surface samples. Reversing thermometers are accurate to within ±0.02°C and the immersion thermometers to 0.1°C.

Salinities were measured on an inductive salinometer for which accuracy to 0.003‰ is claimed (Brown and Hamon, 1961). However, salinities have been rounded off to the nearest one-hundredth and are reported as such.

pH measurements were conducted on all samples during the period September–November 1967 on a pH meter whose accuracy was well within the two decimals reported. However, since no significant variation or correlation of pH with other data was found and no alkalinity measurements were made, these measurements were discontinued for the remainder of the study. Calcium carbonate saturation measurements were made on a number of occasions and it was found that the water was always saturated or supersaturated with respect to calcium carbonate during the September–November period, and thereafter no others were made.

A standard shallow-water bathythermograph was used for subsurface temperature measurements and Nansen bottles were used for collection of subsurface water samples. Usually a 10-m spacing between Nansens was maintained, with the exception that samples were collected as near the bottom as possible, and on occasion the top 10 m of water was sampled at closer intervals.

Meteorological observations included air temperature (wet and dry bulb), wind direction and force, cloud cover, atmospheric pressure, and sea state. In addition, the state of the tide was recorded in order to investigate its relationship to various other properties.

II. Climate

A. Wind Systems Affecting the Queensland Coast

The major wind system operating along the Queensland coast is the southeast trade. This system dominates the entire Queensland continental shelf for about 9 months of the year, or from roughly March through November.

During the summer months two systems operate along the coast. The northwest monsoon in the northern areas reaching occasionally as far south as Cairns, and the southeast trades which continue over the southern remaining areas. The northwest monsoon reaches the Torres Strait about the middle of December and advances southward during January and February. Usually about the middle of February, the monsoon begins to retreat northward again. The northwest monsoon does not have the same dominating effect as the Indian southwest monsoon, but it is sufficiently effective for the northern areas of Australia to term the summer months the "Northwest Season." As the northwest monsoon re-treats, the southeasterlies reestablish their dominance over the shelf.

Hurricanes ("cyclones") are another feature which may be felt on the Queensland coast. On the average, two or three per year may affect some part of the coast. They usually occur during January, February, and March, but have on occasion occurred in December and April as well. They are most frequently felt between Cooktown and Rockhampton, but they are not limited to these areas. Hurricanes are usually generated in an area between lat. 8°S and 18°S with about one-half moving initially in a southwesterly direction, and the other half in a southeasterly direction. Of those which initially move southwesterly and toward Queensland, most of them recurve to the southeast; however some continue moving southwesterly and die out on the Australian conti-nent. Some have crossed the Cape York Peninsula and the Gulf of Carpentaria. In addition to the tropical hurricanes, there is on the average about an equal number of tropical depressions which are similar in type, but less violent, that reach the Queensland coast during these months.

The southeast trades are usually at their strongest in September, but remain at a fairly constant level throughout the year having a Beaufort force of 3–4. Near their northern boundary the southeasterlies become unreliable in late summer due to the presence of the northwest monsoon.

B. RAINFALL

Along the Queensland coast the annual rainfall varies from less than 40 inches to over 150 inches (Fig. 2). There are two areas of the coast where the lower figures prevail. One is near Gladstone and extends northward to the vicinity of Rockhampton. The second extends north-ward from Bowen to the vicinity of Townsville. In these areas, annual rainfall is between 30 and 40 inches. The highest coastal averages are south of Cairns to Innisfail where over 150 inches of rain fall annually.

The high coastal averages in the Innisfail region are largely due to the forced uplifting of moisture laden air moving on-shore and striking the Bellenden Ker range. The orientation of the topographical highs north and south of this general area are more nearly parallel to the southeasterlies and do not cause uplifting to such an extent, and there-fore, less rainfall results.

Two main circulation patterns of the atmosphere are connected with the rainfall. The northwest monsoonal circulation which occurs in the southern hemisphere summer, along with the tropical cyclones and

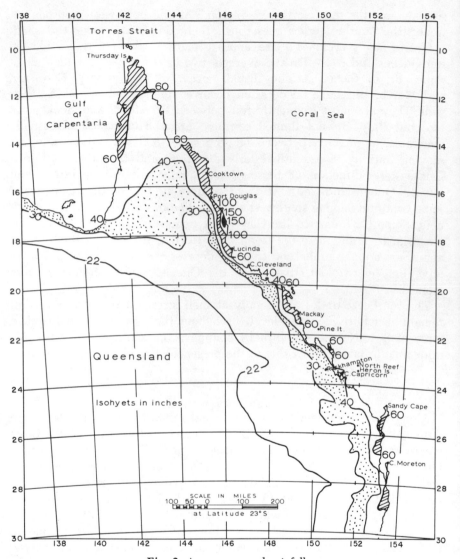

Fig. 2. Average annual rainfall.

storms, accounts for the large amounts of rain which fall over short periods of time. During the winter the rainfall is dependent on troughs and fronts which develop over the continent in the southern portions.

There are well-defined wet and dry seasons in the northern portion of the coast. Table I is a listing of selected coastal stations along the

TABLE I

AVERAGE RAINFALL FROM SELECTED STATIONS ALONG THE COAST OF QUEENSLAND

Location	Number of years	Average monthly rainfall (inches)												Average annual rainfall	Percentage of total rainfall occurring Jan. through March
		Jan.	Feb.	Mar.	Apr.	May	Jun.	Jul.	Aug.	Sep.	Oct.	Nov.	Dec.		
Cape Moreton	95	6.34	6.72	7.86	6.11	7.00	5.28	4.64	3.24	2.92	3.12	3.56	5.12	61.91	33.8
Sandy Cape	87	6.37	6.43	6.47	4.72	4.59	4.29	3.46	2.49	2.22	2.07	2.46	3.98	49.55	38.9
Heron Island	9	5.60	4.96	4.51	3.49	4.01	4.03	1.90	2.21	0.69	1.30	2.77	3.75	39.22	38.4
North Reef	7	5.39	4.92	5.91	4.47	3.89	5.27	2.52	1.97	0.55	1.14	1.92	3.68	41.63	38.9
Cape Capricorn	65	4.81	5.30	4.06	2.68	2.29	2.35	1.76	0.90	1.03	1.41	2.07	3.12	31.78	44.6
Rockhampton	82	7.12	7.67	4.63	2.32	1.62	2.41	1.72	0.79	1.23	1.84	2.42	4.45	38.22	50.8
Pine Islet	34	6.13	7.48	5.22	2.72	3.06	2.42	1.32	0.88	0.63	0.88	1.36	2.34	34.44	54.7
Mackay (Sugar Stn.)	67	14.60	13.63	11.68	5.45	3.27	2.29	1.45	1.05	1.16	1.66	2.79	6.72	65.75	60.7
Bowen	94	9.82	9.72	6.32	2.82	1.39	1.51	1.01	0.70	0.65	0.85	1.30	4.12	40.21	64.3
Cape Cleveland	37	11.58	13.51	9.40	2.96	1.59	1.24	0.79	0.39	0.40	0.77	1.96	3.45	48.04	71.8
Townsville (Rly.)	34	10.50	12.69	8.65	2.34	1.10	1.11	0.78	0.39	0.26	1.07	2.00	3.43	44.32	71.8
Lucinda	67	14.37	18.59	15.80	7.80	4.27	2.77	2.11	1.78	1.28	1.46	3.60	6.32	80.15	60.8
Innisfail	84	21.17	23.52	26.78	19.00	11.61	7.28	4.93	4.59	3.60	3.21	6.00	10.41	142.10	50.3
Green Island	16	18.89	15.89	15.68	8.84	4.97	3.64	2.61	1.99	1.74	0.85	3.62	6.01	84.73	59.6
Cairns (Aero.)	22	16.79	16.63	16.71	6.76	3.61	1.78	1.15	1.07	1.55	1.49	3.16	5.42	76.12	65.9
Low Isles	71	15.16	15.91	16.90	8.74	3.75	2.63	1.36	1.33	1.48	1.67	3.19	7.67	79.79	60.1
Port Douglas	79	15.52	15.99	16.49	8.29	2.61	1.97	1.01	0.95	1.31	1.72	3.97	8.20	78.03	61.5
Cooktown	91	14.35	13.97	14.81	8.17	2.83	1.94	1.02	1.19	0.60	0.89	2.34	6.15	68.26	63.2
Thursday Island	64	17.85	16.10	13.84	8.05	1.84	0.56	0.43	0.22	0.12	0.37	1.78	7.29	68.45	69.8
Booby Island	56	13.66	12.18	11.29	5.20	1.27	0.39	0.39	0.15	0.08	0.24	1.97	6.63	54.45	69.5

entire shelf going from south to north by month for all years of record. The wet season is readily apparent in the northern portions of Queensland. The three months of January, February, and March account for over 60% of the total annual rainfall in these northern areas. The wet season corresponds to the northwest monsoon and the dry season corresponds to the period during which the southeast trades influence the entire coastal area. In the southern areas of the shelf and coast, as the latitude increases the difference between the wet and dry seasons decreases and the rainfall is more equally distributed throughout the year.

The variability of rainfall along the Queensland coast is one of the most significant factors of the climate in this area (Dick, 1958). This variability is largely due to tropical storms which, although short-lived, lead to extremely heavy falls of rain. As might be expected, the highest degree of variability is in the areas where the lowest annual average rainfall occurs, such as Bowen, and the lowest degrees of variability are in the high rainfall areas. Bowen, for example, in January has experienced rainfalls varying from 0.19 inches in 1902 and 1947 to 46.57 inches in 1918.

This latter rainfall period resulted mainly from a tropical cyclone which accounted for 35 inches in 8 days and brought about the destruction of Stone Island Reef (Hedley, 1925). A rainfall as high as 34 inches in 1 day has also been recorded on the coast (Dick, 1958).

Isohyets generally parallel the coast and one is tempted to try to extrapolate the isohyets onto the shelf. Very few data are available pertaining to rainfall on the Queensland shelf proper and no extrapolation at this time could be done with a fair degree of confidence. However, in recent years the Brisbane Office of the Commonwealth Meteorological Bureau has begun to collect some data from the shelf areas. These data, for the most part, come from the few islands where there are still manned lighthouses. Table II is a compilation of these data for Heron, Pine, Green, and Low Isles. The coastal weather stations nearest these islands which have data available for the same years are also listed.

In the southern Great Barrier Reef area, Heron Island is compared with the closest coastal station Cape Capricorn, which is about 38 miles distant. Although there are only 8 years of comparable data available, it can be seen that Heron Island experiences a significantly higher rainfall than does the coast in the area of the Cape. This area of the coast is part of the southern member of the two lower rainfall regions along the Queensland coast. We may conclude from this that the low rainfall

TABLE II

A Comparison between Rainfall on the Great Barrier Reef and Coastal Stations

Location	Number of years compared	Average monthly rainfall (inches)												Average annual rainfall
		Jan.	Feb.	Mar.	Apr.	May	Jun.	Jul.	Aug.	Sep.	Oct.	Nov.	Dec.	
Heron Island	8	5.20	4.36	4.30	3.49	4.01	4.03	1.90	2.21	0.69	1.30	2.77	3.75	38.01
Cape Capricorn	8	5.40	3.77	2.49	1.26	2.46	1.10	1.22	0.68	0.42	0.84	2.12	3.81	23.66
Rockhampton	31	5.20	6.27	4.41	1.48	1.46	1.62	1.36	0.91	0.83	1.82	2.32	3.28	30.96
Pine Islet	31	6.21	7.03	5.15	2.75	2.89	2.46	1.35	0.88	0.67	0.84	1.23	2.27	33.73
Pine Islet	30	6.34	7.20	5.25	2.81	2.93	2.22	1.36	0.88	0.69	0.86	1.14	2.26	33.94
Mackay (Sugar)	30	14.39	16.37	11.39	6.11	2.64	2.10	1.39	1.29	0.77	2.25	2.47	4.84	66.01
Green Island	16	18.89	15.89	15.68	8.84	4.97	3.64	2.61	1.99	1.74	0.85	3.62	6.01	84.73
Cairns (Aero.)	16	17.53	14.70	18.02	7.05	3.53	2.05	0.95	1.17	1.52	1.51	3.58	5.81	77.42
Low Isles	69	14.90	15.54	16.78	8.52	3.65	2.68	1.37	1.34	1.48	1.47	3.05	7.56	78.34
Port Douglas	69	15.86	15.48	16.52	8.25	2.54	1.96	1.02	0.93	1.22	1.71	3.79	7.77	77.05

isohyet does not extend very far off the coast onto the shelf in this area.

Pine Islet is 142 miles northwest of Heron Island and is compared to Rockhampton and Mackay (sugar station). Mackay is much closer to Pine Islet (being about 65 miles distant, in contrast to 105 miles from Pine Islet to Rockhampton), but has about twice the average annual rainfall. Rockhampton and Pine Islet compare very favorably in their average rainfall for the 31 years of relatable data. Therefore the waters in the Pine Islet vicinity receive a lower rainfall than might be suggested by the nearest isohyets, and are more properly considered in the southern dry belt.

Green Island is situated about 15 miles from the Cairns Airport where the rainfall measurements are made. For the 16 years of comparable data, we see that the shelf regions received as much rainfall as the coast. Actually, there appears to be a slightly higher rainfall at sea in this portion of the shelf than that occurring on the adjacent coast. This is reflected in the 69 years of records for Low Isles and Port Douglas. Low Isles is about 33 miles northwest of Green Island and 10 miles from Port Douglas. The total rainfall in this area of the coast and on the shelf is in very good agreement and we can safely conclude that the shelf waters receive a very substantial amount of fresh water in the form of direct rainfall. In addition, runoff from the land must be quite considerable and it would be expected that this condition would find an expression in the total makeup of the waters occupying the Queensland shelf. As we shall see later, this is indeed the case.

North of Low Isles, there are no other weather stations on the shelf itself, and with the exception of Cooktown, no others on the coast. Not until the tip of the Cape York Peninsula and Thursday Island is reached are further rainfall data available. Data from Thursday Island and Booby Island in the western approaches to the Torres Strait are given in Table I.

III. Seasonal Variations in Temperature and Salinity

A. TEMPERATURE

The average minimum water temperatures found along the Great Barrier Reef are well within the limits recognized for coral growth. Kinsman (1964) has stated that corals flourish best in a temperature range of 25°–29°C, but they can withstand limited exposure to lows of 16°–17°C. The northern areas of the shelf fit the optimum temperature range almost perfectly. Booby Island has an average minimum temperature of 24.2°C in August and an average maximum of 29.5°C in December. The average temperatures in this area range from 28.6° to 25.3°C throughout the year.

The northern area of the shelf has prolific reef growth across it, from the outer barrier ribbon reefs, to large reefs on the inner shelf, and fringing reefs along the coast.

The southernmost island composed entirely of coral is Lady Elliott Island. This island is not considered part of the Great Barrier Reef system. Minimum average winter temperatures in this area fall below 20°C, and southward of this area temperatures become the limiting factor in reef growth.

The highest sea surface temperatures along the shelf occur in January in the northern and central regions. In the southern region there is a month lag, and the highest temperatures occur in February. The lowest water temperatures occur in June–July in the northern regions and lag to July–August in the southern areas. Figure 3 is a graph of average water temperature vs latitude by month. December is not shown because of an insufficient amount of data. From the figure, January appears as a rather uniform month over the shelf with temperatures increasing from south to north. February, March, and April compare favorably with the conditions of January with the one exception of lower temperatures existing on the shelf in February between lat. 14°S and 15°S. In April there is a definite lowering of the surface temperatures along the shelf. May temperatures appear to be rather erratic in that temperatures in the central region are lower than in the southern region. How accurate a deduction this is, is not known as May was a month which did not have a large sample density. June, July, and August contain the lower limits of temperature and in September the temperature increases along the entire shelf. Throughout October and November the surface waters continue to warm and by December they compare favorably with the summer maximums. Thus the four months, December through March, mark the high point of the temperature cycle, April and May contain the transitional cooling period to the winter lows in June, July, and August. September, October, and November are the transitional warming months to the summer maximums reached on the shelf commencing in December.

Between lat. 14°S and 15°S there appear two distinct periods when temperatures are lower than those to the south and north. This area contains the narrowest portion of the Queensland shelf and there are a number of relatively wide passages through the outer barrier reefs

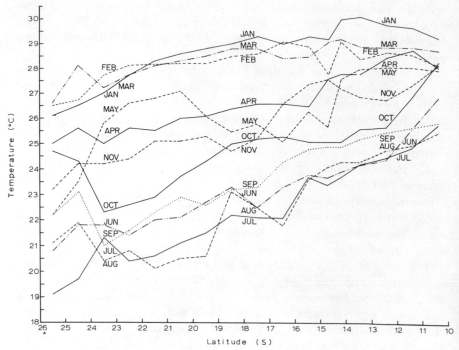

Fig. 3. Average sea surface temperature over the Queensland continental shelf by month and latitude.

at the narrowest width of the shelf. These lower temperatures are more probably a result of surface Coral Sea water having a greater influence on the small volume of shelf water at this point than it does to the south and north. This area of the shelf is unique in another respect as it contains the large, shallow Princess Charlotte Bay indentation of the coast. It is felt that the water temperatures here are a combination of shelf conditions, wind, and weather. This area is somewhat more independent than other shelf areas immediately north or south in that a larger number of factors can influence the water column here.

1. Subsurface Temperatures

The subsurface temperature distribution over the shelf is very similar to the surface temperature distribution. The southeast trade winds mix the water column throughout most of the year so that the vertical temperature gradient is never steep. Only during the summer rainy season are any thermal gradients maintained.

In the southern areas of the shelf where relatively deep water is found, the southeast trades operate over the entire year. In the Capricorn channel, vertical temperature variation is only 1°–2°C on the average, and this range is probably only rarely exceeded. Further north, the water depth decreases and the vertical temperature gradients are less than 1°C during the southeast trade season. Bathythermograph measurements made by the author over the continental shelf during the southeast trade season revealed essentially isothermal water with only a small warming of a few tenths of a degree centigrade generally restricted to the upper few meters.

During the summer rainy season, the vertical temperature distribution is altered. The southwest trades diminish and in the northern regions the northwest monsoon season begins. This season of the year is characterized not only by tropical storms, but also by long periods of light wind conditions. This is particularly the case in the buffer zone between the two systems.

Vertical gradients result from the diminished wind force and the influx of less dense water from direct rainfall and land runoff. The fresh water, being lighter, tends to remain on the surface and undergoes further warming. The surface waters become less dense, the sea becomes more stable, and the vertical temperature gradient is increased. However, a relatively short period of strong winds can readily mix the entire water column and destroy the gradient.

B. SALINITY

The importance of knowledge of the salinity of oceanic waters is well recognized by the marine scientist. Salinity is a basic parameter to the understanding of water movements and systems. Gerard (1966) described salinity as the most useful conservative property of seawater in circulation studies. The salts are moved from area to area by advective processes and from water masses of greater concentration to those of lesser concentration by diffusion processes. Salinity is of extreme importance to the life in the oceans and possibly there may have been no life without it. Pearse and Gunter (1957) have stated that "the most significant property of seawater in relation to the ecology and physiology of organisms depends on the fact that it is a complex solution of salts."

Along the Queensland continental shelf, the surface distribution of salinity is primarily a product of four processes: (1) evaporation, (2) precipitation, (3) continental runoff, and (4) mixing of coastal and oceanic waters. The first, second, and third factors are of primary importance for the northern half of the shelf, whereas the fourth process is dominant in the southern shelf areas.

1. Seasonal Salinity Cycle

The salinity on the portion of the Queensland shelf containing the Great Barrier Reef is seasonal in nature. Figure 4 is fairly typical of the salinities encountered between lat. 11°S and 14°S. Shelf salinities to the north of this region are influenced by the intrusion of West Irian and Gulf of Papua (Fly River) water masses; the area shows lower salinities during the monsoon season. The maximum salinities recorded in the northern regions exceed 36.15‰, and the minimum salinities are of the order of 33.60‰.

Evaporational processes appear to be very significant in these areas. Average salinity values to the south are all lower than on this portion of the shelf. Evaporation in these areas is considered responsible for the formation of a high-salinity coastal water mass which exists during the spring–early summer period of the year. It is bounded to the north by Papuan–Northern Torres Strait water and to the south by lower temperature, less saline waters. This seasonal water mass can be traced eastward where it grades into the surface Coral Sea waters.

The shelf waters between lat. 14°S and 19°S show a much greater range in seasonal salinity values than do those areas to the north and south. Figure 5 is a graph of the monthly salinity values contained in an area bounded by the coastline and the 200-m contour, and between lat. 16°S and 17°S. All salinities recorded in a particular month of the year within this area have been averaged to obtain the average curve. The maximum and the minimum value recorded in a month within this area is also plotted to give the range of values that might be expected. The region between lat. 14°S and 19°S contains the best-developed salinity cycles, and the area in Fig. 5 within this region, the most extreme.

The shelf between lat. 16°S and 18°S borders a coastline which receives the highest rainfall that occurs anywhere in Australia. Rainfalls may total over 150 inches/year off Innisfail and the majority of this rain occurs between January and March. The relatively great annual range of salinity along this portion of the shelf is attributed not only to the high rainfall and runoff during the monsoon season, but also to evaporation which can create localized zones of high-salinity water. It is felt that the maximum value of 36.18‰ in Fig. 5 reflects a localized condition of short duration. Most likely, shelf waters rarely exceed

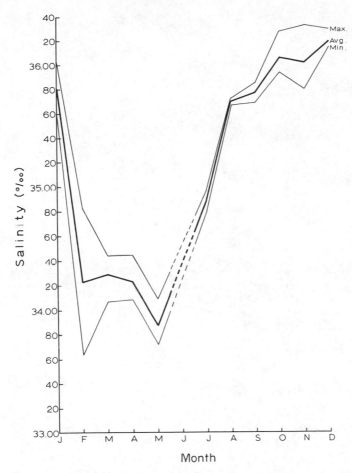

Fig. 4. Monthly salinity values between lat. 11°S and 12°S.

35.80‰ through this area, except very close to shore during the late spring.

The total annual variation of salinity south of lat. 20°S decreases markedly. Salinities greater than 36.00‰ or less than 35.00‰ occur only rarely and for the most part, only near the continent. The outer portions of the shelf in this region have not been sampled so that the above remarks do not apply to this portion of the continental shelf. Figure 6 contains the salinity cycle between lat. 22°S and 23°S and the coastline and long. 152°E. It can be seen that the seasonal cycle is much reduced from that in Fig. 5. The salinity cycle shown in Fig. 6 is representative

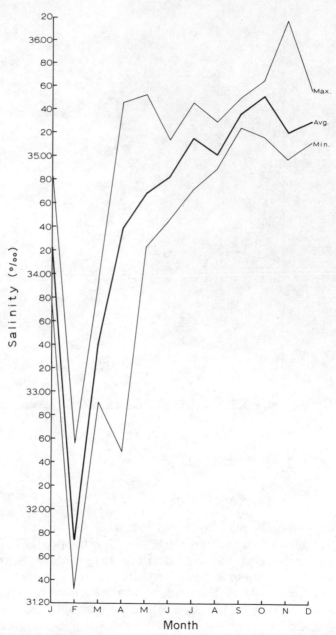

Fig. 5. Monthly salinity values between lat. 16°S and 17°S.

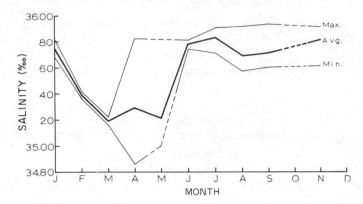

Fig. 6. Monthly salinity values between lat. 22°S and 23°S.

of the seasonal change in salinity on the Queensland continental shelf between lat. 20°S and 27°S. This area is relatively open to mixing with Coral Sea waters and the pronounced monsoon season is absent over this area. The southeast trade winds predominate over the entire year.

2. The Surface Salinity Gradients across the Queensland Continental Shelf

Well-defined salinity gradients exist across the northern portion of the Queensland shelf during two distinct seasonal periods. These periods are the spring–early summer and the northwest monsoon season. It is probable that surface salinity gradients exist as far south as about lat. 20°S. South of this latitude data are not available in the form of traverses across the shelf, but from the previous section we know that seasonal cycles are less distinct and therefore only very minor gradients would be expected.

In the spring–early summer period (about September–December), salinities reach their maximums on the shelf. The salinity is greatest near shore and decreases seaward to the outer barrier reefs. Salinities in the northern region begin increasing at the end of the northwest monsoon season and when land runoff is no longer significant. This is generally around the fall period (March–May), but depends on the local bordering continental climate. As the shelf salinities increase, they reach a point where they are homogeneous with Coral Sea waters and thus no gradients exist. The increase in salinities is due to (1) the lessening of rainfall and runoff, (2) surface Coral Sea water mixing with the shelf waters, and (3) evaporation. Evaporation becomes the significant process after the shelf waters have reached the same salinity level as that of the bordering Coral Sea surface water, which is generally between 35.00 and 35.50‰. Thereafter gradients are established across the shelf and generally increase until the northwest monsoon season begins.

When the northwest monsoon season begins, there is a reversal in the salinity cycle. Salinities across the shelf pass through a point where they are generally homogeneous with bordering Coral Sea waters and then continue to decrease until the low-salinity waters near-shore establish positive salinity gradients across the shelf.

Figure 7 shows a portion of the northern shelf from Cape Grenville to the Gulf of Carpentaria. Shelf salinities during October and November, 1967, have been plotted and contours drawn. It can be seen that very distinct gradients exist across the shelf and that high-salinity water (greater than 36.00‰) borders the Australian continent and extends through the southern Torres Strait into the northeastern part of the Gulf of Carpentaria.

A traverse to Great Detached Reef showed that surface Coral Sea waters bordering the Great Barrier Reef were of the order of 35.50‰. In the Gulf of Carpentaria there was a very sharp boundary between the high-salinity Australian coastal water and Gulf of Carpentaria water. The Gulf of Carpentaria surface water had salinities between 35.60 and 35.69‰, or only slightly higher than Coral Sea surface water. On the eastern Queensland shelf, the salinities grade into the Coral Sea waters, but in the Gulf of Carpentaria there is a distinct boundary between the two masses. Northward, in the Torres Strait, salinities are much lower than those in the Coral Sea, the Gulf of Carpentaria, or on the Queensland shelf from Cape York southward. A rather distinct salinity boundary exists in the southern Torres Strait. Waters with salinities generally less than 35.10‰ border waters with salinities greater than 35.80‰. The Tokyo University of Fisheries training ship *Umitaka Maru*, proceeding southward in the steamer channel down the Australian coast in December 1967, confirmed that the high-salinity water mass (greater than 36.00‰) still existed at that time. She reported salinities up to 36.29‰ between Cape York and Cape Grenville.

During the northwest monsoon season, positive salinity gradients exist seaward across the shelf. These gradients are much larger than those during the spring–early summer period due to the large amounts of rainfall and runoff. During March and April, 1968 surface salinities were measured between the Torres Strait and lat. 15°S. Although by this time the northwest monsoon season was essentially over and the south-

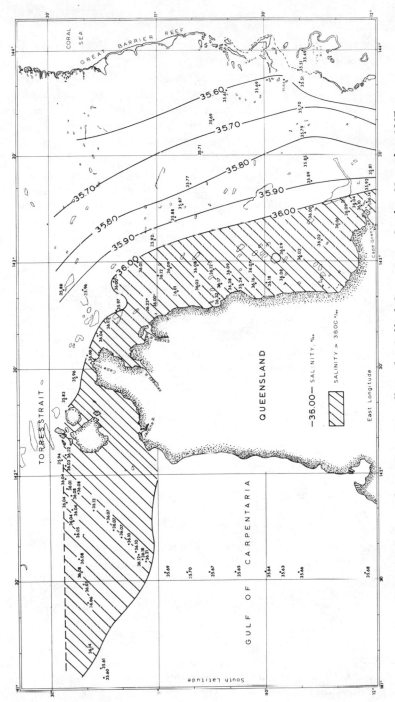

Fig. 7. Salinity gradients, Cape Grenville to the Gulf of Carpentaria, October–November 1967.

205

east trades had begun, the salinity change across the shelf in the vicinity of Lizard Island was on the order of 2.0‰.

Salinities in the Torres Strait were still relatively low at this time being less than 33.00‰, and salinities less than 34.00‰ still extended to lat. 11°S. The lower salinity waters in the Cape York and southern Torres Strait area are a reversal of the conditions found in the spring–early summer period. During that time a tongue of high-salinity water extends westward through the strait and into the Gulf of Carpentaria (Fig. 7), but during the monsoon season and for some time thereafter, a tongue of low-salinity water extends southward down the Australian coast. In addition, near-shore coastal low-salinity water unrelated to the previous low-salinity water can be expected southward due to rainfall and runoff from the adjacent continent. Between Cape Grenville and Princess Charlotte Bay, this low-salinity water may not be significant, but southward from the bay, through the bordering coastal high rainfall areas to about lat. 18°S, it is probably very significant.

In April 1968 surface salinities on the Queensland shelf from Trinity Opening in the outer barrier to lat. 22°30′S were recorded. Surface Coral Sea water with a salinity of 34.88‰ was found just outside Trinity Opening, but near Cairns the salinity was 33.76‰ and southward off the high rainfall areas it was between 32.00 and 33.00‰. In the vicinity of Townsville and southward, salinities were 34.75‰ and greater. Therefore, it is expected that a low-salinity water mass borders the continent from about Princess Charlotte Bay southward to the northern vicinity of Townsville during the northwest monsoon season. The largest salinity gradients are probably found off the Innisfail region and these probably persist until May. Differences in salinity across the shelf during the peak of the season may be as much as 5‰.

In June and July 1968, the coastal salinities north of Cairns to Torres Strait were again determined. Salinities in June from Cairns to Cape York varied between 34.40 and 35.00‰. In July, they had increased so that they varied between 35.00 and 35.40‰. Two traverses across the shelf into the Coral Sea via Cook's Passage and Trinity Opening revealed that the surface water was essentially isohaline, but that very good thermal gradients existed.

3. Subsurface Salinity

The vertical salinity profile on the Queensland shelf is, for the most part, isohaline for the majority of the year. The only exception to this is during the rainy season when the surface salinity may be greatly reduced. Rainfall and land runoff seem to be the only important process whereby any vertical salinity gradients may be established.

From May to the commencement of the rainy season, salinity differences between the surface and bottom waters on the northern Queensland shelf probably rarely exceed 0.10‰. Local variations may occur during this time after a rainfall, but these would be exceptions and of short duration. This is most likely true of the southern shelf also as the southeast trades remain dominant over the year and are capable of keeping the water column mixed.

When the rainy season begins surface salinities are greatly reduced. Depending on the amount of wind at the time, the less saline water may tend to stratify near the surface, or if the winds are sufficient, salinity in the entire water column may be reduced. Orr (1933) has suggested that this effect is also of short duration, and that the less saline bottom water is rapidly replaced by water from outside of the outer barrier. For the most part, when rainfall and runoff cause a stratified layer of less saline water to form, and the wind does not destroy this stratification, the effects may be noted in the water column from 5 to 15 m, although the smallest figure will usually show the sharpest break.

Northward of Low Isles, the outer barrier reefs become more dense and it is reasonable to assume that by lat. 14°S the Coral Sea water may have less effect on modifying the salinities developed in the shelf waters during the rainy season. Again, however, the northern areas receive less rainfall so that the water column does not contain the extreme changes in salinity which may be found southward.

The shelf water north of lat. 11°S should contain substantially reduced subsurface salinity values during the northwest monsoon season. It has been stated that Gulf of Papua or West Irian coastal waters, or both, move into this area during the monsoon season, and their salinity values are significantly lower than shelf waters further south.

IV. The T–S Characteristics of the Water on the Queensland Shelf

In this section the temperature–salinity (T–S) characteristics of the waters on the Queensland shelf are discussed. Data collected by the author in 1967 and 1968 are used to develop the relationships between waters on various portions of the shelf during a particular season of the year, and over the year.

T–S diagrams are widely used in physical oceanography in order to describe water masses in terms of their temperature and salinity values. Helland-Hansen (1916) devised this method by which the two variables, temperature and salinity, can be related. The diagram is instrumental in determining or recognizing water masses at various depths in a vertical column of the ocean, in recognizing the mixing of two or more water masses, and in locating regions of convergence and divergence. In addition, by using the "core method" developed by Wüst (1935) the spreading of water types and masses may be followed.

Normally, T–S diagrams are not used in discussing the surface layers of water. This is due to the fact that the surface temperature and salinity in the ocean cannot be considered as strictly conservative properties of the water. This then normally precludes using T–S diagrams for describing horizontal surface T–S characteristics of the oceans.

There are a number of exceptions in which T–S diagrams have been used constructively in dealing with the surface water. As an example, Iselin (1939) showed that the T–S curve for a column of water in the Sargasso Sea was similar to a horizontal T–S curve for surface waters in the North Atlantic in the winter. This was explained by the sinking of the water mass at convergences, after its formation at the surface, essentially along isopycnic surfaces.

Montgomery (1955) used a T–S diagram for the quantitative representation of the T–S characteristics of the surface water at weather ship "J" in the North Atlantic Ocean. From 3 years of data at an essentially fixed point in the ocean, he assembled a frequency distribution of water temperature and salinity at the surface. The prevalence of certain surface temperatures and surface salinities are easily read from the two-dimensional frequency distribution derived from the T–S diagram.

Cochrane (1956) described the frequency distribution of the surface water characteristics of the Pacific Ocean using T–S diagrams. From charts of mean temperature and salinity, he determined their frequencies of occurrence in six two-dimensional distributions covering the months of February and August, and over the year in both the North and South Pacific.

Two-dimensional frequency distributions of the T–S characteristics of the surface waters on the Queensland shelf have not been attempted in this work. Although it is believed by the author that this would be of value in describing the Queensland waters in a quantitative manner, it is also believed that not enough data are available to undertake this type of analysis at the present time. It would be meaningless to combine all the surface data on the Queensland shelf as it has been shown that location on the shelf is important in understanding the waters which occupy a particular place at a certain time. Therefore, until enough data for a particular location on the shelf can be accumulated, this type of quantitative analysis is not recommended.

In this section T–S diagrams have been used for plotting the surface characteristics of the shelf waters for three time periods. These are: September through November, March–April, and July–August. It should be pointed out that the surface values of temperature and salinity are representative of the entire water column on the Queensland shelf for most of the year. It is only during the rainy season in certain areas, that the surface T–S characteristics are not representative of the water column. It is hoped that the study of the areal distribution of the T–S characteristics on the Queensland shelf will lead to an understanding of the relationships between waters on the shelf and adjacent water masses in the bordering seas.

A. T–S Characteristics from September throuh November

Figure 8 is a T–S diagram containing the T–S values for stations occupied on the Queensland shelf during September and October 1967. These cover the inner shelf waters from lat. 24°43′S in the south to lat. 10°35′S in the north. The data from these stations represent conditions in the inner steamer channel between the continent and the off-lying reefs of the Great Barrier Reef Province, and also the southern portion of the Queensland shelf.

If geographical location is combined with the data plotted on Fig. 8, three general divisions are observable, these are shown (with others) in Fig. 10. The first is a general increase in temperature from the south to the north with the salinity remaining fairly constant, varying only 0.35‰, while the temperature changed 4°C. This covers the shelf roughly from Gladstone to Townsville.

The second is nearly constant temperature accompanied by an increase in salinity from 35.54‰ at about lat. 18°S, to 36.00‰ off Cape Sidmouth. The sea-surface temperature varied within 1°C for nearly 400 miles along the coast and was $25.0 \pm 0.1°C$ from lat. 17°S to 14°S with very few exceptions.

The third division is high-temperature, high-salinity water between Cape Grenville and Cape York. The highest T–S values occur about midway between the two capes.

Figure 9A contains the T–S plots for stations in the Western Torres Strait and the northeastern part of the Gulf of Carpentaria. Very good separation on the basis of salinity may be noted here. Salinities were

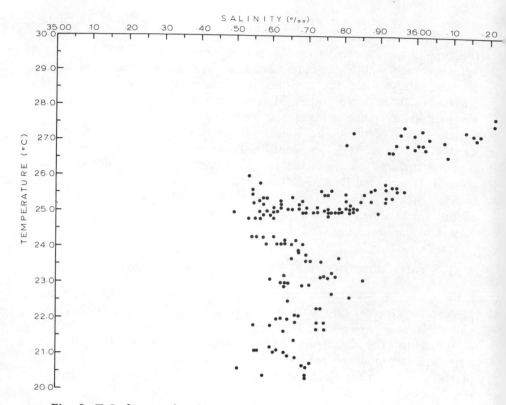

Fig. 8. T–S diagram for the inner Queensland continental shelf, September–October 1967.

36.00‰ or greater in the Western Torres Strait, while those in the Northeast Gulf of Carpentaria were less than 35.70‰.

Figure 9B contains all the stations taken on the Queensland shelf between Cape Sidmouth and Cape York in November 1967. These stations lie mainly on traverses across the shelf. As is readily apparent from the figure, the salinity variation is striking. The high-salinity values (36.00‰ and greater), are found along the coast and as the shelf is traversed toward the outer barrier, the salinities decrease to those of surface Coral Sea water (Fig. 10).

Figure 10 is a composite of the T–S values contained in Figs. 8 and 9A, B. The divisions previously discussed are graphically outlined in Fig. 10 according to their T–S values. The geographical regions on the shelf containing these values are named in, or next to, their respective groupings of T–S characteristics.

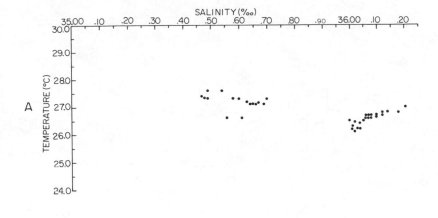

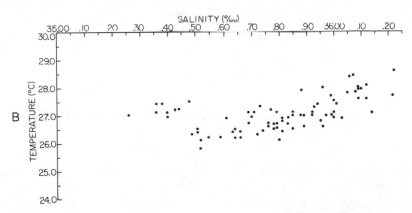

Fig. 9. (A) T–S diagram for the Western Torres Strait and NE Gulf of Carpentaria waters, October 1967. (B) T–S diagram for the Queensland continental shelf from Cape Sidmouth to Cape York, November 1967.

The most interesting portion of Fig. 10 for the early summer period, is the change in salinity values in the northern region. From the outer barrier reefs bordering the Coral Sea, the salinity increases across the shelf until the inner shelf is reached. The increase is linear and mixing ratios between the inner shelf waters and the Coral Sea surface water may be easily obtained. Coral Sea T–S characteristics near the shelf edge are not shown in the figure, but in the southern shelf regions they would correspond to temperatures of about 21.5° ± 0.5°C and

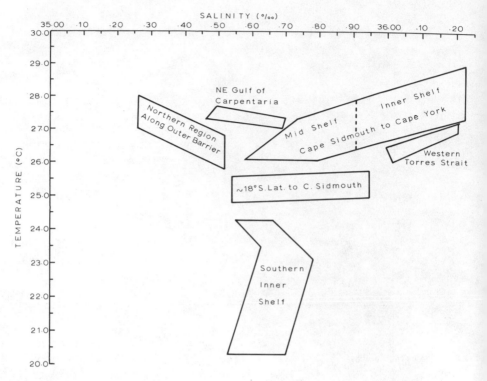

Fig. 10. T–S characteristics on the Queensland continental shelf, September through November 1967.

salinities of $35.65 \pm 0.10\%_o$ for the September–October period. In the northern region, they would occupy or be slightly to the left of the T–S boundary given for the area along the outer barrier in Fig. 10.

The separation between Northeast Gulf of Carpentaria water and that to the east and north in the Western Torres Strait is very distinct in Fig. 10. It would appear that separation between Torres Strait water and northern inner shelf water is implied by the figure, but this is not the case. T–S measurements in the NE Gulf of Carpentaria and the Western Torres Strait were made in October while the majority of northern shelf measurements were made 2–4 weeks later. The temperature increase during this time difference explains the separate boundaries for Western Torres Strait water and those on the inner shelf. As was shown in Fig. 7, the Western Torres Strait waters are an extension of high-salinity inner shelf waters into the Gulf of Carpentaria.

The T–S characteristics for the more northern portion of the Strait and the areas south of Papua are not given in the figure. Indications are that these waters have much lower salinities than any of the waters in Fig. 10 or in the adjacent Coral Sea areas. They more properly represent a separate and distinct water mass in their own right. It seems clear, however, that Northwest Coral Sea surface water which passes through the Torres Strait to the Arafura Sea cannot do so without losing its T–S characteristics.

Wyrtki (1960), in a paper on the surface circulation in the Coral and Tasman Seas, has stated that water masses flowing into the northwestern Coral Sea can pass through the Torres Strait only to a small extent. The water piles up against the shallow shelf and must sink. This convergence of Coral Sea water against the barriers of Australia and New Guinea is not limited only to the northwestern portion of the Queensland shelf but is operative along the entire Queensland shelf. The converging waters in the western Coral Sea result in a general downsinking seaward of the shelf, and the small amount of water which can pass through the Torres Strait is modified as it crosses the shelf, as has been demonstrated for the early summer period.

Convergence of waters in the Coral Sea also gives rise to the East Australian Current. This current is formed in the Coral Sea at about lat. 20°S and between long. 153°E and 158°E. From January to March it is derived from equatorial water masses which have moved west-south-westerly under the influence of the monsoonal winds. From April to December its source is subtropical water masses from the trade drift and the Central South Pacific Ocean.

Wyrtki (1960), in a series of diagrams on surface circulation, shows the East Australian Current on the outer portion of the Queensland shelf moving southeasterly from about lat. 20°S. This current, although still not concentrated at this point, should then have an influence in the vicinity of the Swain Reefs. Maxwell and Swinchatt (1970) state that the marginal zone of the shelf in their southern region (lat. 24°S to 20°S), extends eastward into the path of the East Australian current and this has contributed to prolific reef growth in this area. The current becomes more concentrated, according to Wyrtki, to the north of Great Sandy Island (northernmost point: lat. 24°41.5′S, long. 153°15.5′E) where Coral Sea water masses between the continent and the reefs are integrated into the current. The current then reaches its maximum strength between lat. 25°S and 30°S. Very little data exist for the waters on the outer portions of the shelf between lat. 20°S and 23°S. However, currents on the inner portions of the shelf move northwesterly under the influence of the southeast trade winds and the East Australian Current can have very little to no influence here. The East Australian Current may influence outer portions of the Great Barrier Reef, especially in the Swain Reef complex, but this is not documented. Its greatest effect on the Queensland shelf is south of lat. 25°S.

B. T–S CHARACTERISTICS DURING MARCH AND APRIL

Figure 11 contains the surface T–S values found on the shelf in March and April 1968. The data are generally confined to the inner steamer channel throughout the length of the Great Barrier Reef. Two traverses across the shelf were made. One traverse was across the shelf and into the Coral Sea via Cook's Passage through the outer barrier. The other traverse was made after reentering the shelf through Trinity Opening. T–S measurements were made on the traverses and in the Coral Sea. The Coral Sea surface T–S values are shown in Fig. 11.

As has been discussed previously, the northwest monsoon season begins generally in late December or early January. Heavy rainfall and tropical storms are common during this period and in the northern regions salinities are reduced. In addition, during the northwest monsoon period water movements in the Torres Strait are reversed. Low-salinity water derived from the coastal regions of West Irian moves essentially eastward into the Torres Strait and combines with waters moving south to southeasterly from the south coast of Papua. Transport of water is then eastward through the strait and into the Great Barrier Reef region.

In the northwest Coral Sea, upwelling occurs in January according to Wyrtki (1960). The upwelling is due to the monsoonal winds but is only of short duration and is not present by February. The upwelling is probably limited to the southern areas of the Gulf of Papua and it is not known if this upwelling has any effect on the Queensland shelf.

The T–S values shown in Fig. 11 were not taken during the peak of the northwest monsoon season, but they do reflect some magnitude of the changes in water properties that occurred. Boundaries have been drawn around the T–S characteristics for particular regions in order to illustrate some of the different zones existing on the shelf.

In the top portion of the figure, there is a good break between the waters in the Torres Strait and those southward of Cape York based on salinity differences. For the inner northern shelf, here roughly from Cape Sidmouth to Cape York, water temperatures were fairly uniform and salinities were less than 34.50‰.

Southward of Cape Sidmouth lower temperatures and salinities were

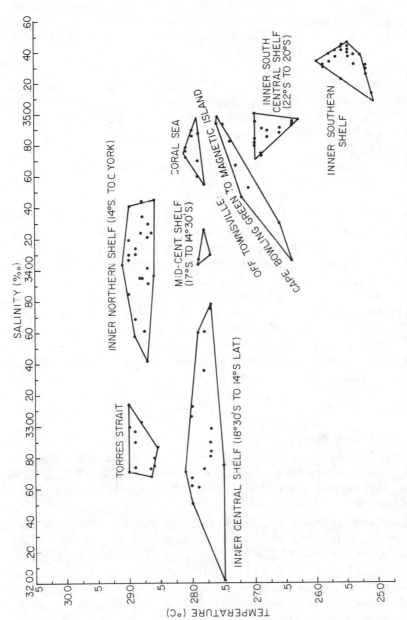

Fig. 11. T–S characteristics on the Queensland continental shelf during March and April 1968.

found. The area of low salinity in this region (labeled inner central shelf in Fig. 11) borders the high rainfall areas of the Queensland coast. Again, very uniform water temperatures were found as was the case during September and October. Temperatures varied only 0.6°C in the steamer channel over 4.5° of latitude.

Salinities during the traverses across the shelf in the central shelf region increased to the values found in the Coral Sea. Temperatures on the shelf and for the Coral Sea surface waters were uniform. The Coral Sea surface salinity just outside of Trinity Opening was 34.55‰, while the salinity inside of the reefs, but still in the exposed passage, was 34.26‰.

Off Townsville between Magnetic Island and Cape Bowling Green, temperatures and salinities decreased. The salinity decrease reflects the crossing of coastal surface salinity gradients.

South of Cape Bowling Green salinity variation decreased as the more uniform waters of the southern shelf areas were approached. Surface temperature continued to fall over the changes in latitude to Brisbane.

The salinities measured on the shelf during this cruise varied from a low of 32.01‰ to a high of 35.47‰. As expected, these values were found in the high rainfall area and the southern shelf area, respectively. Temperatures on the shelf were uniform, varying only 4.2°C over 16.5° of latitude. Coral Sea surface waters and the southeast trades dominated the T–S characteristics of the southern shelf and runoff from the monsoon season was still quite effective in depressing the salinity of the water in the high rainfall areas along the coast.

C. T–S Characteristics during June and July

Figure 12 contains the T–S characteristics of the northern shelf for June and July 1968. June through August are the winter months south of the equator and therefore shelf temperatures should be at their lowest values. Most of the stations were situated in the inner steamer channel and again two traverses across the shelf were made. One traverse was through Cook's Passage into the Coral Sea, and the other was made after reentering the shelf via Grafton Passage.

During June between Cairns and Cape Grenville shelf water characteristics appear to be fairly stable. Salinity gradients for the most part are absent. Surface temperatures on the shelf have a larger range and it is this effect which is most noticeable. For example, surface temperatures between Cairns and Cape York during March and April 1968 had a maximum difference of only 0.9°C, while in June it was 3.0°C and by July it had increased to 4.3°C.

The T–S characteristics shown in Fig. 12 for the Torres Strait in July are enclosed by temperatures of 25.5° ± 0.2°C and salinities between 35.01 and 35.28‰.

The T–S values on the shelf south of Cape Grenville to Princess Charlotte Bay in July are not shown in Fig. 12, but they reflect the trend of gradually increasing salinities from the low values registered during the monsoon season. The T–S characteristics in this region were approaching those values found in the Coral Sea surface waters.

The thermal gradients across the shelf south of Princess Charlotte Bay are very distinct and are probably best developed at this time. Stations 622 through 630 in Fig. 13 represent the T–S characteristics of the traverse from the Coral Sea through Grafton Passage and onto the shelf. The traverse ended at station 630 off Russell Island. Salinities on the traverse were uniform with Coral Sea waters. Station 622 outside of Grafton Passage had isohaline conditions to at least 123 m and, in general, these values were found on the shelf. The temperature at station 622 changed only 0.53°C in the top 123 m

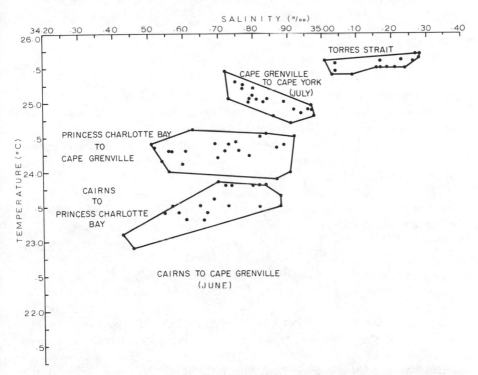

Fig. 12. T–S characteristics on the Queensland continental shelf during June and July 1968.

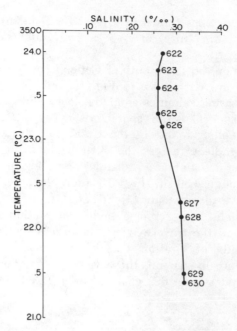

Fig. 13. T–S characteristics across the shelf from Grafton Passage to Russell Island, July 1968.

of water. The surface temperature was 24.0°C, while that at 123 m was 23.47°C. Similarly, at station 630 isothermal, isohaline conditions existed but the temperature was 21.4°C, or 2.6°C lower than that in the Coral Sea. Therefore a thermal gradient of 2.6°C existed across the shelf at this time and the waters on the shelf were denser than those in the Coral Sea.

The T–S characteristics of the Queensland continental shelf are summarized in Section VI. Table III is a summary of the T–S characteristics found on the shelf during 1967–1968 and used in this section.

V. Water Masses in the Coral Sea and Their Effect on the Great Barrier Reef

Maxwell (1968) has pointed out that the Great Barrier Reef can be divided into three regions based on bathymetry, geomorphic changes, and differences in reef density and development. The three regions, according to Maxwell, are: (1) the southern region from lat. 24°S to

TABLE III
SUMMARY OF T–S CHARACTERISTICS ON THE QUEENSLAND SHELF
DURING 1967–1968

	Oct.–Nov. 1967	Mar.–Apr. 1968	Jun.–Jul. 1968
Torres Strait	26.1°–27.0°C	28.5°–29.0°C	25.4°–25.7°C
	36.02–26.21‰	32.68–32.87‰	35.00–35.28‰
NE Gulf of Carpentaria	26.0°–27.6°C		
	35.47–35.70‰		
Northern shelf (Cape York to Cape Sidmouth)			
Inner shelf	27.0°–28.6°C	28.6°–29.1°C	24.1°–25.4°C
	35.90–36.32‰	33.42–34.45‰	34.54–35.11‰
Mid-shelf	26.0°–28.0°C		
	35.25–35.90‰		
Outer shelf	27.0°–27.2°C		
	35.25 35.55‰		
Cape Sidmouth to lat. 18°S	24.8°–26.0°C	27.5°–28.1°C	21.4°–24.7°C
	35.50–36.15‰	32.01–33.76‰	34.54–35.32‰
Lat. 18–20°S	23.7°–24.3°C	26.3°–27.6°C	
	35.55–35.79‰	34.07–34.99‰	
Southern shelf (south of lat. 20°S)	20.4°–23.8°C	25.1°–26.0°C	
	35.51–35.86‰	35.14–35.47‰	

21°S (20°S at the shelf edge); (2) the central region from lat. 21°S to 16°S; and (3) the northern region north of lat. 16°S.

Reef development and growth in the northern and southern regions are classified as being prolific by Maxwell, while reefs in the central region show only weak growth or degeneration. The marked reefal differences between the regions have been described by Maxwell (1968) as being related to differences in "carbonate-enriched" water circulation patterns related to upwelling induced by changes in the Queensland continental slope. Maxwell has noted that all of the nine major reef systems in the southwest Pacific reach their maximum development adjacent to deep troughs and trenches. Thus in the Great Barrier Reef system, the northern region is bordered by the Queensland Trough and the southern region is influenced by the Townsville Trough (called Chesterfield Trough by Maxwell). As the distance from deep water increases, argues Maxwell, the development of reefs decline sharply, and thus reefs are sparse and scattered on the central part of the Queensland shelf.

Granting that the relative abundance of reef growth and development may be correlated with distance from deep water, it is difficult for this author to understand how a change in the gradient of the Queensland continental slope can induce the upwelling as described by Maxwell. The upwelling of deep "carbonate-enriched" water is proposed to occur where the slope is steep. Most coral reefs have steep slopes, but their slopes may or may not extend to oceanic depths (Shepard, 1963).

Maxwell has undoubtedly deduced correctly that the differences in reef growth and development are related to water circulation and chemistry. The application, however, of upwelling of deep, carbonate-enriched waters on steep continental slopes to explain these differences does not seem possible. Carbonate saturation measurements made by the author along the Great Barrier Reef showed that the water was always saturated or supersaturated in carbonates. Considering the region, this is most likely always the case. If at a point waters became undersaturated in carbonate, then carbonate would be taken into solution until an equilibrium was established again. With the large amount of carbonate debris associated with a region as huge as the Great Barrier Reef ecosystem, it seems hard to understand how carbonate depletion or enrichment of the water can occur to a point at which reef development would be affected. As Wells (1957) stated: "availability of $CaCO_3$ in solution does not seem significant in coral distribution, for seawater of normal salinity between 30 and 40‰, to which corals are restricted, is normally supersaturated in $CaCO_3$."

Maxwell and Swinchatt (1970) again correlate regions of prolific reef growth along the Queensland continental shelf with nearness of a steep continental slope. Carbonate enrichment of the shelf waters is not discussed in this paper, but penetration of the shelf by cool, deep Coral Sea water is called upon. Relatively deep troughs bordering the Queensland shelf in some areas certainly allow deeper, higher density Coral Sea water masses to be physically closer to the shelf-edge reefs. It is also true that in those areas where shelf-edge reefs border small gradient continental slopes the deeper Coral Sea water masses are, in an areal fashion, further off-shore. However, the mere fact that deep waters fill troughs with steep slopes and are nearby does not dynamically contribute to an upwelling phenomenon. However, as was stated previously, the differences in reef development in the three regions may still be related to water circulation and nutrient supply as Maxwell has implied.

Orr (1933) suggested upwelling as a possible factor in the nutrient supply for the Great Barrier Reef system, as Maxwell has, but his "upwelling" was restricted to between 50 and 100 m and was wind-induced, whereas Maxwell's is much deeper and is based on deep enriched waters upwelling where the continental slope is steep. Orr proposed that if the surface layers of water outside of the outer barrier were sufficiently disturbed important biological changes might result. His reasoning was that "if the disturbance were sufficiently great, the water which normally lies between 50 and 100 meters, and which contains significant quantities of plant food salts, might be brought up into the better illuminated zone nearer the surface and so give rise to an increase in the phytoplankton." Orr based his conclusions on the results obtained by sampling the water through four different passages in the outer barrier and comparing the waters on the continental slope outside of these passages to that in deeper water further eastward. The four locations examined by Orr are: (1) Cook's Passage, lat. 14°32′S; (2) Papuan Pass, lat. 15°47′S; (3) Trinity Opening, lat. 16°19′S; and (4) Palm Passage, lat. 18°28′S.

These traverses were made during different months and are not strictly comparable one with another. The general results, however, were the same for the four traverses. The upper 50 m of water were fairly homogeneous from the passages to the deep stations, but on all traverses there were colder more phosphate-rich waters at shallower depths outside the passages than at stations further eastward. This indicated to Orr that some small-scale upwelling was going on outside the outer barrier.

Water masses flowing into the western Coral Sea converge against the barrier created by the Australian–New Guinea land mass and only a small portion of this water can escape through the Torres Strait. These water masses are piled up along the coast and must sink (Wyrtki, 1960). Some of the water flows southward along the coast and is integrated into the East Australian Current, but the general character of the area is that of converging water masses and downsinking of water seaward of the outer continental margin. The only upwelling noted by Wyrtki in the western Coral Sea is during January when the northwest monsoonal winds cause upwelling in the northwestern part of the Coral Sea and the Gulf of Papua.

North–south trending currents outside of the outer barrier have been noted by various mariners and some indication of these are noted on various American and English charts. Captain H. G. Chesterman, Master of the Australian Lighthouse Service's ship M. V. *Cape Moreton* has said (1968) that it is not uncommon for a southerly current to be encountered outside of the outer barrier and an opposing current met some distance seaward in the approximate region between Cook's Passage in the north and Trinity Opening in the south. In the British Admiralty "Australia Pilot" Vol. III (1960), it is noted that close to the outer edge of reefs between One- and Half-Mile Opening (14°26'S, 145°26'E) and Flinders Passage (18°45'S, 147°58'E) a current has always been found to set southward from May to November, parallel to the line of reefs. This current does not seem to be affected by the southeast trades blowing in the opposite direction. It is also noted in the "Australia Pilot" that in other areas, northwesterly currents may be encountered in the Coral Sea within 50–60 miles of the outer barrier and these increase as the edge of the reefs is approached.

The presence of currents outside of the outer barrier and parallel to it establishes two possibilities for upwelling to occur near its vicinity. First, if opposite going currents are operable, then the shearing of these currents could cause upwelling. Second, if a southerly current is adjacent to the outer barrier, then Coriolis deflection of this current in the southern hemisphere would be to the left, away from the reefs, and again upwelling would result.

An opportunity to gain some insight into the possibility of upwelling occurring by these means presented itself in December 1967 when the training and research ship *Umitaka Maru* of the Tokyo University of Fisheries carried out oceanographical measurements in the Coral Sea. Stations on three of the traverses she made from the outer barrier into the Coral Sea were selected to see if any indication of upwelling could be found in the top few hundred meters of water. Before these traverses are analyzed, the various water masses present along the outer barrier are described in the next section.

A. CORAL SEA WATER MASSES

Four main water masses occur in the Coral Sea, although at times other masses may be present. Figure 14 is a T–S curve of station UM 6715 occupied by the *Umitaka Maru* in December 1967. This station was located about 32 km from Second 3 Mile Opening in the outer barrier and is illustrative of the four main masses in the Coral Sea. Depths in meters have been located along the curve.

The surface waters are usually composed of the South Equatorial Water Mass. This mass, shown in the upper portion of the T–S curve, occupies the upper hundred meters or so of the water column. The mass has a temperature greater than 24°C and a salinity range between 34.0 and 35.6‰. Surface waters in the Coral and Tasman Seas may

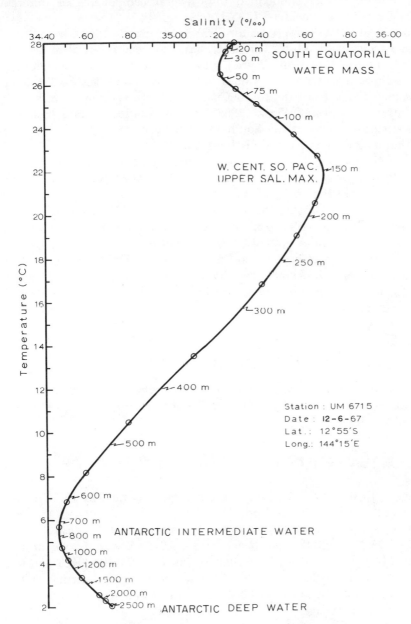

Fig. 14. Water masses in the Coral Sea.

sometime be further subdivided into several components of different origin according to Rochford (1957, 1959).

Underneath the South Equatorial Water Mass is the West Central South Pacific Water Mass. This mass has also been referred to as Subtropical Lower Water by Wyrtki (1962). The mass is characterized by a salinity maximum with values between 35.5 and 36.0‰. Temperatures may range from 18° to 25°C and the core layer may vary from the surface to 250 m (Wyrtki, 1962). Rochford (1968) states that this mass occurs throughout the Southwest Pacific north of about lat. 35°S and that its core layer has a σ_t of about 25.20.

Antarctic Intermediate Water occurs between 650 and 1100 m. It is characterized by a salinity minimum. Wyrtki (1962) lists its properties in the Southwest Pacific as follows: temperature between 4.2° and 6.0°C, salinity between 34.37 and 34.53‰, density near 1.0272. Rochford (1960) in a paper dealing with the $27.20\sigma_t$ surface, also assigned to this water mass a phosphate value of about 1.88 μg-atom/liter.

The bottom waters in the Coral Sea are composed of Antarctic Deep Water. This water is characterized by a weak salinity maximum of about 34.74‰ at a temperature of approximately 1.7°C between 2500 and 3000 m, according to Wyrtki (1962). Wyrtki states that "the maximum is found only in the East Australian Basin, but the Deep water fills the Coral Sea Basin and the Solomon Basin. Defant (1961) listed the temperature range for this water as −1° to 3°C and salinity between 34.6 and 34.7‰.

These four masses were found along the Queensland shelf and in the Coral Sea during December 1967 outward from the Queensland continental shelf.

B. OCEANOGRAPHICAL CONDITIONS BEYOND THE OUTER BARRIER REEFS

Figure 15 shows the location of the stations occupied by the *Umitaka Maru* and used in this section. The data can be considered as three lines of stations outward from the outer barrier reefs. In Figs. 16–18, stations 15 through 19 have been designated (A), stations 20 through 24 as (B), and stations 25 through 30 as (C). Station 15 is located about 32 km from Second 3 Mile Opening in the outer barrier, station 24 is only about 16 km from Grafton Passage, and station 25 is 35 km from Flinders Passage.

Figure 16 is a plot of the density in the upper 250 m for the three lines. Isopycnal surfaces are drawn in at $0.20\sigma_t$ intervals. The geostrophic flow

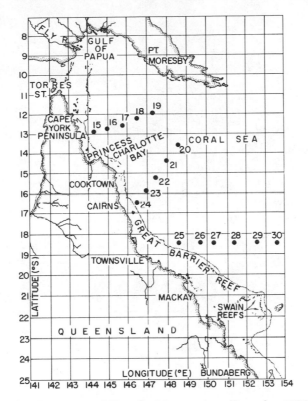

Fig. 15. Location of *Umitaka Maru* stations, December 1967.

has been calculated at 50 m intervals with the direction indicated as southerly by ⊙, and northerly by ⊗. The magnitude of the current is given next to the direction symbol and is in centemeters per second. For the purpose of these calculations, the level of no motion is arbitrarily assumed to be at the base of each 50 m interval.

The strongest development of a southeasterly current is best shown in Fig. 16B, between stations 24 and 23. From the surface to 200 m, a geostrophic flow of 28.2 cm/sec is indicated by the calculations. This current is in keeping with the reports given in the Admiralty's "Australia Pilot" as mentioned previously. Southerly surface flow is also found in (A), but in (C) the flow between stations 25 and 26 appears to be northerly at the surface.

Rochford (1968), in an investigation of the continuity of water masses along the western boundary of the Tasman and Coral Seas during No-

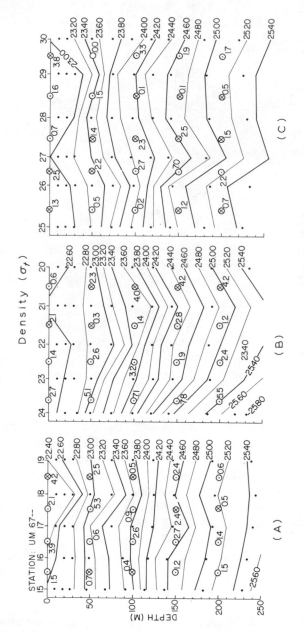

Fig. 16. Density (σ_t) for *Umitaka Maru* stations, upper 250 m, with geostrophic flow calculated at 50-m intervals arbitrarily assuming the level of no motion to be at the base of each 50-m interval. Southerly flow is indicated by $\odot$, northerly flow by $\otimes$. Magnitude of flow is adjacent to the directional symbols and is in cm/sec.

226

vember and December, has shown that water masses along the Queensland continental margin could be traced southward to the New South Wales coast as far as Sydney. Surface waters off the Queensland shelf at this time were of two origins: (1) Gulf of Papua low-salinity water which extended southward for some distance from the gulf, and (2) South Equatorial Waters at about lat. 15°S which were separated from gulf waters by a region of higher salinity water. Gulf of Papua water may be differentiated from South Equatorial water by its lower total phosphorus content. The South Equatorial surface waters which flow into the Coral Sea were traced along the continental margin from lat. 18°S to 30°S.

The core layer of the Upper Salinity Maximum Water Mass was also traced southward along the Queensland and New South Wales continental margins. The geostrophic interpretation between this core layer and a deeper level of no motion by Rochford was that south-flowing currents occurred along the continental margin and north-flowing currents occurred further off-shore. In addition, there was some evidence of vortical exchange between them.

Antarctic Intermediate salinity minimum water was also shown by Rochford to move southward along the continental margin south of lat. 20°S in greater amounts than off-shore. However, it appears that this mass is branched at about lat. 20°S and one component flows northward in a cyclonic flow around the continental margin in the northwest Coral Sea, while the other moves southward as stated.

Assuming then, that there is a southward flow of water at least in the upper few hundred meters along some portions of the Queensland continental shelf we would expect some uplifting of water to occur because of the Coriolis deflection of this movement seaward. Figure 17 is a plot of temperature vs depth for the stations located in Fig. 15. Again, stations on line (B) show the strongest confirmation of this, while it is suggested in (A), but not present or complicated in (C).

Figure 18 also indicates uplifting of water near the continental margin especially in (B), where equal PO_4–P values show marked changes in depth between stations 24 and 23. This uplifting of phosphate is present even in (C) between stations 25 and 26.

The isotherms, isopycnal, and equal phosphate surfaces also suggest that there might be some small-scale cellular motion going on in the upper few hundred meters whereby waters rise along the continental margin, spread out, and sink again. An interpretation of this type is not new.

Sverdrup and Fleming (1941) described cellular motion in the upper

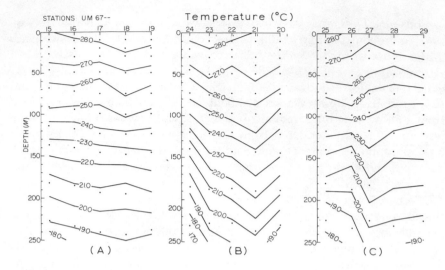

Fig. 17. Isotherms in the Coral Sea, December 1967.

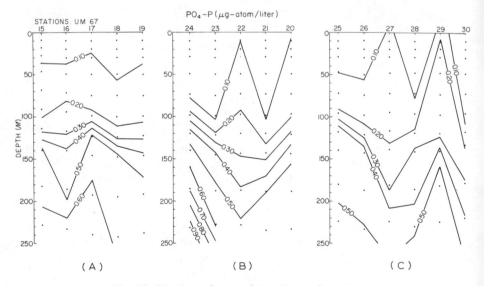

Fig. 18. PO₄–P in the Coral Sea, December 1967.

80 m off the coast of Southern California. These cells developed perpendicular to the currents flowing parallel to the coast and produced patterns of warm and cold surface water.

Off the coast of Peru, Gunther (1936) described an alternating pattern of cold and warm surface waters which were present above 100 m. More recently, Wexler (1959) applied a cellular explanation to temperature differences found across the Antarctic Divergence.

Although a thermal cellular structure is not as obvious in Fig. 17 as it is in the previously mentioned papers, a cellular structure seems to appear when consideration is also given to the isopycnal and equal phosphate surfaces. The cells proposed here are in general agreement with those found by the previous authors, although the present cells are probably not as strong nor as well developed.

The concept of "upwelling" proposed here is of rather limited scale and it is not known if the deeper phosphate-rich water, such as shown in Fig. 18B, actually intersects the surface, or if it makes contact with the living reefs. The probability of contact with living reefs, however, should be fairly good. Orr's basic concept of upwelling between 50 and 100 m seems justified, although not by the means he implied, while Maxwell's much deeper upwelling hypothesis is not believed possible. Coastal upwelling studies have largely shown that the water actually brought to the surface rises from a depth of only about 100–200 m and this would certainly hold for any limited upwelling along the Queensland shelf.

The implications are that some of this nutrient-rich water does get on the Queensland shelf and may be an important factor in reef growth. Correlation of this effect with Maxwell's three regions is not possible at the present time, but further investigations involving nutrient measurements in the outer barrier region should reveal if this mechanism is operable in the area. It does not seem likely that this mechanism would be in effect year round or that nutrient rich water could penetrate very deeply onto the shelf. It is rather thought that this mechanism would have its primary effect on outer barrier reefs.

VI. Summary and Conclusions

The general T–S conditions on the Queensland shelf can be summarized as seasonal in nature with isothermal, isohaline conditions existing over the entire shelf except during the summer rainy season. The southern shelf areas show the smallest variation in salinity with the season and the largest variation in temperature. The northern shelf regions have clearly defined temperature and salinity gradients established across the shelf during spring and summer. Seasonal variation in water temperature is the smallest on the shelf in the northern region, but variation in salinity is significant. The central region has a somewhat complex pattern due to high rainfall and runoff in some portions of the coast, while other parts have an anomalously low amount of rainfall.

The extreme southern portions of the shelf (about lat. 22–27°S) are quite open to Coral Sea waters and stable conditions are found on the shelf. Fairly high salinities may sometimes occur close to shore, but their lateral extent never seems very great.

From lat. 16°S to 22°S, T–S characteristics are more complicated. The high rainfall regions in the northern part of this zone substantially reduce the surface salinity during the monsoon season. High salinities have been found in the central part of this region and they are thought to be related to an anomalous shore climate in part. These high-salinity values have been found even during the rainy season. The T–S values in the southern part of this region grade into those of the southern shelf.

The shelf north of lat. 16°S received the greatest amount of attention. The water over the entire shelf is almost always isothermal and isohaline when the southeast trades are blowing. During the months of October through December, lateral variations in both temperature and salinity are clearly defined. This is also the case during the rainy season and the northwest monsoon. In the first case, evaporation seems to play a decisive role in the creation of a shelf water mass near shore. Waters to the east then take on the characteristics as determined by the diffusion of this mass and the infusion of Coral Sea water which has penetrated the outer barrier. The mixing of these waters along with further evaporation establishes the lateral patterns across the shelf. During the rainy season, reciprocal patterns occur and salinities of the order of 31–32‰ may be found extending seaward.

North of Cape York and the Torres Strait, a different water mass often may be found. There is usually a sharp break in salinity with the lower value always found northward. This division between the two masses migrates depending on the time of the year. The author suspects that this lower salinity water may be originating along the south coast of Papua, and especially the Fly River area, even during the period of the southeast trades. During the northwest monsoon season, this mass moves southward and may extend to Cape York.

Westward of the Torres Strait into the Gulf of Carpentaria, the high coastal salinities of October–December are pinched off and the lower salinity water mass of the Gulf is encountered. In this area also, a migrating line between waters of the Arafura Sea and those to the east is met. It has been proposed by Rochford (1966) that coastal waters from

West Irian are responsible for the low-salinity water mass in the Torres Strait and around Cape York during the northwest monsoon. Rochford has shown that the West Irian coastal water mass moves eastward under the northwest monsoonal winds. It is felt by the author that the water mass in the Torres Strait during the monsoon season is not only West Irian coastal water, but should include a combination of southern Papuan–Fly River coastal waters. It may well be that there is an interplay between the two as their T–S characteristics are very similar.

Finally, Coral Sea waters cannot move across the shelf, through the Torres Strait, and into the Gulf of Carpentaria without losing their T–S characteristics at least during the early summer period. From Fig. 10 it can be seen that waters along the outer barrier as they move onto the shelf continually increase in salinity across the middle and inner shelf portions. Water in the Torres Strait is similar to water on the inner northern shelf and, although the T–S characteristics of water in the Gulf of Carpentaria and those in the Coral Sea are similar (Rochford, 1966), Gulf of Carpentaria T–S characteristics are not derived from Coral Sea water during this period. During the summer, water movement is eastward through the Strait and Coral Sea water can exercise no influence in the Gulf at this time either.

Data obtained in the Coral Sea in December 1967, leads the author to believe that some small-scale upwelling may occur along the outer barrier reefs in the regions where a southerly current is prevalent outside the outer barrier. It is not known whether this uplifted water actually intersects the surface or if contact is made with living coral reefs, but the implications are that some of this water could get on the Queensland shelf. If this water does penetrate the shelf, it is thought that its primary effect would be on outer barrier and outer shelf reefs. This relatively nutrient-rich water is not thought to penetrate the shelf for any appreciable distance.

Correlation of this upwelling with Maxwell's three regions of differential reef growth and development is not possible at the present time, but it is felt that further study, especially nutrient measurements, could determine the importance of this effect.

References

British Admiralty. (1960). "Australia Pilot," 5th ed., Vol. III, Sailing Directions No. 15. British Admiralty, London.
Brown, N. L., and Hamon, B. V. (1961). *Deep-Sea Res.* 8, 65.
Chesterman, H. G. (1968). Personal communication.

Cochrane, J. D. (1956). *Deep-Sea Res.* **4**, 45.

Defant, A. (1961). "Physical Oceanography," Vol. 1, p. 217. Pergamon, Oxford.

Dick, R. S. (1958). *J. Trop. Geogr.* **11**, 32.

Gerard, R. (1966). *In* "The Encyclopedia of Oceanography" (R. W. Fairbridge, ed.), pp. 758–763. Van Nostrand-Reinhold, Princeton, New Jersey.

Gunter, G. (1957). *Geol. Soc. Amer., Mem.* **67**, 159–184.

Gunther, E. R. (1936). *'Discovery' Rep.* **13**, 107.

Hedley, C. (1925). *Gt. Barrier Reef Comm. Rep.* **1**, 35.

Helland-Hansen, B. (1916). "Nogen hydrografiske metoder." Skand. Naturforsker Möte, Kristiania.

Iselin, C. O'D. (1939). *Trans Amer. Geophys. Union* Part 3, p. 414.

Kinsman, D. J. J. (1964). *Nature (London)* **202**, 1280.

Maxwell, W. G. H. (1968). "Atlas of the Great Barrier Reef." Amer. Elsevier, New York.

Maxwell, W. G. H., and Swinchatt, J. P. (1970). *Geol. Soc. Amer., Bull.* **81**, 691.

Montgomery, R. B. (1955). *Deep-Sea Res.* **3**, 331.

Orr, A. P. (1933). *Sci. Rep. Gt. Barrier Reef Exped.* **2**, No. 3, 37.

Orr, A. P., and Moorhouse, F. W. (1933). *Sci. Rep. Gt. Barrier Reef Exped.* **2**, No. 4(a), 87.

Pearse, A. S., and Gunter, G. (1957). *Geol. Soc. Amer., Mem.* **67**, 129–158.

Rochford, D. J. (1957). *Aust. J. Mar. Freshwater Res.* **8**, 369.

Rochford, D. J. (1959). *Aust., CSIRO, Div. Fish. Oceanogr., Tech. Pap.* No. 7.

Rochford, D. J. (1960). *Aust. J. Mar. Freshwater Res.* **11**, 127.

Rochford, D. J. (1966). *Aust. J. Mar. Freshwater Res.* **17**, 31.

Rochford, D. J. (1968). *Aust. J. Mar. Freshwater Res.* **19**, 77.

Shepard, F. P. (1963). "Submarine Geology," 2nd ed. Harper, New York.

Sverdrup, H. U., and Fleming, R. H. (1941). *Bull. Scripps Inst. Oceanogr.* **4**, No. 10, 261.

Wells, J. W. (1957). *Geol. Soc Amer., Mem.* **67**, 609–631.

Wexler, H. (1959). *In* "The Atmosphere and the Sea in Motion" (B. Bolin ed.), pp. 107–120. Rockefeller Inst. Press, New York.

Wüst, G. (1935). *Deut. Atl. Exped. "Meteor" 1925–1927* **6**, Part 1, No. 2.

Wyrtki, K. (1960). *Aust., CSIRO, Div. Fish. Oceanogr., Tech. Pap.* No. 8.

Wyrtki, K. (1962). *Aust. J. Mar. Freshwater Res.* **13**, 18.

GEOMORPHOLOGY OF EASTERN QUEENSLAND IN RELATION TO THE GREAT BARRIER REEF

W. G. H. Maxwell

I. Introduction

The evolution of the Great Barrier Reef Province has been determined by the complex interplay of a number of processes—biological, eustatic, geological, geomorphic, hydrological, and meterological. Some are readily observed operating at the present time, e.g., water circulation, organic growth, and skeletal accumulation. Others are more subtle in their influence, more difficult to observe because of the immeasurably slow rates at which they operate. Nevertheless, these are equally fundamental to the development of the province. Included in this group are those of geologic–geomorphic kind—the processes which led to the formation of the basement on which the reefs are built, the ones which sculptured this basement and enhanced particular zones for reef growth. Tectonism (crustal adjustment), volcanism, weathering, erosion, sedimentation, and eustatism (sea-level change) during the prereefal history of the province largely determined the manner in which the present reef systems have developed, and they continue to exert a controlling influence that is now tempered by biological and hydrological factors.

II. Geological Framework of Eastern Queensland

A. THE TECTONIC FACTOR

Tectonism or crustal adjustment is manifested in the fracturing, warping, folding, and differential movement of segments of the earth's crust. Many of the tectonic patterns evident today have resulted from crustal adjustments that began early in the earth's history and continued spasmodically over long periods of geological time. Many such changes are still in progress and, in comparatively recent time, they have exerted significant influences on the circum-Pacific belt. This belt is dominated by geosynclinal frameworks in which long segments of the crust have experienced differential vertical movement, uplift (tectonic highs) and subsidence (basins or lows), as well as intensive deformation through lateral compression. Because of their elevation, tectonic highs are initially the sites of intensive weathering and thus the main sources of sediment which is supplied to the subsiding basins. In this way the highs feed the basins until an equilibrium is reached either through the normal processes of erosion and deposition bringing both basin and high toward base level or through tectonic or eustatic changes altering base level. Subsequent erosion is frequently concentrated on the softer materials of the filled basins and it is these which then feed newer basins. Sea-level changes during the tectonic evolution of coastal and off-shore regions exert significant influences on the rate of infilling of basins. At times of lower sea level these rates are maximal and the regional topography is drastically modified. When high sea levels prevail, denudation and infilling decline and a tectonic rather than an erosional topography survives. Under such conditions, the tectonic highs are manifested as zones of shallow bathymetry on the continental shelf and it is on these zones that reef growth is favored (Maxwell, 1969a,c). Paradoxically, these are the crustal zones of minimum subsidence, a fact that conflicts with Darwin's theory when it is applied incorrectly to shelf reefs.

In addition to normal geosynclinal trends which, for the circum-Pacific belt, are approximately meridional, cross-flexures (faults, gentle crustal warps, and more complex fracture zones) have been recognized in both eastern and western areas of the Pacific (Maxwell, 1968, 1969a,b,c; Menard, 1955, 1959a,bc, 1964; Menard and Dietz, 1952; Mason, 1958, 1960; Mason and Raff, 1961; Vacquier *et al.*, 1959, 1961; Shepard, 1963). In the west these cross-trends are of smaller magnitude than those of the basins and highs of the geosynclinal belt, but their small vertical dimension is still sufficient to have had important effects on the bathymetry of the off-shore areas and on the drainage patterns, both Recent and pre-Recent.

In summary, the tectonic factor is responsible for the initial relief of the coastal and off-shore region and, depending on the eustatic situation, tectonic relief may be closely reflected in the bathymetric character of the continental shelf. This appears to be the case in the Great Barrier Reef Province. Furthermore, the main reef development has occurred on the tectonic highs. Reef growth in basin areas is quite sparse except where the bathymetric level of such areas has been raised by cross-flexures.

B. BASINS, HIGHS, AND CROSS-FLEXURES

The Tasman Geosynclinal Belt (Fig. 1), which forms the tectonic framework of Eastern Queensland, consists of a series of five basins and five highs trending submeridionally (Hill, 1951; Hill and Maxwell, 1962, 1967; Maxwell, 1968, 1969a,b,c). Four of these elements constitute the continental shelf, viz., Maryborough Basin, Bunker High, Capricorn Basin, and Swains High. Initial evidence of their character and extent was geophysical (Dooley, 1959, 1965; Dooley and Goodspeed, 1959; Hartmann, 1962; Jenny, 1962; Ellis, 1966; Bruce, 1964; Bruce and Thomas, 1964; Affleck and Landau, 1965). Subsequent drilling of the exploration wells Capricorn 1A and Aquarius 1 by Australian Gulf Oil (1968) and the earlier drilling of Wreck Island by Humber Barrier Reef Oil Pty. Ltd. (Traves, 1960; Maxwell, 1962; Lloyd, 1967) confirmed much of the geophysical interpretation. Throughout the history of the geosyncline, beginning in Ordovician time, crustal activity appears to have progressed eastward so that the most recent basin subsidence has been that of the Maryborough and Capricorn Basins (Maxwell, 1968, 1969a,b,c; Ellis, 1966). In the Maryborough Basin, more than 9770 ft of Mesozoic sediment has been established by drilling (Siller, 1955, 1956, 1961) and a thin cover of Cainozoic sediment has been recognized. The Capricorn Basin contains more than 5580 ft of Cainozoic sediment and at least 3000 ft of Mesozoic rock (Carlsen and Wilson, 1968a,b).

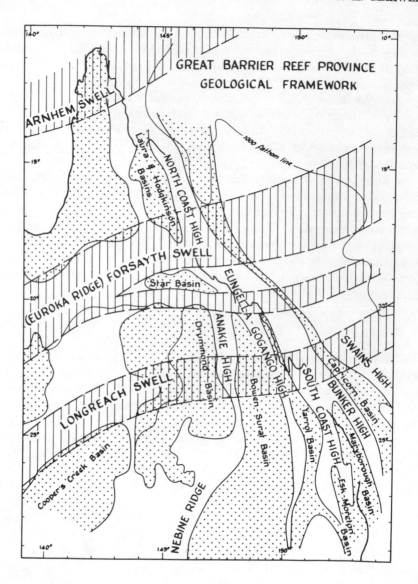

Fig. 1. Geological framework of Queensland. Three framework elements are represented: (a) depositional basins ranging in age from Devonian in the west to late
Mesozoic in the east (stippled); (b) tectonic highs separating the basins (clear); and
(c) cross flexures (vertical lines). The 1000-fm contour line marks the approximate
eastern boundary of the continental mass.

The Bunker High has a veneer of Cainozoic sediment of approximately 1795 ft (Traves, 1960; Lloyd, 1967), while aeromagnetic data suggest that basement in the Swains High lies at approximately 5000 ft and less.

On the land, the highs have survived long periods of erosion since their initial uplift in middle and late Palaeozoic time and in most cases they still remain topographically higher than the infilled basins that derived sediment from them. Furthermore, the softer sedimentary rocks of the basins have yielded more readily to stream erosion and most of the major drainage systems of Eastern Queensland have established themselves on the now stable basins. This also appears to have been the case in pre-Recent time when the present continental shelf was above sea level and the main drainage systems were located on the Maryborough and Capricorn Basins.

In addition to the submeridional basins and highs, three cross-flexures of smaller magnitude have been recognized (Maxwell, 1968, 1969a,b,c; Maxwell and Swinchatt, 1970). They are the Longreach, Forsayth, and Arnhem Swells, all of which have been identified on geological and geomorphological evidence. Geophysical data (unpublished reports of Australian Gulf Oil) support this evidence in the case of the Longreach and Forsayth trends. These swells or cross-flexures have been responsible for the truncation and shoaling of the basins between lat. 23°S and 20°S, as well as for the numerous continental islands in this region. Vertical uplift of the swells has been small by comparison with that of the submeridional highs, but it has been sufficient to affect shelf topography and drainage systems, which in turn have exerted a profound influence on reef development.

C. Sequence of Crustal Development

Seven major tectonic episodes have been recognized in the crustal history of Eastern Queensland (Maxwell, 1968). They are as follows:

1. Initial geosynclinal subsidence, sedimentation, and vulcanicity in early Palaeozoic time, resulting in thicknesses of the order of 20,000 ft of material that is now of metamorphic character

2. Crustal disruption, uplift, igneous intrusion and metamorphism in the Middle Devonian (Hill, 1951) leading to the emergence of highs (Anakie–North Coast High, South Coastal High) and formation of basins (Drummond–Star–Hodgkinson and Yarrol Basins), which rapidly filled with marine and terrestrial sediment and supported reefal zones on the basin margins (Maxwell, 1960, 1961a,b)

3. Downwarp of Bowen, Esk, and Maryborough Basins in late Carboniferous–early Permian time, rejuvenation of adjacent highs (Anakie–North Coast, Eungella–Gogango–Auburn, and South Coastal Highs), and intensive sedimentation

4. Final major deformation of the geosynclinal belt in late Permian–early Triassic time (Derrington, 1961; Webb and McDougall, 1967; Runnegar, 1970), extensive vulcanicity and coal measure deposition; active subsidence of Maryborough and Laura Basins and initial phase of Capricorn Basin

5. Deformation of Maryborough Basin in late Jurassic time, followed by renewed subsidence of this basin and the Capricorn Basin (Ellis, 1964, 1966), and widespread marginal uplift of Eastern Queensland

6. Severe deformation of the Maryborough Basin in late Cretaceous time (Ellis, 1966), followed by uplift and extensive peneplanation in early Tertiary time (Conaghan, 1967; Fairbridge, 1950; Hill, 1956; Watkins, 1967; Whitehouse, 1940, 1951)

7. Marginal upwarping in late Cainozoic time of the Queensland coastal belt, fracturing of the old Tertiary surface and differential vertical movement of the resultant blocks, in many cases along ancient zones of crustal weakness (Andrews, 1902, 1910; Conaghan, 1966, 1967; Cotton, 1949; Daneš, 1911, 1912; Fairbridge, 1950, 1967; Kirkegaard et al., 1966; Stanley, 1928a,b; Sussmilch, 1923, 1930, 1938; Thomas, 1966)

Of the seven tectonic episodes, the most significant in terms of their effects on the framework of Eastern Queensland and on its geomorphological character were those of the Middle Devonian, the late Carboniferous–early Permian and the late Cainozoic. The earlier two established the main basin–high configuration as well as the zones of crustal weakness along which most subsequent adjustments have taken place. Furthermore, igneous intrusion of the structural highs and consequent metamorphism resulted in their increased resistance to erosion so that, in subsequent phases of geomorphological development, the softer sedimentary rocks of the basins have yielded more readily and more extensively to eroding forces. The main effects of the late Cainozoic movement (the Kosciusko Period of Andrews, 1910) were to reactivate old zones of weakness, to cause further block faulting which is reflected in the coastal corridors and plateaus of the present day (Sussmilch, 1938), and to cause regional upwarping of the coastal belt, thereby effecting drastic changes in the drainage systems of Eastern Queensland. Severe erosion of newly uplifted areas occurred with the close of the Kosciusko tectonic phase. Deep incision of stream-beds, particularly during the pluvial periods (Whitehouse, 1940; Gill, 1961a; Watkins, 1967), together with the excessive movement of sediment, resulted in large scale deposition of this material on the continental shelf (which was largely exposed) and on the more landward flood-plains. Fairbridge (1967) has suggested that the last of the late Cainozoic movements occurred less than 50,000 years ago. The persistence of terraces on the shelf and along the continental slope (Maxwell, 1968, 1969a; Phipps, 1967) provides evidence of relatively stable conditions since then. Furthermore, the dating of the sediment on the 65–70-fm terrace (Phipps, 1967) at 13,000 years *B.P.* has established that tectonic stability prevailed throughout the history of reef growth on the Queensland Shelf. Fairbridge (1967) concluded that the main reef development has occurred since 10,000 years *B.P.*

III. Geomorphological Character

A. THE GEOMORPHOLOGICAL FACTOR

Erosion of the land surface, growth of drainage systems, sediment transportation toward the shoreline, deposition of land-derived material in the marine environment, submarine erosion and deposition on shallow shelf areas and reef surfaces, expansion of reefs and shoals by accumulation of both skeletal and terrigenous sediment—all are phases of the geomorphological process. The material involved in the process is determined by the source conditions, composition (igneous, metamorphic and sedimentary rocks, reef types) and nature of the weathering process (dependent on climate, relief, and hydrological situation). In wet tropical zones where igneous and metamorphic rocks are dominant the main sediment type is mud (Maxwell and Swinchatt, 1970). In the more arid zones, coarser sediment increases but the rate of transportation is generally less. In the exposed parts of the reef province, wave and current action concentrate large quantities of skeletal material on the protected sides of reefs. In the deeper shelf waters movement of sediment (both land- and reef-derived) is minimal.

The energy driving the geomorphological process is essentially gravitational. Source regions, elevated initially by tectonic forces, are weathered and, under gravity, their materials move toward sea level. Further movement below sea level depends on hydrodynamic factors, viz., wave action and tidal current flow. Both are less effective at increasing depth so that the main activity occurs in the shallow areas of the shelf or on reef surfaces. The role of tidal currents at greater depths is important as evidenced by deltaic deposition in the Pompey Complex (Maxwell, 1972). Thus the geomorphological process represents a response to tectonic, eustatic, and hydrodynamic influences. The first two determine base level and therefore the zones of active erosion and deposition. The third is more important below base level. When constant sea level is maintained over a considerable period, erosion and deposition tend to eliminate tectonic relief on the land and the near-shore zone. With falling sea level, this tendency is accentuated and previously deep areas of the shelf are brought into the zone of activity and similarly modified. Rising sea level inhibits deposition on the deeper parts of the shelf and reduces the intensity of the geomorphological process on land. Consequently, where an active tectonic and positive eustatic situation prevails, tectonic relief tends to be reflected in the bathymetric character of the shelf. In the case of the Great Barrier Reef Province, there is convincing evidence of tectonic adjustment during the Tertiary and Pleistocene. Furthermore, late Pleistocene and post-Pleistocene fluctuation in sea level was significant and, with the exception of a few stillstands and regressions, there has been a fast net rise in sea level since the Pleistocene, at an average rate of approximately 4 mm/year. Gutenberg (1941), Valentin (1952), and Disney (1955) have shown that the rate of sea level rise over the past 20–50 years averages 1.1 mm/year, which indicates that rates in excess of 4 mm/year were reached during the Holocene. Such rapid eustatic change has been responsible for the preservation of a recognizable tectonic relief in the province and for the minimal accumulation of sediment on the shelf since late Pleistocene time.

B. Physiographic Units of Eastern Queensland

The major physiographic units of Queensland were defined by Bryan (1930), who included the entire eastern region in the *eastern highlands, coastal plains,* and *continental shelf.* These units have been accepted by subsequent workers (Cotton, 1949; David and Browne, 1950; Sussmilch, 1938; Whitehouse, 1951), although divergent views have been advanced as to the relative importance of tectonism and erosion in the development of various features within them. Kirkegaard *et al.* (1966), in their treatment of Central Queensland, recognized high ranges, low ranges, alluvial and coastal plains, and sand dunes. Maxwell (1968) followed Bryan's scheme and used four major divisions of his two land units, viz., the Main Divide, coastal ranges, coastal plain, and intermediate plateau remnants (Fig. 2).

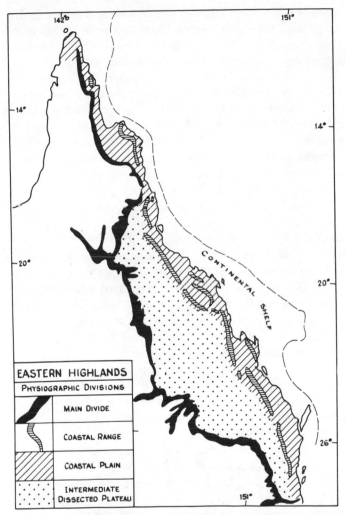

Fig. 2. Physiographic divisions of Eastern Queensland.

1. Eastern Highlands and Coastal Plains

The *Main Divide* forms the western zone of the eastern highlands and separates the Pacific stream systems from those flowing westward into the Gulf of Carpentaria and southwestward into Lake Eyre and the Murray–Darling system. It is extremely varied in relief, ranging from high mountain ranges in the north and south (more than 3000 ft) to low, featureless plainland in parts of the central region. Its very irregular trend from north to south brings it to within 5 miles of the coastline near Cairns and takes it to more than 300 miles from the coast west of Rockhampton. No correlation has been established between the divide and the geological trends within the Palaeozoic rocks which form a major part of this unit. Hedley (1911) noted that "harmony exists between the margin of the continental shelf on the one side and the line of the Pacific watershed on the other." In other words, the line of the Main Divide which is convex westward is almost a mirror image of the line of the continental shelf edge. Bryan (1928) extended this observation, noting that the two lines "are symmetrical about the axis of the coast ranges" which he and Taylor (1911) believed formed the dominant divide of the drainage system preceding the present one. He further suggested that the continental shelf was the foundered counterpart of the Eastern Highlands and, in this, his views have been partly substantiated by the results of recent geophysical and drilling programs. Whitehouse (1941) has presented quite convincing evidence to show that the present position of the Main Divide was established prior to the Miocene laterization, largely as a result of late Cretaceous orogeny (sixth tectonic episode mentioned earlier) and subsequent erosion.

The *Coastal ranges* follow a regular NNW trend, diverging from the Main Divide in the south and approaching it again near Cairns in the north. The highest mountains of the state occur in the northern coastal ranges (more than 5000 ft). Lower Palaeozoic metamorphic rocks and igneous intrusions constitute the bulk of the coastal ranges. Cainozoic volcanics form substantial cappings on some of the older rock masses.

The *coastal plain* extends from the shoreline westward to the Eastern Highlands—the Main Divide, the coastal ranges, or the plateau remnants. It varies in width from almost zero to 120 miles. According to Bryan (1930), it is made up "largely of sand-drift superimposed on soft Mesozoic sediments; of plains of marine deposition recently elevated a few feet above sea level; and of widespread alluvial deposits resting upon various formations." Two features of particular interest in this physiographic unit are the coastal "corridors" (Jardine, 1925b; Sussmilch, 1938) and the extensive sand dunes. Corridors are long, narrow, lowland strips between coastal ranges and plateaus of the Eastern Highlands. They have been variously interpreted as erosion features (Bryan, 1928; Marks, 1924) and as grabens (Andrews, 1910; David, 1914; Stanley, 1928a; Sussmilch, 1938; Taylor, 1911). Cotton (1949), in reviewing the evidence from both schools, favored the tectonic interpretation. Similar corridors are present on the continental shelf, e.g., Whitsunday Passage (Stanley, 1928b; Thomas, 1966). Sand dunes are strongly developed over most of the length of the Queensland coastline. Whitehouse (1963a) defined the various types, viz., stranded beach ridges, storm sands, foredunes, and old parabolic dunes. The main development of the old parabolic dunes in the reef province is restricted to areas north of Cooktown and south of Broad Sound, particularly on Curtis Island. Whitehouse (1963a) has suggested that these dunes were previously extended over the continental shelf during the low sea levels of the Pleistocene and that they have been reworked by currents, waves, and subaerial agencies since that time.

The *intermediate plateau remnants* and low mountain ranges form the largest division of the Eastern Highlands and occupy the area between the coastal ranges and the Main Divide. They include the large tablelands of the Cape York Peninsula, Atherton Tableland, Harvey Plateau, Alice Tableland, Carnarvon, Buckland Tableland, the Upper Burdekin region, and the northern Hughenden area. The plateau region provides the main drainage area for the major river systems of Eastern Queensland—Burnett, Fitzroy, and Burdekin. Geologically the region consists of the Palaeozoic–early Mesozoic basins and metamorphic highs, east of the main Cretaceous formations of the Great Artesian Basin. The general structural grain of the region is controlled by the Palaeozoic formations and trends NNW to N.

2. The Continental Shelf

The continental shelf (Fig. 2) occupies an area of almost 110,000 miles^2, of which 27,000 miles^2 form the main zone of reef growth. Subdivision of the shelf is based on bathymetric variation. The main zones are the near-shore zone (0–5 fm), inner shelf (5–20 fm), marginal shelf (20–50 fm), and Southern Shelf Embayment. Beyond the shelf proper are the Queensland Trench and the Coral Sea Platform (Maxwell, 1968).

The *near-shore zone* is essentially an extension of the coastal plain where active movement of sediment is now occurring. Topographically, the zone is shallow, extremely varied in relief, and changing continually. In most areas, the near-shore zone steepens abruptly between 2 and 5 fm, due to the progradation of coastal sediment (sand in the more arid regions, mud in the more humid regions).

The *inner shelf* (5–20 fm) is relatively constant in width (23 mi) and quite varied in topography, depending on the nature of the underlying rock formations and on the development of sediment deposits. Two persistent changes in bathymetric gradient occur at 10 and 16 fm. These changes are suggestive of ancient strandlines. Reef development, particularly in the northern region, appears to have been favored by these bathymetric features.

The *marginal shelf* (20–50 fm) is almost nonexistent in the northern region, but it widens progressively southward and in the southern region it is separated into Western and Eastern Shelves by the Southern Shelf Embayment (Capricorn Channel). This dichotomy occurs near lat. 20°S. The Western Shelf varies from 5 to 35 miles, and only in the far south does it support reef growth (Capricorn and Bunker Groups). The Eastern Shelf averages 45 miles in width, with a maximum of 80 miles. It forms a broad platform of comparatively uniform depth (32–35 fm) and unlike its western counterpart, it supports the greatest density of reefs of the entire province. Important gradient changes occur between 20 and 22 fm, and at 32 and 35 fm. Features at these depths are associated with strong reef development. They have been interpreted as old strandline features (Maxwell, 1968, 1969a,b,c).

The *southern shelf embayment* is defined by the 35-fm contour. It reaches a maximum width of 55 miles in the south, and wedges out northward near lat. 20°S. An isolated depression further north as well as a reentrant in the central region provide evidence of its previous extension northward, possibly into the Queensland Trench.

Geologically, the various shelf units are related to the basins and highs described earlier. The most obvious relationships are between the Swains High and the Eastern Marginal Shelf in the south, the Capricorn Basin and the Southern Shelf Embayment, the Bunker High and the Western Marginal Shelf, the Maryborough Basin and the Inner Shelf in the south, the Yarrol Basin and the Inner Shelf in the central region. In the northern and central regions there are less obvious relationships between shelf topography and tectonic framework, largely because of the interference caused by the Forsayth Swell to the main meridional trends. North of Princess Charlotte Bay where the shelf widens, the Inner Shelf appears to correspond to the seaward extension of the Laura Basin. The shelf margin in this region is probably a submarine continuation of the North Coastal High.

C. DRAINAGE SYSTEMS—MODERN AND ANCIENT

Eastern Queensland is drained by three major systems (Fig. 3) developed mainly to the west of the coastal ranges (Burnett, Fitzroy, Burdekin) and numerous smaller rivers which rise in the coastal ranges and flow across narrow coastal plains to the sea. The average annual discharge of the three major streams has been calculated at approxi-

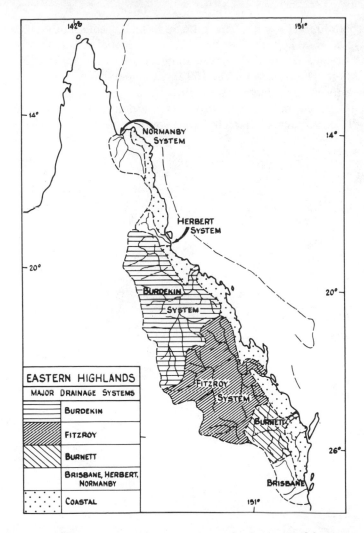

Fig. 3. Major drainage systems of Eastern Queensland. The Burdekin, Fitzroy, and Burnett catchments (hatched) are the largest. The Normanby, Herbert, and Brisbane River systems (clear) extend back into the plateau remnants beyond the coastal plains. Short, fast-flowing rivers drain the coastal area (stippled).

mately 3.6 miles³ (Maxwell, 1968) and that of the coastal streams at approximately 0.8 miles³, thereby giving a total annual discharge of 4.3 miles³ of water. By far the largest discharge is that of the Burdekin system which has a drainage basin of approximately 53,500 miles² (David and Browne, 1950) much of which falls in a rainfall belt of more than 30 inches/year. The Fitzroy basin, of comparable size (55,600 miles²), is in a belt of less than 20-inch annual rainfall. Most of the coastal streams north of the three major systems are located in the high rainfall belt ranging from 60 to 150 inches/year.

The intermediate plateau remnants form the physiographic unit on which the three major drainage systems are located. This unit changes markedly in geological character from south to north. The Burnett system

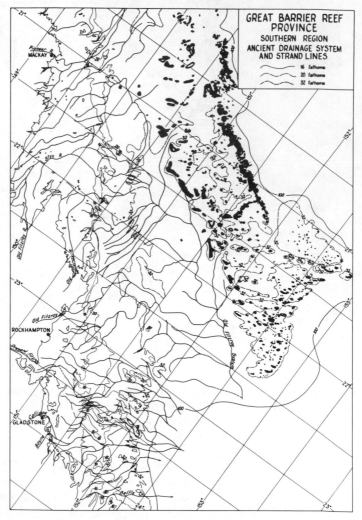

Fig. 4A. Southern region.

Fig. 4. Great Barrier Reef Province, showing reef location, 16-, 20-, and 32-fm lines and axes of maximum depression of the shelf floor reflecting the ancient drainage pattern.

drains predominantly granitic terrain of the Auburn High, as well as the sedimentary formations (Carboniferous to Jurassic) of the Yarrol Basin and the metamorphics of the South Coast High. The Fitzroy system is located mainly on the Bowen Basin which contains Triassic–Jurassic sediment in the southern half, Permian sediment and Tertiary basalt in the north and west, while Carboniferous sediment and Lower Palaeozoic metamorphics of the adjacent Yarrol Basin and Gogango High

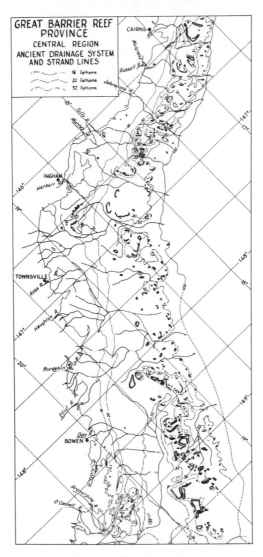

Fig. 4B. Central region.

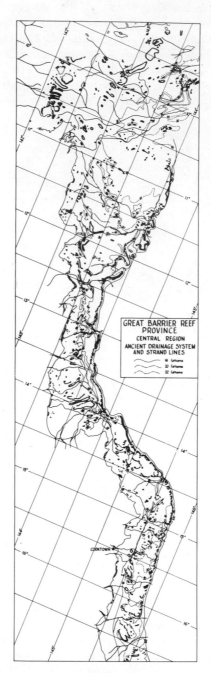

Fig. 4C. Northern region.

account for approximately one-third of the Fitzroy drainage area. The Burdekin system occupies the Drummond and Star Basins (Palaeozoic sediment) as well as draining large areas of igneous and metamorphic basement which forms the core of the Forsayth Swell and the North Coastal and Anakie Highs. Extensive basalt terrain occurs in the north. There are almost no Mesozoic formations in the Burdekin drainage area.

Each system has its major stream direction trending NNW to N, approximately coinciding with the geological grain. Each river breaks through the coastal ranges in an east–west direction. All enter the sea between lat. 19°30′S and 25°S, which represents a coastline distance of almost 400 miles. There is quite convincing evidence of the Fitzroy River having changed its course several times during the Holocene (Maxwell, 1968). Broad Sound, Shoalwater Bay, and Waterpark appear to have been earlier mouths for this river.

The other coastal streams are short and of steep gradient. Their courses, as noted by earlier writers (Bryan, 1930; Daneš, 1912; David and Browne, 1950; Jardine, 1925a,b; Whitehouse, 1951), are irregular to the point of angularity, many follow north–south trends along the floors of corridors, others flow easterly, still others have cut deep gorges into the scarps forming the eastern margin of the Eastern Highlands. The majority of the coastal streams head in the coastal ranges and the edge of the plateau remnants. The dominant rocks of these areas are fine-grained metamorphics and granites. In the high rainfall belt, the source rocks provide mud loads, in the drier areas particularly where granites are extensive, sand load is more significant. On the coastal plain, the streams are eroding alluvial deposits formed in earlier periods. Crustal upwarping in the late Cainozoic has led to the diversion of large parts of river systems to the west and in some cases the present east-flowing streams represent very reduced remnants of the original system. Whitehouse (1963b) clearly demonstrated that the rivers of the Mackay region are of this kind, "misfit streams which, in their present form, could never have carved these big estuaries." Similarly, evidence of larger eastern streams is found in the now submerged deltaic deposits off the mouth of the Pioneer and Fitzroy Rivers (Maxwell, 1968).

The Holocene drainage patterns of the shelf have been considered by a number of authors (Hedley, 1925a,b; Jardine, 1925a, 1928; Jukes, 1847; Maxwell, 1968; Richards and Hedley, 1923; Saville-Kent, 1893; Steers, 1929), but sufficient bathymetric information has been available only recently to permit reasonable interpretation of palaeodrainage. Maxwell (1968) produced maps on which the lines of maximum depression of the shelf were drawn to indicate possible courses of ancient streams (Fig. 4). In some cases, the lines followed actual channels that could

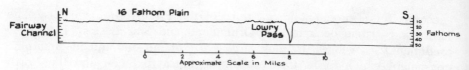

Fig. 5. N–S *Echogram Profile* inside the shelf edge, showing the deep incision of Lowry Pass (northern region).

be interpreted as old stream beds. In most cases, the depressions were broad, vaguely defined features. Nonetheless, in almost every case, these "stream" lines connected the mouths of existing rivers to deeply incised reentrants in the shelf edge. If the paleodrainage pattern represented by these "stream" lines is correct, two significant conclusions may be drawn. The first relates to the drainage pattern in the south and the nature of the Swains High. All the southern streams discharged into the sea after flowing along the axes of the Maryborough or Capricorn Basins. The main Fitzroy drainage changed from north to southward along the Capricorn Basin. Rivers to the south (Burnett, Mary) flowed into the Maryborough Basin and cut deep valleys through the shelf edge formed by the Bunker High. There is no evidence of stream incision on the eastern margin of the Swains High, thus suggesting that this unit formed a significant barrier in late Cainozoic time. The second conclusion relates to the changing course of the Fitzroy and apparent isolation of the Pioneer, O'Connell, and Don River system. It appears that the gradual uplift of the Longreach Swell resulted in a watershed forming in a zone through Sarina and Mackay, and with progressive uplift, the Fitzroy system was deflected southward. Initially, the Fitzroy may have joined the Burdekin system near the southern end of the Queensland Trench.

Further north, the coastal streams cut deeply into the shelf edge and these old valleys today form the major passages through the reefs. Possibly the best example is found near Princess Charlotte Bay where the very deep, precipitous Lowry Pass (Fig. 5) can be traced shoreward to the Normanby River system. The absence of reefs in such passages is due, not to sediment pollution, but to the prohibitive depth of the channels.

IV. Eustatism

A. TERTIARY DEVELOPMENT—TECTONIC, GEOMORPHIC, EUSTATIC

Two major episodes in the geological evolution of eastern Australia during the Tertiary have had a profound influence on the course of subsequent development of this region. The first relates to the extensive peneplanation that climaxed a long period of tectonic stability, probably in Miocene time. After the orogeny at the end of the Cretaceous when the Eastern Highlands were elevated, intensive erosion of this belt led to the recession of the Main Divide to its present position, widespread planation of higher areas, lacustrine and fluviatile sedimentation in lower regions, and extensive extrusion of basalt over much of the landscape.

The second episode terminated the stable conditions of the Tertiary. Toward the end of the Pliocene and possibly during early Pleistocene time, the Kosciusko uplift, an epeirogenic movement, resulted in the rejuvenation of the eastern belt. Differential movement of fault blocks, tilting, and warping led to the formation of plateaus, coastal ranges, corridors, and submarine grabens. Older drainage systems were truncated and, in many cases, their eastern remnants were redirected toward the Pacific. Severe erosion and deep incision of the old disrupted Miocene surface by rejuvenated and new drainage systems accelerated the broad geomorphological cycle that persists today.

During these preglacial episodes, sea level apparently stood some 150 ft above the present level. Evidence for this higher level has been published by a number of authors (David and Browne, 1950; Maze, 1945; Raggatt, 1936), and David and Browne (1950) have suggested that the terraces between 100 and 150 ft may reflect both preglacial and interglacial strandlines. The chronology of Tertiary eustatic fluctuation has not been reliably established, but it is quite obvious that the geomorphic development prior to the Pleistocene was strongly influenced by sea level change.

B. PLEISTOCENE AND HOLOCENE SEA LEVELS

In a comprehensive review of the problem of sea-level changes in the Quaternary, Guilcher (1969) clearly demonstrated the uncertainties which confront workers involved in strandline correlation and chronology. While most authors accept the basic principles of glacioeustatic theory (Maclaren, 1842; Tyler, 1872; Delamothe, 1918; Depéret, 1918, 1926; Umbgrove, 1930; Daly, 1934; Fairbridge, 1961) and recognize the complications of tectonic and geodetic factors, neither the evidence available nor the numerous opinions expressed have permitted a broadly acceptable chronology to evolve.

Fundamental to the problem are certain assumptions concerning the mode of formation of coastal terraces, the relative heights of terraces, beach deposits and lagoon floors, the significance of material from these features in terms of its age relative to the age of formation of the particular feature, and, possibly most important of all, the datum to which all strandline features should be referred—mean sea level, low-tide mark, or high-tide mark. Remarkably few authors concerned with strandline chronology have discussed the actual levels which may develop in a shoreline profile and their position relative to high- and low-tide marks. On hard rock the profile may consist of a high-water platform resulting from contemporary storm-wave action or previously high sea level, a lower wave-cut bench at or below low-tide mark, and a marine-built terrace extending seaward beyond this bench (Johnson, 1919; Kuenen, 1950; Thornbury, 1954). Where a soft formation overlies a harder basement, erosion may be entirely differential and the ancient surface topography of the harder formation may be the only evidence of the sea level,—neither terraces nor benches, but irregular undulation. With beaches, the width and gradient vary with the energy conditions of the particular area. Except for storm beaches above the high-tide mark, the greatest width of beach occurs just above low-tide mark. Lagoons may be totally marine or they may form through the entrapment of fresh water in depressions behind coastal dunes. In most cases their floor levels are below low-tide mark and therefore below the level of contemporaneous beaches, wave-cut benches and high-water platforms. Should any of these features be preserved after regression, then the topographic difference between them could be of the order of 5 m, depending on the size of tidal range, the intensity of prevailing wave attack and the nature of the coastal material, hard rock or soft sediment. Fairbridge (1961) and Fairbridge and Gill (1947) emphasized this aspect of the Quaternary problem. If such features are dated, then accurate assessment of the ancient sea level will depend on recognition of their relative positions, e.g., peat formed in a lagoon should be considerably lower than a shell bed on a wave-cut bench or a shell gravel of a storm beach. Furthermore, if these differences are recognized in one area what common datum should be used for correlation between all areas? Fairbridge (1961) advocated mean low-tide level rather than mean sea level on the grounds that it is the true base level. If this is accepted as the datum, then it is essential that all measurements be accompanied by the tidal ranges of the areas from which they are taken since there will be considerable divergence of low and high-water mark features between areas of greatly differing tidal ranges.

The final area of uncertainty relates to the material selected for dating. In the case of material into which a terrace has been cut, e.g., a coastal limestone or an emergent reef, dating of that material gives the maximum possible age of the terrace; its minimum possible age could approach zero. In the case of encrusting skeletal material on such a terrace (e.g., shells, secondary coral growth, worm tubes, calcareous algae), dating of the encrustation would give the minimum possible age of the terrace since encrustation could only occur at or after the time of terrace formation. With ridges and platforms of cemented gravels (coral, shell, etc.), dating would give a maximum age for ridge formation since the gravel actually predates the ridge. In the case of peat, its age coincides with that of a particular sea level, but with a continuously regressive sea, peat of widely ranging age may result. Thus dating of material for strandline chronology can only lead to a maximum or minimum possible age of a strandline feature; rarely can it give an exact age. Many of the emergent reefs of the Pacific have been dated by Veeh (1966) and others, and ages of 110,000 to 160,000 years BP have been obtained. Veeh's conclusion that these dates provide evidence of a higher sea level 120,000 ± 2000 years BP is valid, but the present surface of these emergent reefs may have formed at any time between then and the present.

Because of the uncertainties involved in the recognition and dating of old sea-level features, conflicting opinions have been the rule rather than the exception. Two opposing schools of thought have emerged, particularly with respect to late Holocene strandline changes. The first, typified by the writings of Fairbridge (1947, 1958, 1960, 1961), postulates a fluctuating but overall advance in sea level during the Holocene, i.e., over the past 10,000 years. This view has been challenged by the second school (Shepard, 1964; Shepard *et al.*, 1967; Scholl, 1964a,b; Thom *et al.* 1969) which favors a steady rise in sea level and rejects the idea of higher sea levels in late Holocene time.

Fairbridge (1961) examined in considerable detail the many hypotheses advanced to explain shoreline displacement and he assembled the voluminous data that has been used to support these varied and, at times, conflicting ideas. He developed his "integrated theory" of shoreline displacement, based on a wide range of geomorphological, palaeoclimatic, and palaeontological data which enabled him to correlate Quaternary oscillation features (Table I) and to construct a table of late Quaternary eustatic events (Table II). It is the latter which has drawn strongest criticism from authors of the opposing school. In particular, attempts have been made to discredit the significance which Fairbridge attached to the Older Peron, Younger Peron, Abrolhos, and Rottnest

TABLE I

CORRELATION OF PLEISTOCENE OSCILLATION FEATURES[a]

| Glacial | Interglacial | | Mean altitude (m) |
	Europe	North America	
Würm–Wisconsin			−100
	Eemian (Monastrian)	Sangamon (Pamlico/ Chowan)	18, 8, 3
Riss–Illinoian			Very low? −55
	Hoxnian or Holstein (Milazzian/ Tyrrhenian)	Yarmouth (Surry/ Wicomico/ Sunderland)	55, 30
Mindel–Kansan			Low? −5
	Cromerian (Sicilian)	Aftonian (Coharie/ Brandywine)	+100, 80
Gunz–Nebraskan			Relatively high, +55
Donau–"pre-Nebraskan"	Tegelen (Calabrian)	Citronelle	200, 130

[a] After Fairbridge (1961).

Terraces, all representing middle and late Holocene sea levels higher than that of the present.

Shepard (1960, 1963, 1964) in denying the possibility of these higher sea levels, based his opinion on the lack of higher shoreline features in a number of areas—Gulf of Mexico, Californian coast, and the Netherlands. He dismissed much of Fairbridge's evidence from Western Australia as unreliable, suggesting that dated material from this region had been derived from kitchen middens and not from natural shoreline deposits. An examination of 33 islands of the Central Pacific was made by Shepard et al. (1967), who concluded that no emerged reefs or other direct evidence of postglacial high-stands of the sea could be recognized. Rubble conglomerates (age range from 4500 to 1890 years BP) forming ridges and platforms were not admitted as evidence because such features could have formed from storm-wave action. Thom et al. (1969), using peat samples from eastern Australia, published dates which they claimed contradicted those of Fairbridge (1961), Gill (1961b), and Ward (1965). Scholl (1964a,b) also reached conclusions opposed to the hypothesis of higher Holocene sea level. In dismissing Fairbridge's interpretation, these authors based their arguments largely on negative evidence, i.e., no terraces or strandline features of appro-

TABLE II

MAJOR/LATE QUATERNARY EUSTATIC EVENTS[a]

Identification	Elevation		Transgression submergence factor (m)	Regression emergence factor (m)	Dates
	Metric	British			
A Late Monastirian	6–8	20–25 ft	+		ca. 95,000–90,000
B Epi-Monastirian	3–5	10–15 ft	+		ca. 80,000–75,000
C Main Würm Emergence	−100	55 fm		103	30,000–17,000
D Masurian Submergence	−45–66	25–35 fm	±50	15	17,000–12,500
E Allerod Submergence	−32–40	18–22 fm	±20	8	12,000–10,800
F Yoldia Submergence	−15–24	7–13 fm	±15	8	10,300– 8,900
G Hydrobia Submergence	−10	5.5 fm	14	2	7,500– 6,700
H Older Peron Submergence	3–5	10–15 fm	15		6,000– 4,600
I Bahama Emergence	−4	2 fm		6	4,600– 4,000
J Younger Peron Submergence	3	10 ft	7		4,000– 3,400
K Crane Key Emergence	−2	1 fm		5	3,400– 3,000
L Peltham Bay Emergence	−3	1.5 fm		3	3,000– 2,600
M Abrolhos Submergence	1.5–2	5–6 ft	5		2,600– 2,100
N Florida Emergence	−2	1 fm		4	2,100– 1,600
O Rottnest Submergence	60 cm	2–3 ft	3		1,600– 1,000
P Paria Emergence	−2	1 fm		3	1,000– 600

[a] After Fairbridge (1961).

priate age were found in the areas which they examined, and therefore higher sea levels could not be postulated. The areas used by Shepard (1960, 1963, 1964), viz., Gulf Coast, California, and the Netherlands, are all recognized as having suffered substantial subsidence (Fisk, 1956; Gutenberg, 1941; Disney, 1955; Edelman, 1954; Van Veen, 1954; Vening Meinesz, 1954; Dechend, 1954), and therefore are less likely to have terraces or other such features preserved. Of the Central Pacific islands examined by Shepard *et al.* (1967), most are located on the Darwin Rise (Menard, 1964), which is regarded by many workers as representing a currently subsiding or recently subsided zone. The third criticism concerning the reliability of radiocarbon dates from Western Australia was not supported by evidence in Shepard's publication and furthermore, additional dates from other parts of Australia are in accord with Fairbridge's interpretation (Rubin and Suess, 1955; Rubin and Alexander, 1958). The more recent paper by Thom *et al.* (1969) is inconclusive. Of the eight peat samples dated by them, only three fall in the age range of the Younger Peron submergence, (4000–3400 years BP) and these three are at elevations (0.0 and 1 ft above H.W.M. in an area where tidal range is of the order of 6 ft) which are consistent with Fairbridge's interpretation of a 10-ft higher sea level for that time. The remaining five peat dates of Thom *et al.*, fall into the Hydrobia, Crane Key, and Bahama Emergences, $5\frac{1}{2}$, 4, and 2 m below present sea level. Furthermore, as Fairbridge (1961) and Scholl (1964a,b) clearly illustrated, peat is more typical of regressive sedimentation and as such is to be expected in the emergent rather than the submergent phases. It would seem that the evidence of Thom *et al.* (1969) confirms the emergent phases of Fairbridge's chronology; no evidence has been presented that disproves the submergent phases.

In view of the obvious weaknesses in the criticisms levelled at Fairbridge's interpretation and in view of more recent evidence that supports this interpretation, one can but accept the probability that higher Holocene sea levels did exist and that these have had a significant influence on geomorphic development, both of the shelf and of the coastal region (Table III). At the same time, one cannot ignore the inherent difficulties associated with strandline interpretation. Two of the major difficulties relate to the near-coincidence of certain interglacial surfaces with those of possible mid-Holocene age, and to the possible formation of raised ridges and banks of old material by processes of more recent origin. Fairbridge (1961), in discussing the Epi-Monastirian terrace, emphasized that "great confusion has been caused in the geological literature by the coincidence in height of this pre-Wisconsin terrace and the mid-Recent 10-ft (3-m) level." He drew attention to similar difficulties relating to the Older Peron and Younger Peron surfaces. Shepard *et al.* (1967) were quite explicit in their rejection of gravel ridges on reefs as old sea-level markers because of their possible formation by storm waves. Both difficulties are very real but the reliability of surfaces and dates can be gauged when data from many distant areas are assessed. If a persistent correlation emerges for particular surfaces and particular dates, then the surface–age relationships may be accepted.

C. QUATERNARY STRANDLINES AND REEF LOCATION

Evidence of Holocene strandlines on the Queensland shelf is found in the bathymetric anomalies that occur at 56, 48, 36, 32, 20, 16, and 10 fm (i.e., 92.5, 88.5, 66, 59, 36.5, 29.5, and 18.3 m) The anomalies are manifested in a number of ways: gradient changes from wide flat surfaces to steep scarps, low ridges, and broad depressions (Maxwell, 1968, 1969b). Particularly well marked are the features at 16, 20, and 32 fm. Furthermore, these features persist for the greater length of the shelf. The difficulties involved in the dating of such features are even greater than those related to emerged strandlines and on the Queensland shelf accurate dates have not been obtained. Data on the New South Wales shelf and the Gulf of Carpentaria have been published by Phipps (1967, 1970), and limiting age values have been obtained for the 65–70-fm terrace (119–128 m) which occurs at the shelf edge. Sediment from this terrace has given a date of 13,000 years (Phipps, 1967), while a shelly sandstone from the same depth has given a date of 17,900 years, and shells from the sandstone a date of 24,600 years (Phipps, 1970), thus suggesting that the terrace was cut at least 24,000 years *BP*, during the Würm Glacial. Shallow water conditions possibly prevailed at 13,000 years *BP* when the younger sediment was deposited, possibly in the late Würm.

In the initial interpretation of the Queensland shelf features (Maxwell, 1968), it was suggested that the time sequence of strandline development was 36, 16, 32, 20, and 10 fm. This interpretation inferred a major transgression and regression to and from the 16-fm terrace, the evidence being found in the deep dissection of the terrace in the northern region and the extensive distribution of sediment beyond the 16-fm contour. However, reappraisal of the Queensland data in conjunction with information from other parts of the world suggests that the major oscillation at 16 fm was from 32 fm and not 36 fm. Possible ages of the various

TABLE III
DATA SUPPORTING HOLOCENE SEA-LEVEL CHRONOLOGY

Sea levels				Evidence		Age range
Level (m)	High	Low	Area	Author	Date	Years BP
+6 to 8	Late Monastirian					ca 90,000–95,000
+3 to 5	Epi-Monastirian					ca 75,000–80,000
−100		Würm	Pacific Coast	Emery (1958)	17,000–24,000	30,000–17,000
			Mississippi	Broecker and Kulp (1957)	28,000; 31,850	
−50 to −65		Late Würm	Gulf of Mexico	Brannon et al. (1957)	11,200; 11,700	ca 17,000–12,500
−32 to −40		Allerod	Guiana, S.A.	Nota (1958)	8,000	12,500–10,800
−15 to −24		Yoldia	Texas, U.S.A.	Shepard (1956)	9,800; 9,300; 8,900	10,300– 8,900
		Texas	Poole Harbour, U.K.	Godwin et al. (1958); Godwin and Willis (1959)	9,298	
			Melbourne, Aus.	Suess (1954)	8,780	
			Melbourne, Aus.	Deevey et al. (1959)	8,300	
			Perth, Aus.	Dury (1966)	8,270; 8,340	
			Sydney, Aus.	Dury (1966)	8,360	
			Saunders Is., Greenl.	Suess (1954)	8,570	
−10		Hydrobia	Christchurch, N.Z.	Suggate (1958)	8,000– 6,700	7,500– 6,700
			Netherlands	de Vries and Barendson (1954)	7,200	
+3 to 5	Older Peron		Netherlands	van Straaten (1957)	7,485– 6,905	6,000– 4,600
			Texas, U.S.A.	Shepard (1956)	5,150	
			Greymouth, N.Z.	Fergusson and Rafter (1954)	4,600	

		Location	Reference	Date	Range
		Rottnest, W. Aus.	Deevey et al. (1959)	5,120	
		Rottnest, W. Aus.	Dury (1966)	5,660; 4,950	
		Victoria, Aus.	Dury (1966)	5,560; 4,910; 4,750	
		Victoria, Aus.	Rubin and Suess (1955)	4,830	
		Curacao Is., Qld., Aus.	Hopley (1968)	5,250; 5,070	
		Torres Strait, Aus.	Maxwell (unpublished)	5,120	
		New South Wales, Aus.	Thom et al. (1969)	4,600	
−4	Bahama	Bimini, Bahamas	Broecker and Kulp (1957)	4,370	4,600– 4,000
		Velsen, Neth.	van Straaten (1957)	3,970	
		Hutt, N.Z.	Stevens (1956)	4,470	
		Hutt, N.Z.	Fergusson et al. (1954)	4,400	
+3	Younger Peron	Algeria	Broecker and Kulp (1957)	3,990	4,000– 3,400
		Alaska	Broecker et al. (1956)	3,600	
		Saipan, Marianna Is.	Libby (1952)	3,479	
		Guam	Rubin and Alexander (1958)	3,400	
		Moreton Bay, Aus.	Rubin and Alexander (1958)	3,710	
		Dennington, Aus.	Dury (1966)	3,980	
		Crescent Hd., Aus.	Thom et al. (1969)	3,600	
		Crescent Hd., Aus.	Thom et al. (1969)	3,600	
		Delicate Nobby, Aus.	Thom et al. (1969)	3,600	

TABLE III (*Continued*)

Level (m)	Sea levels High	Sea levels Low	Area	Evidence Author	Evidence Date	Age range Years BP
+3	Younger Peron		Rottnest, W. Aus.	Deevey et al. (1959v)	3,950	
			Eclipse Is., Qld., Aus.	Gill (1968)	4,100	
−2		Crane Key	Crane Key, Florida	Rubin and Suess (1955)	3,300	3,400– 3,000
			Zuider Zee, Neth.	de Vries et al. (1958)	3,315; 3,505	
			W. Freisland, Neth.	de Vries et al. (1958)	3,240	
			Venezuela	Deevey et al. (1959)	3,570; 3,400	
−3		Pelham Bay	New York, U.S.A.	Libby (1954)	2,830	
−3		Pelham Bay	Maine, U.S.A.	Rubin and Alexander (1958)	2,980	3,000–2,600
			N. Carolina, U.S.A.	Broecker et al (1956)	2,680	
			New Orleans, U.S.A	Broecker and Kulp (1957)	2,900	
+1.5 to 2	Abrolhos		Bimini, Bahamas	Broecker and Kulp (1957)	2,300	
			Andros, Bahamas	Olson and Broecker (1959)	2,300– 2,660	2,600– 2,100
			Velsen, Neth.	van Straaten (1957)	2,420– 2,195	
			Raroia Atoll	Broecker et al. (1956)	2,680	
			Hauraki, N.Z.	Fergusson et al. (1957)	2,270	
			Christchurch, N.Z.	Suggate (1958)	2,402	
			Townsville, Qld., Aus.	Gill (1968)	2,460	
			Balgal, Qld., Aus.	Gill (1968)	2,180	

		Location	Reference	Date	Range
		Bohle R., Qld., Aus.	Gill (1968)	2,350	
		Curacao Is, Qld., Aus.	Gill (1968)	2,620	
−2	Florida	Texas, U.S.A.	Shepard (1956)	2,100	2,100– 1,600
		Florida, U.S.A.	Broecker *et al.* (1956)	1,700	
		Oregon, U.S.A.	Rubin and Alexander (1958)	1,730	
		Mississippi, U.S.A.	Broecker and Kulp (1957)	2,050	
+.6 to 1	Rottnest	Maine, U.S.A.	Suess (1954)	1,050	1,600– 1,000
		Mississippi, U.S.A.	Brannon *et al.* (1957)	1,520– 1,150	
		Andros, Bahamas	Rubin and Alexander (1958)	1,675– 1,025	
		Waitotara, N.Z.	Fleming (1957)	1,020	
		Facing Is, Qld., Aus.	Dury (1966)	1,500	
		Bohle R., Qld., Aus.	Gill (1968)	1,320	
−2	Paria	Twin Cay, Qld., Aus.	Maxwell (unpublished)	1,110	1,000– 600

strandlines may be deduced when they are plotted on the Holocene strandline curve (Fig. 6) of Fairbridge (1960). In most instances they coincide with significant trend changes on the curve. The 56-fm level occurs at a minor peak of 18,000 years *BP*, the 48-fm at a strong low at 15,100 years, the 36-fm at a minor gradient change at 13,400 years, and the 32-fm at a minor oscillation at 13,000 years. A minor oscillation at 11,400 years and a major low at 10,300 years correspond closely with the 20-fm level. The 16-fm level is represented by a strong peak at 10,900 years, between the two ages of the 20-fm level. The 10-fm level

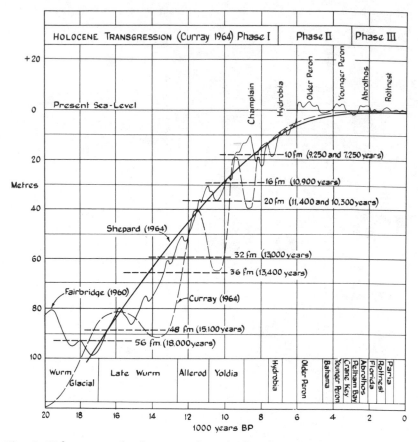

Fig. 6. Holocene sea level curves of Fairbridge (1960), Shepard (1964), and Curray (1964). Old strandline features of the Great Barrier Reef Province at 56, 48, 36, 32, 20, 16, and 10 fm have been plotted against Fairbridge's curve and their possible ages indicated.

coincides with lows at 9250 and 7250 years *BP*, the younger possibly being more significant.

The relationship between the ancient strandlines (particularly those at 16, 20, and 32 fm) and reef location is reflected clearly on the bathymetric charts. In the southern region the reefs of the Bunker Group, the eastern members of the Capricorn Group and the reefal shoals to the north all lie on a line just inside the 32-fm contour. The Swains and Pompey Complexes on the eastern marginal shelf rise from a dissected 32-fm floor which falls away sharply beyond the reef zone. In the central region, the 32-fm level is recognizable as far north as Cairns (lat. 17°S) but, within the main reef zone, the 20-fm level is also well developed, particularly between lat. 19°S and 20°S. In the northern region, the most extensive shelf surface is the 16-fm plain which is deeply incised near the shelf margin (e.g., Lowry Pass) and closely dissected in the back-reef areas. Most of the reefs of the northern region rise abruptly from this 16-fm surface. The lack of 20- and 32-fm features on the northern shelf probably implies that most of this area was emergent until 8000 years *BP* and that reef growth prior to the 16-fm sea was restricted to a narrow zone along the shelf edge.

The significant implication of the reef–strandline relationship is that reef development is controlled by the initial bathymetry and substrate. Because of the nature of shoreline processes, positive relief tends to form in linear zones, and with transgression these topographic highs (off-shore bars, beaches, and dunes) are preserved to varying degrees. This positive relief remains within the range of effective wave and current action long after the adjacent lower floor has been covered by deeper water and as such it provides a favorable site for organic growth. Furthermore, because of its mode of origin and the more vigorous environment, the substrate tends to be predominantly sand and gravel which are more conducive to coral–algal colonization than is mud. Breaks in strandline trends are caused mainly by rivers and by abrupt changes in the geological character of the coastal region. In the case of rivers, deep valleys carved through the shelf during the low sea levels of the Pleistocene have been preserved as channels, particularly near the shelf edge, and these have not been bridged by later reef development. Where hard rock formations have survived the normal weathering processes, islands project above the normal level of the shelf floor and their margins serve as favorable sites for reef colonization. The majority of reefs more than 40 miles distant from the shelf-edge, and not related to one of the three major strandline levels, are associated with islands and island remnants.

Evidence of higher sea levels during the Holocene is found in the terraces and cemented shell beds of the coastal zone of Queensland. Few radiocarbon dates have been obtained but those that are available confirm the sea levels postulated by Fairbridge (1960, 1961). They have been included in Table III. The effect of the middle and late Holocene oscillations is found in the inner shelf reefs of the northern region—the "high" reefs (Maxwell, 1968)—and in the sediment patterns that have developed along the shoreline and around some reefs. In these instances, erosion of old reef surfaces after the higher sea levels of the Holocene has led to dispersal of cemented material over the newly forming surfaces and of recrystallized detritus into the sediment facies adjacent to these reefs. Reefs in the more protected areas of the shelf have retained more of the old surface material than those near the shelf edge (Maxwell, 1968).

V. Reef Development and the Geomorphic Factor

The influence of the geomorphic factor on reef development has been exerted in two phases, a prereefal and a contemporaneous phase. During the prereefal history of the province a considerable thickness of Lower Tertiary sediment of marine and nonmarine character was deposited in the Capricorn Basin, but the rest of the shelf appears to have been emergent for the greater part of this time (Fig. 7). Of the five bores drilled in the province, only Capricorn 1A and Aquarius 1 (Carlsen and Wilson, 1968a,b), southwest of the Swains Complex, penetrated Lower Tertiary rock. In Capricorn 1A, 2135 ft of predominantly quartzose sandstone with minor lignite and interbedded claystone, molluscan limestone, and arkose rest unconformably on pre-Tertiary basement. Thirty miles to the east, Aquarius 1 contains 1500 ft of similar material resting on pre-Tertiary conglomerates. At Wreck Island on the Bunker High, no Lower Tertiary section was encountered (Lloyd, 1967) between the Miocene limestones and the Cretaceous basement.

Extensive peneplanation and laterization of the land followed the Lower Tertiary, together with mass movement of terrigenous sediment across the partly emergent shelf. This phase was terminated by epeirogenic movements which began in Upper Tertiary time and extended into the Pleistocene, and by a generally transgressive sea. Miocene marine sediments are represented in four of the bores drilled on the shelf, while Pliocene limestone is present in two. In Capricorn 1A, 2215 ft of foraminiferal limestone with minor claystone have been dated as Pliocene–Miocene. A similar sequence of 3055 ft occurs in Aquarius 1.

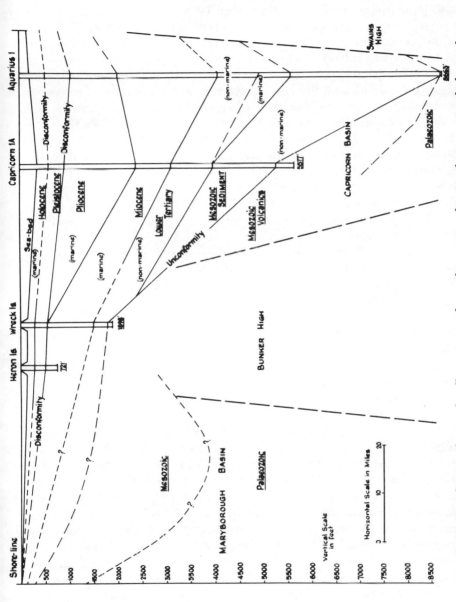

Fig. 7. Stratigraphic correlation of well sections in the southern region showing the increasing thickness of pre-Pleistocene formations to the east and west of the Bunker High.

Only Miocene limestone and calcareous siltstone are represented in the Heron Island and Wreck Island bores, where a maximum thickness of 1240 ft was penetrated. Thus the Upper Tertiary witnessed the progressive submergence of the shelf and deposition of carbonate sediment of nonreefal origin. At the same time geomorphic processes on the land probably led to extensive alluviation of older, deeper valleys and basins which subsequently provided a ready source of sediment for dispersal across the shelf during the low sea levels of the Pleistocene. Emergence of the shelf and present coastal belt began with the onset of the Pleistocene and during the Riss and Würm glacials (120,000–105,000 years BP and 70,000–12,000 years BP) the greater part of the shelf was exposed. In the bores of the province, this emergence is reflected in the disconformity between the Miocene and Quaternary limestones of Heron Island and Wreck Island, and the probable break in sequence above the Pliocene marls of the Capricorn 1A and Aquarius 1 wells. Lloyd (1967) suggested that this break occurs at 445 and 530 ft in the Heron and Wreck Island bores, respectively, basing his view on the absence of recognizable Tertiary faunas above these levels. The lithological evidence from the Heron Island bore suggests that major depositional changes occurred at 506 ft, 308 ft, and 289 ft (Richards and Hill, 1942; Maxwell, 1962). At 506 ft quartzose sands are overlain by foraminiferal limestone which is succeeded from 308 to 289 ft by quartzose sandstone and this in turn by the first algal–coral limestone of reefal character. Thus, it is probable that the Miocene–Pleistocene disconformity caused through regression associated with the onset of glaciation is at 506 ft, and that short-lived submergence during interglacial periods led to carbonate deposition prior to the second quartzose sand deposition reflected in the interval 308–289 ft. Reef development on the Bunker High almost certainly dates from this time, since the first occurrence of abundant coral and algal detritus appears in the limestone above 289 ft. A major buildup of the coral component is found in the interval 183–213 ft and a major decline in the interval 152–183 ft (Maxwell, 1962). Correlation of major faunal and lithological changes in the various reef bores with generally accepted sea-level stands in the Cainozoic leads to possible interpretation of the prereefal palaeogeography of the province. The major disconformity between Pleistocene foraminiferal limestone and Holocene coralline limestone is marked in the Heron Island bore by the quartzose sands of the interval 289–308 ft (Fig. 8). This interval probably postdates the 56-fm sea level (18,000 years) and possibly correlates with the 48-fm strandline (15,100 years). The buildup of coral in the interval 183–213 ft corresponds with the 32-fm sea level (13,000

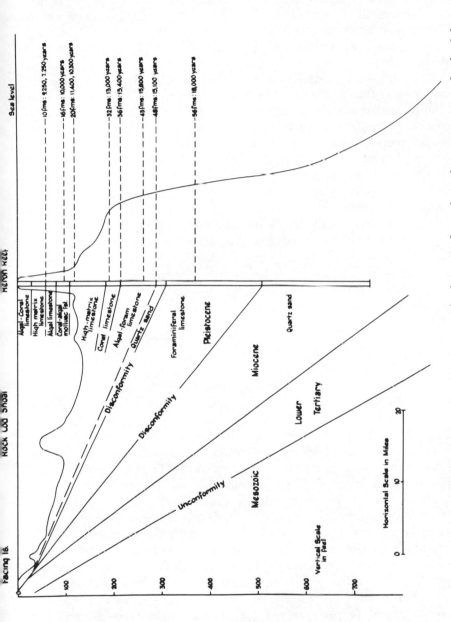

Fig. 8. The *Stratigraphic Succession* in the Heron Island Bore and its relationship to ancient sea levels of the Quaternary.

years) and the decline of the coral component immediately above is possibly related to rapid reef growth and minimal dispersal of reef debris during the rapid transgression from 32 to 16 fm. Oscillations from 16 to 20 fm and again at 10 fm probably account for the second buildup of coral detritus in response to exposure of reefs at 10,300 years and 7250 years BP.

Sedimentological evidence suggests that erosion of reefs by marine processes and the dispersal of reef and terrigenous sediment are minimal at the present time. The reason for this low rate of sedimentation may be found in the transgressive character of the present sea. Under transgression reef growth is the main process. With regression, reefs are exposed to more severe subaerial weathering, the land mass is subjected to rejuvenated stream action, shelf seas are shallower and thus more effectively worked by waves and currents, and consequently sediment dispersal and accumulation become the dominant process. The variation in the sedimentary character of the reef bores provides clear evidence of alternating regressive and transgressive phases in the Quaternary. With each change of sea level the geomorphic character of the shelf floor was modified, generally to the advantage of subsequent reef growth. Even during the regressive phases, the truncation of reef tops led to the expansion of reefal deposits which could serve as the sites for later reef colonization.

The importance of sea-level change in reef development has rarely received proper recognition. Because of the Darwinian theory in which crustal subsidence is accorded the major role in reef growth, and because of the substantiation of this theory for reefs of the oceanic province, many workers have assumed, quite wrongly, that crustal subsidence has played a similar role in the reef provinces of the continental shelves. In the Great Barrier Reef Province, the main reef development is restricted to the tectonic highs, i.e., crustal belts which have risen rather than subsided. Very few reefs are located on the basins where the main subsidence has occurred. Thus, the situation on shelves is virtually the reverse of that of the oceans proper. Reef growth on the shelves has been induced by rising sea level, reef destruction has been the result during lower stands of sea level. Consequently, the thickness of shelf reefs has been determined by the maximum range of sea level since the Tertiary, whereas those of the oceanic province have responded to both sea level rise and floor subsidence and greater thicknesses of reef have formed. During lower stands of the sea, geomorphic processes have sculptured the reefs and shelf and laid the foundations for renewed reef expansion during subsequent transgressive phases.

Acknowledgements

This paper is based on research originally sponsored by the American Petroleum Institute and the American Chemical Society (Petroleum Research Fund). Considerable information was provided by the Australian Gulf Oil Co. and, in particular, by the late Dr. T. C. Wilson and by Mr. E. Ericson.

References

Affleck, J., and Landau, J. F. (1965). Unpublished report for Australian Gulf Oil Co.

Andrews, E. C. (1902). *Proc. Linn. Soc. N.S.W.* 27, 146–185.

Andrews, E. C. (1910). *Proc. Roy. Soc. N.S.W.* 44, 420–480.

Brannon, H. R., Simons, L. H., Perry, D., Daughtry, A. C., and McFarlane, E. (1957). *Science* 125, 919–923.

Broecker, W. S., and Kulp, J. L. (1957). *Science* 126, 1324–1334.

Broecker, W. S., Kulp, J. L., and Tucek, C. S. (1956). *Science* 124, 154–165.

Bruce, I. D. (1964). Unpublished report for Shell Development (Aust.) Pty. Ltd.

Bruce, I. D., and Thomas, J. S. (1964). Unpublished report for Shell Development (Aust.) Pty. Ltd.

Bryan, W. H. (1928). *Rep. Gt. Barrier Reef Comm.* 2, 58–69.

Bryan, W. H. (1930). *In* "Handbook of Queensland," pp. 17–22. Australasian Ass. Advance. Sci., Brisbane.

Carlsen, C. T., and Wilson, T. C. (1968a). Unpublished report for Australian Gulf Oil Co.

Carlsen, C. T., and Wilson, T. C. (1968b). Unpublished report for Australian Gulf Oil Co.

Conaghan, P. J. (1966). *Pap. Dep. Geol. Univ. Queensl.* 6, 1–52.

Conaghan, P. J. (1967). Unpublished Ph.D. Thesis, University of Queensland, Brisbane.

Cotton, C. A. (1949). *J. Geol.* 57, 280–296.

Curray, J. R. (1964). *In* "Papers in Marine Geology–Shepard Commemorative Volume" (R. L. Miller, ed.). Macmillan and Co., New York.

Daly, R. A. (1934). "The Changing World of the Ice Age," pp. 1–271. Yale Univ. Press, New Haven, Connecticut.

Daneš, J. V. (1911). *Soc. Roy. Sci. Boheme Proc., Prague, 2nd Ser.* pp. 1–18.

Daneš, J. V. (1912). *Ann. Geogr.* 21, 344–363.

David, T. W. E. (1914). *In* "Handbook of the Commonwealth of Australia," p. 576. Brit. Ass. Advance Sci.

David, T. W. E., and Browne, W. R. (1950). "The Geology of the Commonwealth of Australia," 3 vols. Arnold, London.

Dechend, W. (1954). *Geol. Mijnbouw* 16, 195–200.

Deevey, E. S., Gralenski, L. J., and Hoffren, V. (1959). *Amer. J. Sci., Radiocarbon Suppl.* 1, 144–172.

Delamothe, L. (1918). *Bull. Soc. Geol. Fr.* 18, 3–58.

Depéret, C. (1918). *C. R. Acad. Sci.* 100, 400.

Depéret, C. (1926). *C. R., Congr. Geol. Int., 13th, 1923* Vol. 3, pp. 1409–1426.

Derrington, S. S. (1961). *A.P.E.A. J.* pp. 18–21.
de Vries, H., and Barendson, G. W. (1954). *Nature (London)* 174, 1138–1141.
de Vries, H., Barendson, G. W., Waterbolk, H. T. (1958). *Science* 127, 129–137.
Disney, L. P. (1955). *Proc. Amer. Soc. Civil Eng.* 81, 666.
Dooley, J. C. (1959). *Rec. Bur. Min. Resourc. Aust.* pp. 1–4.
Dooley, J. C. (1965). *Rep. Bur. Min. Resourc. Geol. Geophys. Aust.* 73, 1–26.
Dooley, J. C., and Goodspeed, M. J. (1959). *Rec. Bur. Min. Resourc. Aust.* pp. 1–5.
Dury, G. H. (1966). *Aust. J. Sci.* 29, 158–162.
Edelman, T. (1954). *Geol. Mijnbouw* 16, 209–212.
Ellis, P. L. (1964). *In* "Maryborough Basin Field Conference Guide Book," p. 12–22. Geol. Soc., Australia.
Ellis, P. L. (1966). *A.P.E.A. J.* pp. 30–36.
Emery, K. O. (1958). *Bull. Geol. Soc. Amer.* 69, 39–60.
Fairbridge, R. W. (1947). *Geogr. J.* 109, 157.
Fairbridge, R. W. (1950). *J. Geol.* 58, 338–401.
Fairbridge, R. W. (1958). *Trans. N.Y. Acad. Sci.* [2] 20, 471–482.
Fairbridge, R. W. (1960). *Sci. Amer.* 202, 70–79.
Fairbridge, R. W. (1961). *Phys. Chem. Earth* 4, 99–185.
Fairbridge, R. W. (1967). *In* "Land Form Studies from Australia and New Guinea" (J. N. Jennings and J. A. Mabbutt, eds.), pp. 386–417. Australian National University Press, Canberra.
Fairbridge, R. W., and Gill, E. D. (1947). *Aust. J. Sci.* 10, 63–67.
Fergusson, G. F., and Rafter, T. A. (1954). *N.Z. J. Sci. Technol., Sect. B* 36, 371–374.
Fisk, H. N. (1956). *Proc. Tex. Conf. Soil Mech. Found. Eng., 8th, 1955* pp. 1–36.
Fleming, C. A. (1957). *N.Z. J. Sci. Technol., Sect. B* 38, 726–733.
Gill, E. D. (1961a). *In* "Descriptive Palaeoclimatology" (A. E. M. Nairn, ed.), p. 332–353. Wiley (Interscience), New York.
Gill, E. D. (1961b). *Z. Geomorphol.*, Suppl. 3, 73–79.
Gill, E. D. (1968). *Aust. J. Sci.* 31, 106–111.
Godwin, H., and Willis, E. H. (1959). *Amer. J. Sci., Radiocarbon Suppl.* 1, 63–75.
Godwin, H., Suggate, R. P., and Willis, E. H. (1958). *Nature (London)* 181, 1518–1519.
Guilcher, A. (1969). *Earth-Sci. Rev.* 5, 69–97.
Gutenberg, B. (1941). *Bull. Geol. Soc. Amer.* 52, 721–772.
Hartmann, R. R. (1962). Unpublished report for Australian Oil & Gas. Corp. Ltd.
Hedley, C. (1911). *Proc. Linn. Soc. N.S.W.* 36, 1–39.
Hedley, C. (1925a). *Rep. Gt. Barrier Reef Comm.* 1, 63–65.
Hedley, C. (1925b). *Rep. Gt. Barrier Reef Comm.* 1, 69–72.
Hill, D. (1951). *In* "Handbook of Queensland," pp. 13–24. Australian and New Zealand Ass. Advance Sci., Brisbane.
Hill, D. (1956). Unpublished report for Australian Mining and Smelting Co. Ltd.
Hill, D., and Denmead, A. K. (1960). *J. Geol. Soc. Aust.* 7, 1–474.
Hill, D., and Maxwell, W. G. H. (1962). "Elements of the Stratigraphy of Queensland," 1st ed. Univ. of Queensland Press, Brisbane.
Hill, D., and Maxwell, W. G. H. (1967). "Elements of the Stratigraphy of Queensland," 2nd rev. ed. Univ. of Queensland Press, Brisbane.
Hopley, D. (1968). *Aust. J. Sci.* 31, 122–123.
Jardine, F. (1925a). *Rep. Gt. Barrier Reef Comm.* 1, 73–110.
Jardine, F. (1925b). *Rep. Gt. Barrier Reef Comm.* 1, 131–148.

Jardine, F. (1928). *Rep. Gt. Barrier Reef Comm.* **2**, 88–92.
Jenny, W. P. (1962). Unpublished report for Gulf Interstate Overseas Ltd.
Johnson, D. W. (1919). "Shore Processes and Shoreline Development." Wiley, New York.
Jukes, J. B. (1847). "Narrative of the Surveying Voyage of HMS Fly," Vols. 1 and 2. Boone, London.
Kirkegaard, A. G., Shaw, R. D., and Murray, C. G. (1966). *Rec. Geol. Surv. Queensl.* p. 1 (unpublished).
Kuenen, P. H. (1950). "Marine Geology." Wiley, New York.
Libby, W. F. (1952). *Science* **116**, 673–681.
Libby, W. F. (1954). *Science* **119**, 135–140.
Lloyd, A. R. (1967). *Bur. Min. Resourc. Aust. Bull.* **92**, 69–113.
Maclaren, C. (1842). *Amer. J. Sci.* **42**, 346–365.
Marks, F. O. (1924). *Proc. Roy. Soc. Queensl.* **36**, 1–18.
Mason, R. G. (1958). *Geophys. J.* **5**, 320–329.
Mason, R. G. (1960). *Liverpool Manchester Geol. J.* **2**, 389–410.
Mason, R. G., and Raff, A. D. (1961). *Bull. Geol. Soc. Amer.* **72**, 1259–1266.
Maxwell, W. G. H. (1960). *J. Geol. Soc. Aust.* **7**, 1–474.
Maxwell, W. G. H. (1961a). *J. Palaeontol.* **35**, 82–103.
Maxwell, W. G. H. (1961b). *Palaeontology* **4**, 59–70.
Maxwell, W. G. H. (1962). *J. Geol. Soc. Austr.* **8**, 217–238.
Maxwell, W. G. H. (1968). "Atlas of the Great Barrier Reef." Elsevier, Amsterdam.
Maxwell, W. G. H. (1969a). *In* "The Future of the Great Barrier Reef," Spec. Publ. No. 3, pp. 5–14. Australian Conservation Foundation.
Maxwell, W. G. H. (1969b). *In* "Stratigraphy and Palaeontology" (K. S. W. Campbell, ed.), pp. 353–374. Australian National University Press, Canberra.
Maxwell, W. G. H. (1969c). *Australas. Oil Gas Rev.* **15**, 15–22.
Maxwell, W. G. H. (1969d). *Aust. J. Sci.* **31**, 85.
Maxwell, W. G. H. (1972). *Deep Sea Res.* **17**, 1005–1018.
Maxwell, W. G. H., and Swinchatt, J. P. (1970). *Bull. Geol. Soc. Amer.* **81**, 691–724.
Maze, W. H. (1945). *Proc. Linn. Soc. N.S.W.* **70**, 41–46.
Menard, H. W. (1955). *Bull. Geol. Soc. Amer.* **66**, 1149–1198.
Menard, H. W. (1959a). *Bull. Geol. Soc. Amer.* **70**, 1491–1496.
Menard, H. W. (1959b). *Colloq. Int. Cent. Nat. Rech. Sci.* **83**, 95–108.
Menard, H. W. (1959c). *Experientia* **15**, 205–213.
Menard, H. W. (1964). "Marine Geology of the Pacific." McGraw-Hill, New York.
Menard, H. W., and Dietz, R. S. (1952). *J. Geol.* **60**, 266–278.
Nota, D. J. G. (1958). *Meded. Landbouwhogesch. Wageningen Ned.* **58**, 1–98.
Olson, E. A., and Broecker, W. S. (1959). *Amer. Sci.* **257**, 1–28.
Phipps, C. V. G. (1967). *A.P.E.A. J.* pp. 44–49.
Phipps, C. V. G. (1970). *Aust. J. Sci.* **32**, 329–330.
Raggatt, H. G. (1936). *J. Proc. Roy. Soc. N.S.W.* **70**, 100–174.
Richards, H. C., and Hedley, C. (1923). *Queensl. Geogr. J.* **38**, 105–109.
Richards, H. C. and Hill, D. (1942). *Rep. Gt. Barrier Reef Comm.* **5**, 1–111.
Rubin, M., and Alexander, C. (1958). *Science* **127**, 1476–1487.
Rubin, M., and Suess, H. (1955). *Science* **121**, 481–488.
Runnegar, B. N. (1970). *J. Geol. Soc. Aust.* **16**, 697–710.
Saville-Kent, W. (1893). "The Great Barrier Reef of Australia " Allen, London.
Scholl, D. W. (1964a). *Mar. Geol.* **1**, 344–366.
Scholl, D. W. (1964b). *Mar. Geol.* **2**, 343–364.

Shepard, F. P. (1956). *J. Geol.* **64**, 56–69.

Shepard, F. P. (1960). In "Recent Sediments, Northwest Gulf of Mexico" (F. P. Shepard, F. B. Phleger, and T. H. van Andel, eds.), pp. 338–344. Amer. Ass. Petrol. Geol., Tulsa, Oklahoma.

Shepard, F. P. (1963). "Submarine Geology," 2nd ed. Harper, New York.

Shepard, F. P. (1964). *Science* **143**, 574–576.

Shepard, F. P., Curray, J. R., Newman, W. A., Bloom, A. L., Newell, N. D., Tracey, J. I., and Veeh, H. H. (1967). *Science* **157**, 542–544.

Siller, C. W. (1955). Unpublished report for Lucky Strike Drilling Co.

Siller, C. W. (1956). Unpublished report for Lucky Strike Drilling Co.

Siller, C. W. (1961). *A.P.E.A. J.* pp. 30–36.

Stanley, G. A. V. (1928a). *Rep. Gt. Barrier Reef Comm.* **2**, 1–51.

Stanley, G. A. V. (1928b). *Amer. J. Sci.* **16**, 45–50.

Steers, J. A. (1929). *Geogr. J.* **74**, 232–270.

Stevens, G. R. (1956. *N.Z. J. Sci. Technol.* **38**, 201–235.

Suess, H. (1954). *Science* **120**, 467–473.

Suggate, R. P. (1958). *N.Z. J. Geol. Geophys.* **1**, 103–122.

Sussmilch, C. A. (1923). *Pan-Pac. Sci. Congr. Proc.* **1**, 721–726.

Sussmilch, C. A. (1930). *Proc. Pac. Sci. Congr., 4th, 1929* Vol. 2B, p. 657.

Sussmilch, C. A. (1938). *Rep. Gt. Barrier Reef Comm.* **4**, 105–134.

Taylor, G. T. (1911). *Commonw. Bur. Met. Bull.* **8**.

Thom, B. G., Hails, J. R., and Martin, A. R. H. (1969). *Mar. Geol.* **7**, 161–168.

Thomas, B. M. (1966). Unpublished Honours Thesis, University of Queensland, Brisbane.

Thornbury, W. D. (1954). "Principles of Geomorphology." Wiley, New York.

Traves, D. M. (1960). *J. Geol. Soc. Aust.* **7**, 369–371.

Tyler, A. (1872). *Geol. Mag.* **9**, 393–399 and 485–501.

Umbgrove, J. T. F. (1930). *Proc. Pac. Sci. Congr., 1929* Vol. 2A, pp. 105–113.

Vacquier, V., Raff, A. D., and Warren, R. E. (1959). *Prepr. Int. Oceanogr. Congr.* p. 60.

Vacquier, V., Raff, A. D., and Warren, R. E. (1961). *Bull. Geol. Soc. Amer.* **72**, 1251–1258.

Valentin, H. (1952). *Petermanns Geogr. Mitt.* **246**, 1–118.

Veeh, H. H. (1966). *J. Geophys. Res.* 3379–3386.

van Straaten, L. M. J. U. (1957). *Verh. Kon. Ned. Geol. Mijnbouwk. Genoot., Geol. Ser.* **17**, 158–183.

Van Veen, J. (1954). *Geol. Mijnbouw* [N.S.] **16**, 214–219.

Vening Meinesz, F. A. (1954). *Proc. Kon. Ned. Akad. Wetensch., Ser. B* **57**, 142–155.

Ward, W. T. (1965). *J. Geol.* **73**, 592–602.

Watkins, J. R. (1967). *J. Geol. Soc. Aust.* **14**, 153–168.

Webb, A. W., and McDougall, I. (1967). *Earth Planet. Sci. Lett.* **2**, 583–588.

Whitehouse, F. W. (1940). *Pap. Dep. Geol. Univ. Queensl.* **2**, 1–74.

Whitehouse, F. W. (1941). *Proc. Roy. Soc. Queensl.* **53**, 1–22.

Whitehouse, F. W. (1944). *Aust. Geogr.* **4**, 183–196.

Whitehouse, F. W. (1951). In "Handbook of Queensland," pp. 5–12. Australian and New Zealand Ass. Advance Sci., Brisbane.

Whitehouse, F. W. (1963a). *Queensl. Natur.* **17**, 1–10.

Whitehouse, F. W. (1963b). "Notes on the Geomorphology of the Mackay Region," pp. 1–4. Mackay High School (unpublished manuscript).

9

STRUCTURAL AND TECTONIC FACTORS INFLUENCING THE DEVELOPMENT OF RECENT CORAL REEFS OFF NORTHEASTERN QUEENSLAND

E. Heidecker

I. Introduction

Structural studies of reefs were pioneered by geologists of the caliber of C. Lyell, C. Darwin, J. Murray, J. D. Dana, and A. Geikie. Their efforts led to recognition of the tectonic implications of reef development.

Interest in structural matters subsided late in the 19th century when it became apparent that development of this subject would have to wait upon advances in oceanographic and geophysical technologies. Scientists and amateur naturalists turned to biological studies of reefs. Lay writers such as E. J. Banfield, the loquacious beachcomber of Queensland's Dunk Island, aided this process by generating public interest in the biological spectacle of reefs. Indeed Banfield's idyllic "gardens of coral" now appear to be part of Australia's cultural and emotional heritage.

Over the last decade spectacular advances in geotectonics, oceanography, and marine geophysics have provided a modern basis for structural studies of reefs.

Conditions conducive to such studies prevail off the coast of Queensland. Well-defined structural units have been revealed by bathymetric and marine geophysical work. These indirect, but vital, studies from ships can be backed to an unusual degree by direct observations on-shore. Tectonic events in the Coral Sea are clearly recorded in the structure, geomorphology, and volcanicity of the peripheral continent. On-shore continuations of structures affecting reef development are also well exposed.

A feature of the following discussion then, will be liaison of on-shore and off-shore information.

II. Evidence for Tectonism and Structural Control

A. THE DURICRUST DATUM

Development of the Great Barrier Reef was associated with revived tectonism along the northeastern margin of Australia. Prior to this resurgence, Australia was, and elsewhere still remains, in a state of geological quiescence. Mountain ranges developed in a previous era had wasted away to monadnocks in a vast peneplain. Mechanical erosion almost ceased, so that a surface zone suffered protracted pedogenesis.

Countless seasonal water-table fluctuations produced highly differentiated soil profiles. Siliceous and ferruginous hard-pans or crusts developed within the profiles. The term "duricrust" (Woolnough, 1927) is applied collectively to the various profiles containing crusts or armour-plate cappings. Siliceous duricrusts are referred to as silcrete or "billy." Ferruginous crusts are referred to as ferricrete or laterite (Grant and Aitchison, 1970).

Duricrust which developed before tectonic resurgence is readily distinguished from much thinner later developments. Duricrust is then a valuable marker for studies of Late Cainozoic tectonism. Its initial configuration was simple, and it is readily recognized. Dissected remnants are common enough west of Townsville for a reconstruction of the duricrust surface. Contours in Fig. 1 reveal a depression west of Townsville (coinciding with the Burdekin River in Fig. 2) from which the surface rises both toward the interior and the coast. Behind the coastal escarpment northwest of Townsville duricrust remnants lie at more than 650 m above sea level. Further east on the coastal escarpments and islands (e.g., the Cardwell Ranges and Hinchinbrook Island in Fig. 2) the capping has been removed. However, summit plateaus indicate that uplift reached a maximum of about 1000 m in this area.

The depression in the duricrust surface represented in Fig. 1 terminates along a northeasterly arch which lies along the elongate Owenee granite pluton. Chains of youthful volcanic vents are aligned along this direction (Wyatt and Webb, 1970). The Broken River Embayment and Euroka Ridge in Fig. 1 have similar trends. The Euroka Ridge is a subsurface feature of the Great Artesian Basin (Whitehouse, 1954). Recent arching effects are also evident on the surface over the ridge. The upper waters of the Diamantina River once flowed northeast to the Flinders River and thence to the Gulf of Carpentaria (Twidale, 1966). Uparching along the Euroka Ridge diverted these southern tributaries of the Flinders River southward into the Diamantina River as shown in Fig. 2.

Progressive rise of the duricrust marker from west to east is reversed beyond the present coastline. An extensive planar reflector underlies unconsolidated sediments of the Coral Sea Plateau. This reflector is considered to be the duricrust surface which subsided some 1500 m after Upper Oligocene times (Ewing *et al.*, 1970).

B. ANCIENT DRAINAGE PATTERNS

Ancient drainage patterns provide perhaps the best evidence that the Great Barrier Reef has developed on a foundered terrestrial high. These ancient drainage systems have been studied intensely as they contain alluvial tin and gold deposits. Changing drainage systems, and thus the tectonic events causing changes, have been dated from the radiometric ages of associated basalts.

The picture that has emerged is perhaps unexpected. Substantial earth movements have occurred quite recently. Until late Miocene or even early Pliocene, rivers radiated from a domal high on the Queensland

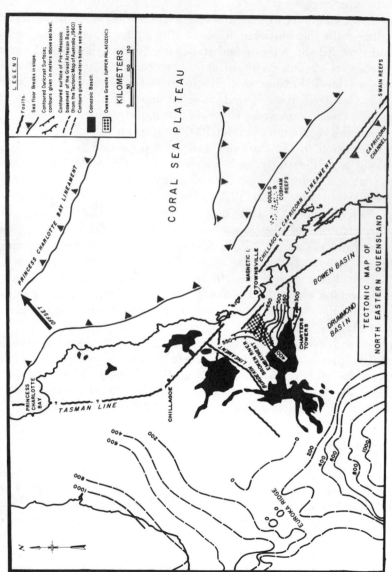

Fig. 1. Tectonic map of Northeastern Queensland.

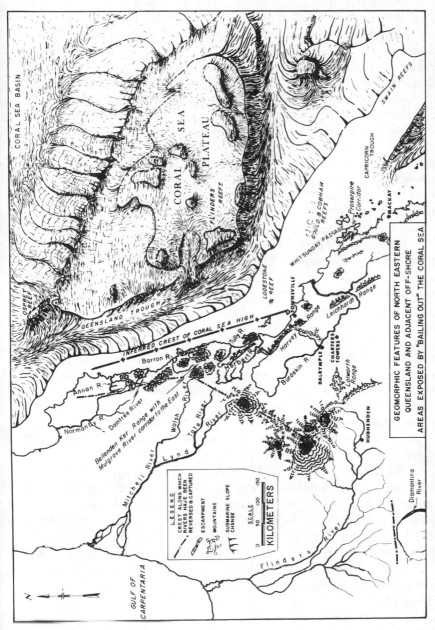

Fig. 2. Geomorphic features of Northeastern Queensland and adjacent off-shore areas exposed by "bailing out" the Coral Sea. (After Heidecker, 1972.)

Shelf. Westward-flowing rivers traversed areas now drained by streams flowing eastward to the coast, crossed the area now occupied by the Great Dividing Range, and emptied into the Gulf of Carpentaria. The ancestral Burdekin and Normanby Rivers flowed southward and northward off the crest of a high (the Coral Sea High in Fig. 2) and hooked around its limits into the Coral Sea.

Numerous abandoned channels and captured river courses are a legacy of the Coral Sea High. For instance gravels mark abandoned channels which once carried the waters of the Annan River westward, rather than eastward, into the Normanby River (Best, 1962). The contiguous Annan and Normanby Rivers shown in Fig. 2 were parts of a single stream. Collapse in the region of its headwaters and uplift across its midcourse tipped the eastern section of this stream, the Annan River, backward into the Coral Sea.

The upper Herbert River is also made up of streams which once flowed westward off the Coral Sea High along the Tate River into the Gulf of Carpentaria (Best, 1962; de Keyser *et al.*, 1964). The Upper Barron River is also a captured part of the west-flowing Mitchell River (Poole, 1909; Jardine, 1925). These reversals are clearly the consequence of collapse of a high in the Coral Sea, with uplift further west along the present Great Dividing Range (Figs. 2 and 3).

The prominent hook in the upper waters of the Burdekin River in Figs. 2 and 3 is the captured upwaters of the Lynd River which flowed from the Coral Sea High into the Gulf of Carpentaria. The Black Burdekin and the White Burdekin Rivers, headwaters of the Burdekin River, rise within sight of the Coral Sea, yet these streams flow in the opposite direction perilously close to the coastal escarpments. It is hardly surprising that upper parts of the White Burdekin which once approached even closer to the present coastline have been captured and short-circuited into the sea by Garrawalt and Stony Creeks.

The Black Burdekin is a notably sluggish stream, particularly in view of the nearby relief depicted in Fig. 3. Further downstream it enters an area of lacustrine gravels and Pleistocene diatomites in the Wairuna–Lake Lucy area. In this area the river appears to have been ponded by the development further west of the McBride basaltic highlands (see Fig. 3). Here, it abandoned its older course westward and turned southward in a great loop to join the ancestral Burdekin River, a river which had worked its way around the southern margin of the Coral Sea High.

South of the Burdekin upwaters there is less evidence of river capture. The upper reaches of some rivers have been truncated as their up-waters

Fig. 3. Panoramic view from over the Coral Sea to the head of the Burdekin River. (After Heidecker, 1972.)

subsided into the Coral Sea. The Keelbottom River (shown in Fig. 4) rises a few miles from the coast in a broad valley which terminates abruptly on the Harvey Range Escarpment. The truncated valley forms the Thornton Gap, one of the few routes across the escarpment. The Keelbottom River near the escarpment is a disorganized creek which does not match its broad, mature valley. At one time the valley at this point was matched by a major stream drawing water from a large catchment to the east in the region of the present Queensland Shelf and Coral Sea. Collapse of the Coral Sea High truncated the Keelbottom River in midcourse and left it near the point of truncation a deprived misfit.

C. Volcanism Associated with Peripheral Doming, Following Collapse of the Coral Sea High

The distribution of Cainozoic basalts (see Fig. 1) is spatially related to the Coral Sea Plateau. Most of the vents also lie along the Great Dividing Range (Fig. 2). Arching along the Great Divide caused river reversal (discussed above) after the early Pliocene. Isotopic dating of basalts has indicated brief periods of volcanism from 4.5 millions of years ago to very recent times (Wyatt and Webb, 1970). Studies of river entrenchment rates and basalt flow directions confirm an implication of this dating work, that the Divide was already actively rising when basalt extrusion commenced.

Collapse of the Coral Sea High to provide the Great Barrier Reef's foundations, the initiation of arching along the Great Divide, and volcanicity appear to have been sequential.

D. The Corridors

1. General Features of the Corridors of Northeastern Queensland

The coastal fringe opposite the Coral Sea displays rugged "square tooth" profiles. An escarped interior plateau is bordered by plains and channels which alternate with steep-sided ridges and islands. The term "corridor" is applied to the sea channels and to the strips of plains set between imposing, inaccessible highs (Sussmilch, 1928). The term is an appropriate one as the principal communication lines follow the corridors.

Individual corridors are from 25 to 150 km long and 5 to 25 km wide. Corridors commonly occur in parallel groups separated by escarped highs. Figure 4 is a view westward across a group of corridors

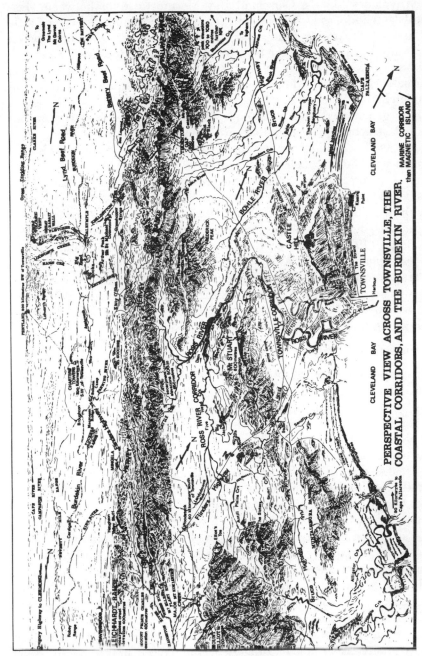

Fig. 4. Perspective view across Townsville, the Coastal Corridors, and the Upper Burdekin River. (After Heidecker, 1972.)

which includes (1) the Cleveland Bay Marine Corridor between Magnetic Island and Castle Hill; (2) the Townsville Corridor between Castle Hill and Mt. Stuart–Mt. Elliot; and (3) the Ross River Corridor between Mt. Stuart–Mt. Elliott and the escarped margin of the Interior Plateau.

Other corridors familiar to tourists and travellers are Hinchinbrook Channel (Fig. 3), the Mulgrave Corridor, the Proserpine Corridor, and Whitsunday Channel (Fig. 2).

Many generalized studies of the corridors have been undertaken, but detailed structural considerations are conspicuously absent from these studies. As the corridors are essentially structural phenomena, it is hardly surprising to find that their mode of origin is unsettled.

Three factors, faulting, differential erosion, and warping, have been considered important in the development of the corridors.

Andrews (1902) and Sussmilch (1928) considered the corridors to be grabens depressed between faults. Marks (1924), Jardine (1928), and Bryan (1925) considered them to be products of differential erosion and escarpment retreat. King (1962) also favored differential erosion, but after upwarping of an ancient land surface. Since then much relevant geomorphic, structural, and geophysical data has come to hand. This information will be reviewed in detail as the corridors hold important implications for tectonism in the adjacent Queensland Shelf.

2. Structural Implications of Geophysical Studies Across the Corridors

An important regional observation is that positive Bouger gravity anomalies increase steadily out to sea across the Queensland Shelf. These changes have been attributed to wedging of the crust and its replacement by denser subcrustal material. The anomalies plotted do not suggest the presence of masses of sediment of low density, such as would be expected in association with rift valleys, but suggest a more regional deep-seated origin (Goodspeed and Williams, 1959; Dooley, 1965).

Studies of particular corridors and trenches provide conflicting evidence for structural or geomorphic origins.

The Paluma Escarpment is the northwesterly continuation of the Harvey Range Escarpment shown in Fig. 4. It is one of a family of linear, apparently structurally controlled corridors. These structures undergo complex changes over a distance of 100 km in the Townsville area. Had they been down-faulted graben structures, then it is likely that some parts of this complex area would be faulted more than others. In some areas the basement would be expected to stand well above sea level, and in others to lie under masses of sediments. However, gravity studies have failed to reveal accumulations of low-density sediments (Dooley, 1965). Drilling has confirmed these geophysical observations. Bedrock lies at a remarkably consistent level under not much more than 100 ft of sediments (Stephenson, 1970). It may be argued then that the corridors in the Townsville area follow linear features, not because of faulting, but because differential erosion has followed weaker fault zones. However, there is still the possibility that a structural mechanism other than graben faulting could have rent the crust to produce corridors.

South of Townsville in the Proserpine Corridor (Fig. 2), magnetic, gravity, and seismic studies have revealed a structural control (Ampol Australian Exploration Proprietary Limited, 1964; Darby, 1969). A thickness of up to 7000 ft of Tertiary sediments underlies the Proserpine Corridor. The pile of sediments thickens toward the northeast margin of the corridor. On the surface the corridor floor is sympathetically tilted. The Hinchinbrook Channel in Fig. 3 has a similar surface expression, with swampy lowlands on the west inclining to a marine channel on the east.

The arcuate Queensland Trough shown in Fig. 2 slopes down to the north and east from a shallow point east of Townsville. The Capricorn Trough and its continuations affect areas to the south of the Queensland Trough. Although the troughs are larger than the corridors, they bear what appear to be genetic spatial and temporal relationships to them. Thus the troughs and corridors of Northeastern Queensland will be considered together.

The Queensland and Capricorn Troughs are morphologically similar to troughs which separate many inner continental shelves from outer subsidiary segments. Troughs on shelves off Florida, Labrador, Scandinavia, and Antarctica are considered to be fault-controlled (Holtedahl and Holtedahl, 1961).

Semicircular parts of the shelf along the Great Australian Bight and elsewhere appear to have subsided along similar fault structures (von der Borch, 1967). Morphologically there is a strong suggestion that the Coral Sea Plateau (Fig. 2) has been detached and subsided along the arcuate Queensland Trough.

Aeromagnetic and seismic studies support the suggestion that the Queensland Trough is a rent marginal to the Coral Sea Plateau. Profiles reveal a narrow infilling of sediments extending to unknown depths between vertical structural discontinuities (Ewing *et al.*, 1970).

3. Seismicity

The Coral Sea is aseismic (Bryan, 1944; Doyle *et al.*, 1968). Yet the evidence reviewed above is for considerable recent tectonism and volcanism. Subsidence and growth of the Great Barrier Reef appear to be proceeding normally.

This apparent contradiction holds important mechanistic implications. "Stick and slip" faulting mechanisms are ruled out in the subsiding shelf areas and in the corridors. Instead tensional renting and stretching mechanisms are implied. Strain release in the course of renting and stretching can be a continuous process, without the spasmodic "stick slip" bursts associated with shearing.

4. Fabrics and the Mechanism of Subsidence in the Corridors

Preliminary fabric studies have been undertaken along the margins of several coastal and marine corridors. Fracture planes have been sampled at random. Their attitudes have been determined, and they have been classified according to fillings, slickensiding, and other surface characteristics. Poles to fracture planes have been plotted on equal-area nets to give diagrams such as Fig. 5. These diagrams fail to reveal fracturing parallel to the inclined normal faults which would be expected in areas of graben subsidence. Grabens are usually bordered by normal faults (with associated lesser fractures) dipping inward at about 60° (Illies, 1969). Secondary antithetic normal faults may also be present. They dip outward at similar angles. Note then the lack of appropriately inclined discontinuities among those measured along the Cardwell Range (Fig. 5). If the adjacent Hinchinbrook Channel (Fig. 3) were a graben it might be expected that strong sets of fractures would strike northwest parallel to the channel and dip eastward at about 60°. As it is, most of the fractures observed strike across the Hinchinbrook Channel (showing rather a relationship to trends in the Macalister Ranges further west) and dip steeply northward.

Another observation is that very few of the discontinuities examined show any sign of slickensiding, and thus faulting. Most have developed as the result of tensional gaping. Those that are fault structures display strike-slip slickensiding rather than the dip-slip slickensiding of normal faults.

Thus fabric studies have failed to reveal evidence of conventional graben rifting. However, though simple graben subsidence can be ruled out in the corridors and the margins of the Queensland Shelf, all fracturing mechanisms have not been ruled out. This matter will be developed later. Corridors and the Coral Sea troughs will be interpreted in terms of tensional renting rather than normal faulting and graben subsidence.

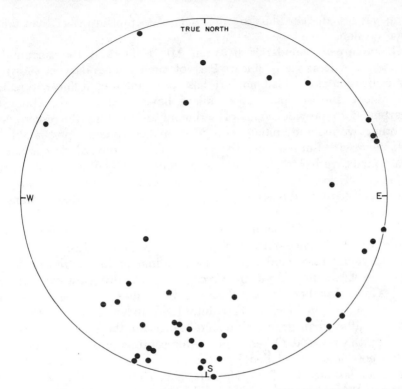

Fig. 5. Poles to fractures plotted on a Schmidt or equal-area net, fractures observed in a road cutting 7 km south of Cardwell on the western margin of Hinchinbrook Channel.

5. *Lithologic and Geomorphic Evidence for Tectonism in the Corridors*

Highs between corridors in the Townsville area tend to have cores of resistant igneous rock (Sussmilch, 1928). Mt. Elliot, Mt. Stuart, and Castle Hill in Fig. 4 have such cores. These structures and associated corridors could be attributed to differential erosion which has left igneous residuals after the eroding away of weaker Permian sediments.

Elsewhere, however, corridors and corridor highs terminate across the geological grain and disregard changes in rock type. Thus it might well be that a tectonic rather than geomorphic mechanism has left resistant blocks at high levels. The crustal thinning mechanism discussed later would leave upstanding blocks which resist thinning and plastic deformation.

Other geomorphic features of the corridors and their boundary escarpments suggest the corridors are products of tectonism rather than differential erosion.

The escarped boundaries between Mt. Elliot and the surrounding corridors are remarkably linear. Part of such a boundary is shown in Fig. 6. Erosion along this boundary has not maintained a linear retreating escarpment. Instead, pocketlike valleys have been eroded behind the escarpment to provide cones of sediment in front of the escarpment. In other words, the initial state was an escarpment, presumably of tectonic origin. Subsequent erosion has worked toward destruction of escarpments, rather than their establishment. Likewise, river capture behind the Harvey Range Escarpment just to the west of Mt. Elliot provides evidence that this escarpment is a tectonic rather than a transient erosive feature.

The upper part of Humpybong Creek in Fig. 7 once flowed into Cattle Creek. Erosive retreat of the Harvey Range Escarpment (see Fig. 4) would have modified the longitudinal profile of the creek but would not have affected the creek's course drastically. However, Humpybong Creek has been reversed just behind the escarpment and now turns in a hairpin bend to flow into Reid River shown in Fig. 4. A tectonic rather than geomorphic significance for the escarpment is indicated by this reversal. Other evidence for continuing tectonic adjustment in the corridors is provided by asymmetric drainage patterns. Several rivers, for instance the Ross, Herbert, and Tully Rivers, have migrated to asymmetric drainage positions hard against marginal escarpments and highs. Migration of this type is displayed by the Herbert River in Fig. 3. The Herbert has swung from its former central course along Trebonne Creek to a channel hugging an escarpment to the northeast. Apparently continuing tectonism has tilted the coastal plain and forced the Herbert River into a new position.

E. LINEAMENTS LIKELY TO CONTROL CORAL SEA TECTONICS

Failure in compressed brittle materials follows, or is refracted toward, suitably oriented weakness zones. Under tensile conditions a greater degree of freedom increases this tendency for structurally controlled failure. Even unfavorably oriented zones are followed in zigzag fashion.

In plan, a section of the earth's crust undergoing tensional renting would separate into a jigsaw mosaic with sutures determined by pre-existing structures. In a vertical sense, zigzag failure is inhibited by friction. Failure will tend to take advantage of *vertical* first-order structures (lineaments) that pass deeply into the crust.

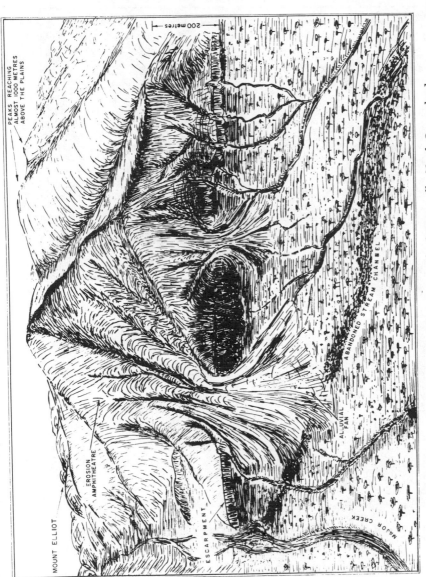

Fig. 6. Geomorphic features, Mt. Elliot area south of Townsville, North Queensland.

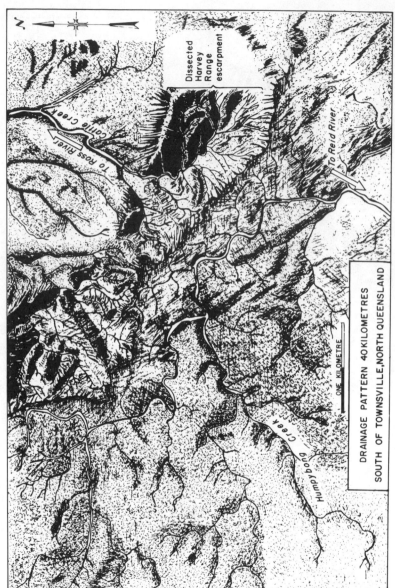

Fig. 7. Drainage pattern 40 km south of Townsville, North Queensland.

In the following section, geotectonic arguments will be advanced for post-Mesozoic stretching and tensional fragmentation of eastern Queensland. Under these circumstances it would be expected that failure would be controlled by major vertical lineaments.

Lineaments which controlled fragmentation of the Queensland Shelf continue onshore. The lineament network onshore is a complex one. This does not mean that any trend in the Coral Sea can be correlated plausibly with one of a wide range of control directions. Lineaments have the property of position as well as direction. Lineament control of a linear off-shore feature will only be considered plausible if the feature is parallel to a lineament direction *and* if appropriately positioned on continuations on-shore of a lineament with such a trend.

Fragmentation of northeastern Queensland and adjacent shelf areas then can be expected to be controlled by the following:

1. Meridional Lineaments

Most of the Palaeozoic basins followed meridional lineaments. The northern part of the "Tasman Line" (Hill, 1967) in Fig. 1 is such a lineament. It separates Palaeozoic sediments from a western Precambrian massif.

Disturbed zones, isopachs, and geophysical anomalies reflect a lineament defining the eastern margin of the Bowen Basin (Malone, 1964; Darby, 1969). This lineament lines up with the coast extending northward from Townsville.

2. Southeasterly Lineaments

a. Chillagoe–Capricorn Lineament [in Part the Palmerville Fault of de Keyser (1963)]. A highly disturbed zone separating Precambrian and Palaeozoic rocks near Chillagoe is followed southeastward by a great swarm of dykes. At the point at which this zone reaches the coast (north of Townsville) the coast turns in sympathy with it (Fig. 1). Along this section of the coast a southeasterly tectonic grain is conspicuous. For instance, swarms of dykes are parallel to this direction near Townsville and prominent linears follow it in aerial photographs of nearby Magnetic Island. In the Proserpine area detailed mapping has revealed major faults and basins trending in this direction (Clarke *et al.*, 1968). Further to the southeast, the coast swings back to its normal meridional course, and the lineament passes offshore as a remarkably linear boundary between Capricorn Trough and Swain Reefs (Figs. 1 and 2).

Parallel to the Chillagoe–Capricorn Lineament are other less-known structures. The break in slope along the seaward ramparts of the Great Barrier Reef between Gould-Cobham Reefs and Swain Reefs appears to be associated with such a structure. Another possible lineament is the disturbed zone which has controlled the generally linear course of the Burdekin River near Charters Towers and Dalrymple in Fig. 2, and which limits the southwestern end of the Owenee Granite in Fig. 1.

b. Princess Charlotte Bay Lineament. The shelf margin adopts a linear, southeasterly course opposite Princess Charlotte Bay (Figs. 1 and 2). The escarped linear margin of the Coral Sea Plateau facing the Coral Sea Basin also trends southeast. These linear features are likely to be defined by another southeasterly lineament which has been offset by gaping of the Queensland Trough. This lineament is likely to have been an important suture in the fragmentation of New Guinea discussed below.

c. Torres Strait Lineament. Another southeasterly lineament is placed tentatively in Torres Strait (some 400 km north of Figs. 1 and 2). The principal evidence is an abrupt change from the dominant meridional fracture pattern of Cape York Peninsula (Richards and Hedley, 1925). The southeastern extensions of this lineament are likely to have been removed in the course of fragmentation discussed below.

3. Northeasterly Lineaments

In the hinterland of Townsville (Fig. 1) the meridional trends which dominate eastern Australia are overshadowed by northeasterly trends, as well as less-conspicuous southeasterly and easterly trends. Northeasterly lineaments follow the Broken River Embayment, a Palaeozoic rift (White, 1965). The Burdekin Lineament along the northwest margin of the Embayment is a major fault which has controlled serpentinite intrusion and Cainozoic volcanicity. The southeastern margin of the Embayment is poorly exposed, but it is evidently a major structure which has controlled emplacement of the northeast-trending Owenee Granite and minor serpentinite masses.

4. Easterly Lineaments

The Burdekin Shelf and the Drummond–Bowen Basins terminate on the northern and southern flanks of a granite high trending through Charters Towers in Fig. 1 (Hill, 1967). This high is aligned with a set of major faults (Heidecker, 1960, 1970) which include the Inkerman Shear Zone (Paine et al., 1966) and the Mosgardies Shear Zone (Clarke,

1969). Broad zones of intense cataclasis and metamorphism follow these structures. Offshore, Gould and Cobham Reefs (Fig. 2) appear to be aligned on an extension of the Charters Towers High and its associated lineaments.

Further north there is evidence of other easterly lineaments. One is placed tentatively along the Macalister Range escarpment (Fig. 3). On line to the west are linear portions of the Herbert River and lines of volcanic vents in the McBride basalt province. To the east are aligned ridges on Hinchinbrook Island and portion of the Queensland Trough trending eastward.

F. GLOBAL TECTONIC SETTING OF THE QUEENSLAND SHELF

To this point, structural and tectonic factors influencing reef development have been considered along the usual scale progression of scientific analyses. Detailed local observations have been collected and analyzed to provide a basis for inferences on a larger scale. For instance, fractures have been used to elucidate the mechanism of corridor development. The corridors have in turn been used to gain an understanding of the broader tectonics of the Queensland Shelf.

Unfortunately, this up-scale progression of structural studies is beset by ambiguities. Rapid advances in geotectonics now permit some reversal of the usual progression. Detailed information can be fitted directly into a working hypothesis derived from considerations on a global scale. According to the fit of the data, the hypothesis can be scrapped or, more hopefully, modified to fit the details more closely.

On a global scale northeastern Queensland is a marginal segment of one of several plates of crust, the fundamental geotectonic units (Dewey and Bird, 1970). These plates are generated at sites of sea-floor spreading on oceanic ridges and are conveyed or "drift" toward oceanic trenches where they are consumed. This scheme of plate tectonics is simply the sophisticated progeny of the earlier concept of continental drift (Wegener, 1924; Carey, 1958; Rod, 1966).

There is growing agreement that the overall pattern for the Southwest Pacific is fragmentation, stretching, and subsidence (Cullen, 1970). New Zealand, the Lord Howe Rise, the Norfolk Ridge, New Caledonia, and New Guinea are considered to be strips of sialic crust that have spread apart like the fingers of a hand.

As a working hypothesis it is suggested that northeastern Queensland suffered relaxation and stretching during expansion of the Coral Sea Basin. These effects can be expected to have diminished as they propogated westward. The most profound effects would have been in the east where marginal areas became detached along a belt of tensional renting and then foundered into the Coral Sea Basin.

This hypothesis will be used in the following synthesis as a framework on which to arrange the observations set out in previous sections.

III. Synthesis

The tectonic sequence given below is a synthesis of data in Section II. This synthesis considers reef structures in a wide geotectonic, geomorphic, and igneous context. In this context it is possible to gain powerful insights which would not have arisen if the Great Barrier Reef had been considered as an isolated marine structural phenomenon.

A. STAGE 1—STRESS RELAXATION

By Miocene times the deep Coral Sea Basin had opened up along Australia's northeastern margin, depriving the continental plate of lateral support. Compressive crustal disturbances, which may have provided some physiographic relief, relaxed. Geomorphic processes settled even further into stagnation. Conditions ideal for duricrust accretion continued.

B. STAGE 2—DOMING AND VULCANICITY

Initial elastic relaxation was followed by continuing extension toward the deep Coral Sea Basin. This extension is likely to have been by flow in the deeper roots of the continental plate (Sproll and Dietz, 1969). The brittle upper crust, unable to respond as readily, tended to gape along lineaments. Gaping and dilation was pronounced in an area north of Townsville. This area lies on the intersection of the Chillagoe–Capricorn Lineament, the Broken River–Owence Trends, and the lineament extending north from the Bowen Basin (see Section II,E). A broad dome (the Coral Sea High) developed in this region, with rivers draining westward across the present coast (see Section II,B).

Development of the Coral Sea High is expected according to several popular models for subcrustal regions.

If the Ringwood (1962) model is adopted, then it could be argued that pressure release beneath dilating crust would be accompanied by upper-mantle phase changes. The crust would dome over a pillowlike zone of expanded material in the upper mantle.

If the Hess (1962) model is favored, doming can again be expected.

Heat loss through a disturbed lineament intersection zone would be facilitated by more efficient fluid convective mechanisms. Increased heat flow to the surface would depress temperatures in the upper mantle to 500°C. Serpentinization would then commence. Water in the mantle would migrate to the zone of serpentinization to reduce chemical potential imbalances (King, 1962, p. 645). This expending upper mantle "pillow" would dome the overlying crust. Mobile material collected in the "pillow" zone might also be expected to pass up fractures to erupt on the surface.

C. STAGE 3—FOUNDERING TO GIVE REEF FOUNDATIONS

Growth of a tumor off the present coast of Queensland, north of Townsville, increased crustal instability. Extension toward the Coral Sea Basin (the initial cause of doming) was aided by a tendency for gravitational slump off the tumor.

Extension toward the Coral Sea Basin was by yield and in the subcrust. The surface subsided as its subcrustal roots extended laterally and thinned. The Coral Sea High foundered. A lateritic landsurface became a steadily subsiding marine platform ideal for reef development.

D. STAGE 4—RENTING OF TROUGHS AND CORRIDORS

The responses of the mobile subcrust and the brittle upper crust failed to match. In analogous fashion the upper, brittle, perished layer and deeper deformable zones of a rubber rod are incompatible. If the rod is loaded, the inner zones yield, but the brittle surface perished zones break into small scabs. These scabs cover only part of the stretched surface. In between the steep-sided margins of the scabs are flat-bottomed rents in which is exposed deeper, yielding material.

Similar effects may be seen in failing soil slopes. Failure in the sole of a slide may be by flow and thinning. Less mobile surface zones disintegrate into a crackled mosaic of interlocking blocks separated by vertical tensional rents.

Likewise eastern Queensland fragmented along lineaments. A major fragment, the Coral Sea Plateau moved subparallel to an easterly lineament zone (discussed in Section II,E) defining its southern margin. This margin received continuing lateral support. In contrast the northwestern section of the plateau parted from the continent along the Queensland Trough and subsided directly toward the deep Coral Sea Basin. Thus the plateau tilted northward as it sank. Subsidence of more than 1500 m outpaced reef developments which contracted into atoll-like

structures. Closer to shore, on the Inner Shelf, reef growth, and subsidence were almost ideally matched. Under these optimum tectonic
conditions the Great Barrier Reef developed.

E. STAGE 5—PERIPHERAL ISOSTATIC UPLIFT

Subcrustal thinning effects tapered out west of the present coast. In
the absence of subcrustal "thinning" there was little tendency for subsidence. The coastal belt nonetheless suffered considerable corridor renting
as it lay between stable areas to the west and foundering crustal slabs
further east. Deep rents (e.g., the lower part of the Herbert River
Gorge) developed close to the stable western plateau. On the coast
itself scablike highs (e.g., Hinchinbrook Island) separated to leave intervening corridors.

The overall effect was to decrease loading imposed by the upper
crust in an area in which the lower crust had suffered very little thinning.
The crust thus underwent isostatic rise analogous to the rise of an iceberg
which has suffered losses from its upper surface. The margin of the
western plateau and coastal "scabs" such as Mt. Elliot and Hinchinbrook
Island rose in this fashion as much as 1000 m.

F. STAGE 6—A NEW CYCLE OF DOMING AND VULCANICITY FURTHER WEST

Disengagement of the Coral Sea Plateau sent a wave of relaxation
further west. The early effects now to be seen along the Great Dividing
Range are particularly interesting. They are a repetition of Stage 2,
which was experienced further east soon after the removal of New
Guinea. River drainage systems and dated volcanic events (Sections II,B
and C) permit detailed analysis of effects along the Great Dividing
Range. Relaxation produced swelling and doming effects at lineament
intersections.

A meridional belt of negative gravity anomalies follows the domed
regions of the Great Divide (Marshall and Narain, 1954). This indicates
that the land surface is standing higher than might be expected if the
crust floated in isostatic equilibrium. A dilated upper mantle "pillow,"
as suggested in Stage 2, would give such an anomaly (King, 1962, p.
642).

With continuing doming, basalts poured from hundreds of focii on
the tumors. These focii fall on lineament intersections, e.g., between
lineaments associated with the Broken River Embayment and the easterly lineaments discussed above. The basalts themselves are rich in fugi-

tive elements such as potassium and water (Morgan, 1968). This is as would be expected for the tumor mechanism proposed in Stage 2.

IV. Conclusions

A. TECTONIC SETTING OF THE GREAT BARRIER REEF

Eastern Queensland has suffered subcrustal thinning and stretching toward the Coral Sea Basin vacated by New Guinea. Doming, vulcanicity, and changes in drainage patterns preceded foundering of the stretched regions. The Queensland Trough, Capricorn Trough, coastal corridors, and coastal escarpments developed as subsidence progressed. Segments of the shelf disengaged along major lineaments. Subsidence outstripped reef development on certain of the segments, e.g., the Coral Sea Plateau. Along the inner shelf, steady foundering at an optimum rate provided an ideal tectonic setting for growth of the Great Barrier Reef.

B. STRUCTURAL FEATURES OF TROUGHS AND CORRIDORS MARGINAL TO THE BARRIER REEF

Troughs and corridors are of particular interest as some have trapped piles of sediments. Otherwise they control the pattern of reef development.

Troughs and corridors along the Queensland Shelf are considered to be products of tensional renting in brittle crust overlying thinned mobile subcrust.

Narrow, structurally controlled depressions in the crust are usually attributed to rifting. In the case of rifting, blocks of crust slip down along marginal faults, as in Fig. 8. These faults dip beneath the depressed blocks and are striated by dip-slip shearing.

Corridors and troughs produced by renting are bounded by near-vertical tensional partings (Fig. 8). These partings need not continue beneath the bottom of the intervening corridor or trough. The floors of these rents, in contrast to those of rifts, are not underlain by broad slabs of crust which once stood at higher levels.

The depth–width–age relationships, and thus sedimentation patterns, of rifts and rents are markedly different. A rift widens and deepens as deformation progresses. In contrast rents are initiated as ravines. The initial depth, dictated by rheologic factors, is maintained with widening. There may be some secondary renting in mature corridors, particularly

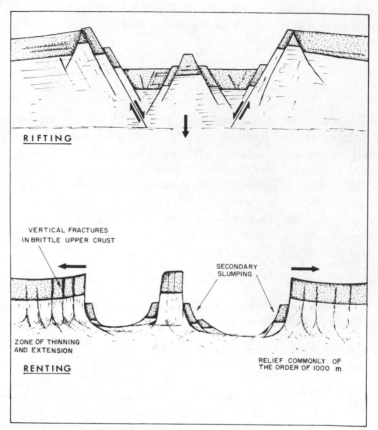

Fig. 8. Cross sections contrasting rifting and renting mechanisms.

if there is an inflow of sediments and isostatic uplift. However, this deepening is a secondary effect. Thus nearby corridors developed under similar rheologic conditions tend to be remarkably similar in depth, though perhaps markedly different in age and width.

C. Palaeogeographic Implications

Large horizontal displacements of crustal blocks (Continental Drift) is widely admitted. Dispersal also attends crustal thinning and fragmentation as postulated in this study.

Regions suffering drift and dispersal (e.g., the Queensland Shelf) provide tectonic environments suited for reef growth. Thus distortions due to drift and dispersal can be expected to be a common problem in palaeogeographic reconstructions of reef provinces.

References

Ampol Australian Exploration Proprietary Limited (1964). "Authority to Prospect 93-94 P; Airborne Magnetometer Survey—Mackay Area." Unpublished company report.

Andrews, E. C. (1902). *Proc. Linn. Soc. N.S.W.* **27**, 146–185.

Best, J. G. (1962). *Bur. Min. Res. Aust. Rec.* **12**, 1–5.

Bryan, W. H. (1925). *Proc. Roy. Soc. Queensl.* **37**, 42–46.

Bryan, W. H. (1944). *J. Roy. Soc. N.S.W.* **78**, 42–62.

Carey, S. W. (1958). "Continental Drift: A Symposium." Geology Dept., University of Tasmania, Hobart.

Clarke, D. E. (1969). *Bur. Min. Res. Aust. Rec.* **117**, 48–50.

Clarke, D. E., Paine, A. G. L., and Jensen, A. R. (1968). *Bur. Min. Res. Aust. Rec.* **22**, 94–98.

Cullen, D. J. (1970). *N.Z. J. Geol. Geophys.* **13**, 72.

Darby, F. (1969). *Bur. Min. Res. Aust. Rep.* **138**, 1–5.

de Keyser, F. (1963). *J. Geol. Soc. Aust.* **10**, 273–278.

de Keyser, F., Fardon, R. S. H., and Cuttler, L. G. (1964). *Bur. Min. Res. Aust. Rec.* **78**, 8–9.

Dewey, J. F., and Bird, J. M. (1970). *J. Geophys. Res.* **75**, 2625–2646.

Dooley, J. C. (1965). *Bur. Min. Res. Rep.* **73**, 1–23.

Doyle, H. A., Everingham, I. B., and Sutton, D. J. (1968). *J. Geol. Soc. Aust.* **15**, 297–306.

Ewing, J. I., Houtz, R. E., and Ludwig, W. J. (1970). *J. Geophys. Res.* **75**, 1963–1971.

Goodspeed, M. J., and Williams, L. W. (1959). *Bur. Min. Res. Aust. Rec.* **70**, 1–11.

Grant, K., and Aitchison, G. (1970). *Eng. Geol.* (Amsterdam) **4**, 93–120.

Heidecker, E. J. (1960). Unpublished Honours Thesis, Geology Dept., University of Queensland, Brisbane.

Heidecker, E. J. (1970). *In* "Handbook, 1970 Field Conference; Townsville, Charters Towers Area," pp. 34–40. Queensland Div., Geological Society of Australia, Brisbane.

Heidecker, E. J. (1972). "Sketch-Guide to Coral-Sea-Coast Geology: Goldfields, Volcanic Sanctuaries and Coralline Gorges of the Burdekin River" (in press).

Hess, H. (1962). *In* "Petrologic Studies: A Volume in Honour of A. F. Buddington" (A. E. Engel, H. L. James, and B. F. Leonard, eds.), pp. 603–610. Geological Society of America, New York.

Hill, D. (1967). *In* "Elements of the Stratigraphy of Queensland" (D. Hill and W. G. H. Maxwell, eds.), Fig. 1. Univ. of Queensland Press, Brisbane.

Holtedahl, O., and Holtedahl, H. (1961). *Bull. Geol. Inst. Univ. Uppsala* **40**, 183–187.

Illies, J. H. (1969). *Tectonophysics* **8**, 5–29.

Jardine, F. (1925). *Rep. Gt. Barrier Reef Comm.* **1**, 131–148.

Jardine, F. (1928). *Rep. Gt. Barrier Reef Comm.* **2**, 70–87.

King, L. C. (1962). "Morphology of the Earth." Oliver & Boyd, Edinburgh and London.

Malone, E. J. (1964). *J. Geol. Soc. Aust.* **11**, 264–282.
Marks, E. O. (1924). *Proc. Roy. Soc. Queensl.* **36**, 1–18.
Marshall, C. E., and Narain, H. (1954). *Univ. Sydney Dep. Geol. Mem.* **2**, 1–98.
Morgan, W. R. (1968). *J. Geol. Soc. Aust.* **15**, 65–78.
Paine, A. G. L., Gregory, C. M., and Clarke, D. E. (1966). *Bur. Min. Res. Aust. Rec.* **68**, 55–59.
Poole, W. (1909). *Aust. Asso. Advan. Sci. Rep. 12th Meet.* pp. 316–317.
Richards, H. C., and Hedley, C. (1925). *Rep. Gt. Barrier Reef Comm.* **1**, 1–28.
Ringwood, A. E. (1962). *J. Geophys. Res.* **67**, 857–867.
Rod, E. (1966). *Eclogae Geol. Helv.* **59**, 849–883.
Sproll, W. P., and Dietz, R. S. (1969). *Nature (London)* **222**, 345–348.
Stephenson, P. J. (1970). *In* "Handbook, 1970 Field Conference; Townsville, Charters Towers Area," p. 42. Queensland Div., Geological Society of Australia, Brisbane.
Sussmilch, C. A. (1928). *Rep. Gt. Barrier Reef Comm.* **4**, 105–134.
Twidale, C. R. (1966). *J. Geol. Soc. Aust.* **13**, 491–494.
von der Borch, C. C. (1967). *J. Geol. Soc. Aust.* **14**, 309–316.
Wegener, A. (1924). "The Origin of Continents and Oceans." Methuen, London.
White, D. A. (1965). *Bur. Min. Res. Aust. Bull.* **71**, 52–62.
Whitehouse, F. W. (1954). *In* "Artesian Water Supplies in Queensland," Appendix G. Report Department of Co-ordinator of Public Works, Brisbane.
Woolnough, W. G. (1927). *Proc. Roy. Soc. N.S.W.* **65**, 1–53.
Wyatt, D. H., and Webb, A. W. (1970). *J. Geol. Soc. Aust.* **17**, 39–51.

10

SEDIMENTS OF THE GREAT BARRIER REEF PROVINCE

W. G. H. Maxwell

I. Introduction

The Great Barrier Reef Province may be defined as the region of the Queensland continental shelf that is occupied by organic reefs and reefal sediment or has come under the influence of reefs since the end of the Tertiary. Both existing and relict reefs and sediment are used as criteria in delineating the province. As such, its limits may be drawn at lat. 24°10′S and 9°20′S (Lady Elliot Island to Bramble Cay) and at long. 141°55′E and 152°50′E (i.e., the western edge of Torres Strait and the eastern margin of the Swains Complex). Total area approximates 103,630 miles², maximum length is 1250 miles. The province narrows progressively southward from its northern boundary for a distance of 320 miles and reaches a minimum width of 13 miles between Cape Melville and Tydeman Reef. It then expands southward for 600 miles to its maximum width of 182 miles between Cape Manifold and the eastern margin of the Swains Complex. From here south, it narrows abruptly to a near-constant width of 50 miles. The main reef development occurs in the reef zone which forms a band approximately 35–45 miles wide bordering the shelf edge and occupies an area of 41,000 miles^2. Reefs outside this zone (less than 25% of the total number) occur in areas of abnormally high tidal current activity such as Torres Strait and Whitsunday Passage. Approximately half the province lacks living reefs. Much of this barren area occurs on the wide, southern shelf. There are more than 2425 reefs in the province, most of which are less than 1 mile^2 in area. Only 75 reefs have surfaces greater than 20 miles^2. Total surface area of all reefs is between 4300 and 5000 miles^2.

The province is divisible into three natural regions, each distinguished by its bathymetric character and reef development. The *northern region,* between lat. 9°20' and 16°S (Papua to Cooktown), narrows southward from 180 miles to 13 miles, with maximum depths generally less than 16 fm (deeper channels penetrate from the shelf edge). Abundant reefs are dispersed across the entire shelf and a near-continuous line of narrow shelf-edge reefs extends along the entire length of the region.

The *central region,* between lat. 16°S and 21°S–20°S (shoreline and shelf-edge) widens southward to a maximum 100 miles, with maximum depths less than 35 fm. No true shelf-edge reef system exists, fringing reefs occur on many of the island margins but reef density on the shelf is less than elsewhere in the province. The "resorbed reef" (Maxwell, 1968) is common in this region.

The *southern region,* between lat. 20–21°S and 24°10'S, reaches a maximum width of 182 miles, maximum depths exceed 35 fm, and the largest shelf-edge reef system of the entire province is developed in its northern sector. Fringing reefs are almost entirely absent. More than half the region is without reefs.

It is in this geographic setting that the sediments of the province have been developed and are in process of developing. Not unexpectedly, the resultant sedimentary pattern is extremely variable and complex. In addition to the sharp differences in bathymetry, reef type, reef concentration, and shelf area, the unusually active hydrological condition of the province contributes to the sedimentary complexity. The prevailing southeast trades (Maxwell, 1968, 1969a–e, 1972; Maxwell and Swinchatt, 1970; Australia Pilot, 1962; Wyrtki, 1960) result in the trade wind drift which diverges into a northward-flowing current in the northern region and the East Australian current which sweeps into the southern region, as well as surface drift within the province (generally moving westward and northwestward). Monsoonal winds cause a southward flow through the northern region in the summer months. The trade winds and cyclonic disturbances (mainly in the late summer) give rise to a persistent and strong southeasterly and easterly swell which results in heavy surf action along reef margins and in the near-shore zone. The tidal regime is equally vigorous (maximum tidal range varies from 9 to 35 ft) and strong tidal streams develop throughout the province, but particularly in the southern and northern regions. In Torres Strait, a maximum tidal stream velocity of 8 knots has been recorded (Australian National Tide Tables, 1970).

The effect of vigorous water circulation is most marked in the reef zone and the near-shore zones of the southern and northern regions where large volumes of sediment are moved. Over the remainder of the shelf, dispersal of sediment is inhibited by excessive depth, and many of the facies are relict. On the reef surfaces, active erosion and sedimentation is occurring in response to organic and physical processes. The reef masses serve as large-scale sediment traps for material eroded from themselves. Recognizable facies zones are developed on the reefs and a narrow fringing zone of reef-derived sediment is present around the bases of most reefs.

In order to appreciate the facies development of the shelf floor and reef surfaces, it is necessary to consider the distribution patterns of the major components which contribute to the sediment mass. These include, in addition to quartz, skeletal detritus from Foraminifera, molluscs, corals, bryozoans, echinoderms, and the calcareous algae *Halimeda* and *Lithothamnion*. Equally important is the recognition of the carbonate–noncarbonate proportions and the mud–sand proportions.

II. Component Distribution

A. Mud–Sand Distribution

Mud, largely of terrigenous origin, is dominant along the axial zone of the province (Fig. 1a,b,c) in all three regions and in the near-shore zone of the high rainfall belt in the northern part of the central region. Very little is derived from carbonate material except for that of the marginal shelf zone in the southern region. In this respect, the province differs from others such as the Persian Gulf (Sugden, 1963; Kendall

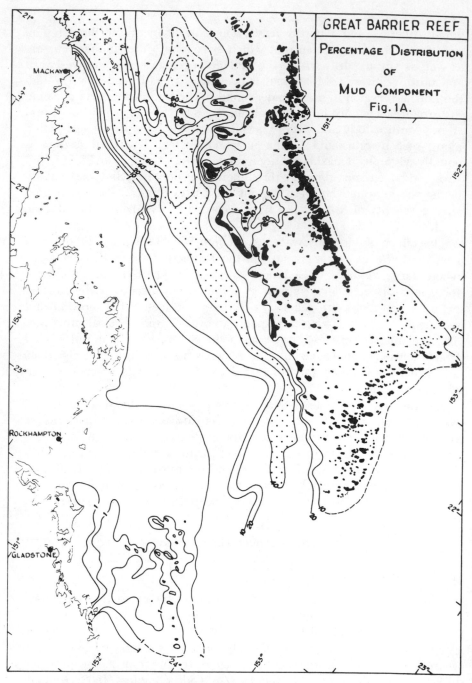

Fig. 1. (A, B, and C) Mud:sand distribution map. (For part C, see foldout.)

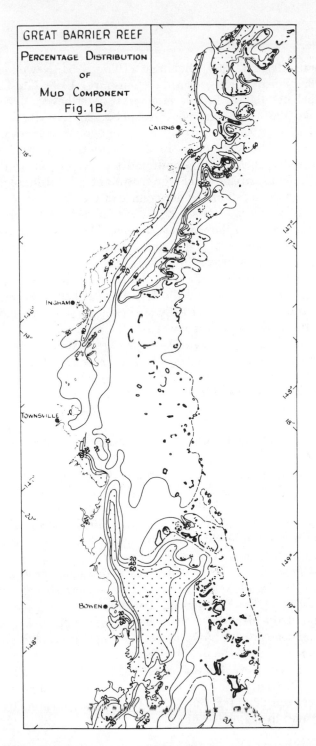

and Skipwith, 1968a,b, 1969), Bahamas (Illing, 1954; Newell *et al.*, 1960; Cloud, 1962; Purdy, 1963), Florida (Ginsburg, 1956; Ginsburg *et al.*, 1963; Swinchatt, 1965), British Honduras (Matthews, 1965; Tebbutt, 1967; Ebanks, 1967) and Bermuda (Neuman, 1963).

Mud distribution varies markedly as one progresses northward through the province. Controlling factors in its distribution appear to be wave and tidal activity, shelf width, depth and topography, proximity of high land areas, rainfall, stream discharge, nature of coastal lithology, coastal vegetation, and the location and abundance of reefs.

The *mud to sandy mud facies* (greater than 40% mud) forms an axial belt 7–45 miles wide in the southern region. The belt is displaced westward in the Whitsunday area (Thomas, 1967) and it bifurcates in the central region north of Bowen, the main development occurring on the inner shelf and near-shore zone until the northern region is approached. North of Port Douglas (beyond the high-rainfall belt), the facies moves away from the shoreline toward the shelf margin and becomes a minor element in the sedimentary framework of the shelf.

The *very muddy sand facies* (20–40% mud) is extensive and persistent throughout the entire province. It covers a larger area than the previous facies which it encloses in the southern and northern regions and which it flanks on the seaward side in the north-central region. In the south it occurs most frequently in the depth range 30–70 fm; in the northern and central regions it ranges from 5 to 20 fm, and in the high-rainfall belt its main development is in the near-shore zone in depths less than 5 fm.

The *muddy sand facies* (10–20% mud) forms the transitional zone between the main sand facies which is extensively developed in northern and southern regions, and the mud-dominated facies. It varies from 2 to 5 miles in width (7–10 miles in the Hinchinbrook area, central region) and occurs in depths of 30–36 fm and 8–14 fm. In the northern and southern regions it is very poorly represented.

The *sand facies* (0–10% mud) includes both the *slightly muddy sand* and the *sand facies* of Maxwell (1968). It is most strongly developed in the southern region (Conaghan, 1967) where it is represented by extensive quartzose deposits (to the north and south of Broad Sound) which extend seaward for 35–75 miles. Extensive carbonate sands occur through the reef zone of the marginal shelf of the southern, south-central, and northern regions (Maxwell, 1968; Maxwell and Maiklem, 1964; Maiklem, 1970). They are poorly developed on the narrow section of the shelf between lat. 14°S and 18°S. This is due, in part, to excessive dilution by mud in the high-rainfall belt.

B. Carbonate–Terrigenous Distribution

The carbonate:noncarbonate ratio is possibly the most significant parameter of shelf sediment as it clearly differentiates the broad facies pattern of the province. Comparatively abrupt transitions occur from the two end members so that facies zones are sharply delineated. The abruptness of change from carbonate-rich to carbonate-poor sediment is related to the relatively deep bathymetry and the consequent ineffectiveness of dispersal processes except in the near-shore zone, to the survival of relict sediment and to the changing sea levels of the Holocene. Four main carbonate–terrigenous zones may be recognized (Fig. 2a,b,c) separated at carbonate levels of 40, 60, and 80%.

The *high-carbonate facies* (greater than 80% carbonate) occupies the reef zone and a long, narrow, discontinuous belt (more than 70 miles) extending northward from the Capricorn group almost to the latitude of Mackay (21°5′S). This belt appears to represent relict reef sediment intermixed with younger bryozoan and molluscan material. The main development of the facies occurs in depths characteristic of the reef zones of the three main regions—32 and 16–32 fm in the southern region, 20–32 fm in the central region, and 16 fm in the northern region. The facies is virtually restricted to the reef zones in the northern and central regions, but it extends for 5–25 miles westward of the reef zone in the southern region where depths are generally greater (more than 32 fm on the eastern shelf and more than 16 fm on the western shelf). Thus, the expansion of the southern facies may seem anomalous if considered in terms of energy and dispersal processes in the deeper water. However, the greater development of reefs in this region, the strong tidal current activity on the eastern shelf and the possibility of the eastern reefs being older than others in the province counteract the effect of depth on facies expansion. By contrast, in the northern region (particularly its southern sector), shallow bathymetry, narrow shelf, and small reef zone contribute to a restriction of the high-carbonate facies because energy levels are sufficient to move terrigenous sediment seaward while the supply of carbonate material is limited by the size of the reefs.

The *impure carbonate facies* (60–80% carbonate) is quite extensive in the deeper axial part of the southern and central regions, but only weakly developed in the northern region. It is irregularly distributed in all three regions. The depression of the carbonate percentage is due mainly to the incursion of terrigenous mud, particularly along the eastern margin where strong currents sweep through interreef passages, carrying mud into the reef zone itself.

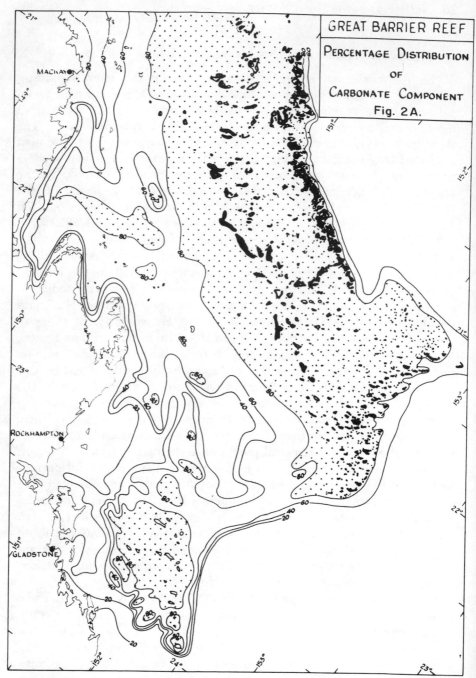

Fig. 2. (A, B, and C) Carbonate:noncarbonate distribution map. (For part C, see foldout.)

306

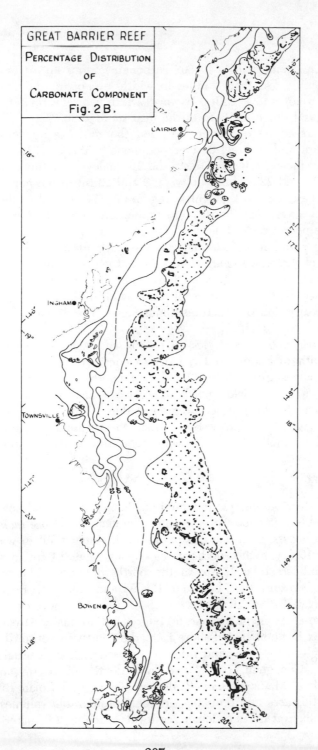

GREAT BARRIER REEF
PERCENTAGE DISTRIBUTION
OF
CARBONATE COMPONENT
Fig. 2 B.

The *transitional facies* (40–60% carbonate) forms an extremely narrow zone generally less than 4 miles in width, except for the central region where it expands to 11 miles. Its small width is a reflection of the ineffective dispersal processes beyond the near-shore zone. Very little intermixing appears to be occurring between the main terrigenous materials on the west and the reef-derived sediment to the east. In the southern region the facies is located near the 16-fm level, in the central region between 14 and 22 fm, and between 9 and 12 fm in the northern region. Its greater expanse in the central region may be related to the discharge of large river systems (Pioneer–Proserpine, Burdekin–Haughton, and Herbert), the poorer protection of reefs against the prevailing easterly swell, and the generation of fast tidal currents particularly in the Whitsunday area—all of these factors are conducive to the dispersal of the terrigenous fraction further across the shelf into zones normally dominated by reef-derived sediment. The effect of ancient strandlines (10 and 16 fm) on the localization of the transitional facies was discussed by Maxwell (1968).

The *terrigenous facies* (less than 20% carbonate) is best developed in the southern region, and it narrows considerably northward. In the south it ranges from 15 to 29 miles in width (Gladstone region) but is less than 1 mile wide north of Princess Charlotte Bay. Its greater development in the southern region is due to the abundant supply of sediment from the Fitzroy River since the beginning of the Holocene and to the remoteness of the main reef zone of the eastern shelf.

C. Quartz

Quartz distribution, plotted on Fig. 3a,b,c, at intervals of 1, 10, 20, 40, 60, and 80%, is almost predictable. The *main quartzose facies* (greater than 60%) occurs, with one exception, on the inner shelf in water depths less than 16 fm in the southern region, less than 14 fm in the central region, and less than 10 fm in the northern region. The exception is the high concentration between Bushy and Chauvel Reefs, east of Mackay (lat. 21°S) in depths of 28–32 fm. The most extensive areas of quartzose sediment appear to be related to major stream systems, and are six in number. Off-shore from the Gladstone–Bustard Head sector, strongly developed sand ridges composed of fine material of aeolian character (Conaghan, 1967), rise from depths of 10 fm and parallel the shoreline. Major streams—Calliope, Boyne, and Kolan Rivers—converge on this area and probably brought considerable volumes of quartz on to the shelf at times of lower sea level. Much of this material appears to have been reworked subaerially and later submerged by transgressive seas. The second area of strong quartz concentration spreads eastward from Waterpark and possibly represents deposition from the old Fitzroy River mouth. It is separated from the first area to the south by a terrigenous–carbonate facies that extends seaward from the present Fitzroy mouth. To the north, in Broad Sound, rich molluscan material has diluted the quartz facies of an earlier Fitzroy alluvium, and only the shallow ridges which are typical of this area have retained high quartz values. In the Cape Palmerston–Mackay–Repulse Bay sector, the third development of quartzose sediment extends over a length of 60 miles, parallel with the shoreline and averaging 15 miles in width. This facies appears to be derived from the Pioneer, O'Connell, and Proserpine Rivers, the first-named being characterized by extensive sand banks in its bed and very sandy alluvial flats. From Bowen to Fitzroy Island (just south of Cairns) a narrow, continuous zone of quartzose sediment, extending almost from the shoreline to depths of 12 fm, constitutes the fourth facies area. It is quite variable in width (1 mile near Townsville to 17 miles off Hinchinbrook Island), and in the high rainfall area between Ingham and Cairns it is separated from the shoreline by a substantial mud zone. As with the previous areas, its most extensive developments appear to be related to major stream systems—Elliot, Burdekin, Haughton in the south, Herbert in the north. The fifth area of quartz concentration extends from Cooktown to Princess Charlotte Bay. Unlike the previous zones, this one is extremely narrow (maximum width $2\frac{1}{2}$ miles). However, it borders a sector in which a number of streams drain large sandy alluvia (Annan, Endeavour, McIvor, Starcke, Normanby) and it is possible that these coastal streams continue to supply substantial sand loads to the area. The narrow width of the zone is related to the narrow shelf and the close proximity of reefs which tend to dilute the quartzose sediment as it moves seaward. Two minor quartz concentrations north and south of Lloyd Bay (lat. 12°40′S) constitute the sixth area, and this appears to have been localized by the Lockhart, Claudie, and Pascoe Rivers. Except for very small isolated patches in the Torres Straits, no quartz-rich sediment occurs north of the Lloyd Bay facies. The absence of the facies in the north is due to the greater abundance of reefs in the near-shore and inner shelf zones with consequent contribution of carbonate to the surrounding sediment.

Sediment with 10–60% quartz content has been named the *moderate to high quartz facies* and it represents the narrow zone where mixing of the main quartz and carbonate facies takes place. Only in the southern region does it exceed 5 miles in width, and for the greater part of

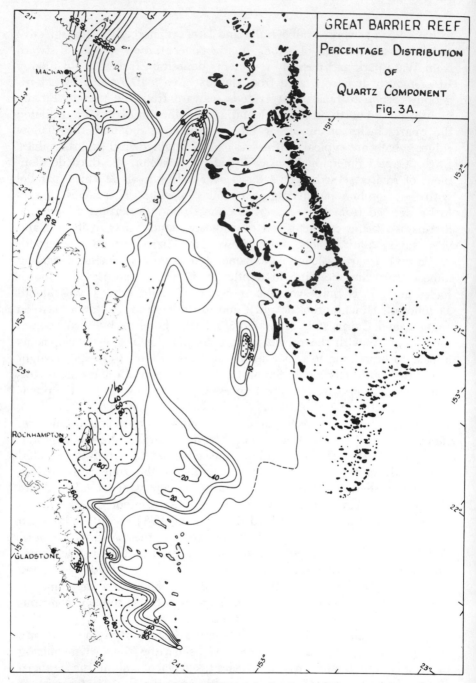

Fig. 3. (A, B, and C) Quartz distribution. (For part C, see foldout.)

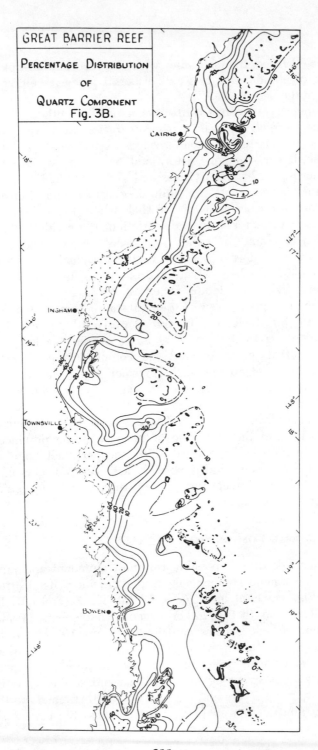

GREAT BARRIER REEF

PERCENTAGE DISTRIBUTION
OF
QUARTZ COMPONENT
Fig. 3B.

CAIRNS

INGHAM

TOWNSVILLE

BOWEN

the province it is less than 3.5 miles wide. It reflects an abrupt transition which in turn is indicative of weak dispersal processes either from the west or from the ocean side.

The *low quartz sediment* (1–10%) occurs in the southern shelf embayment to the west of the main carbonate facies, in a wide area north of Broad Sound, across the broad reefless western part of the marginal shelf in the Bowen–Ingham sector, and as a narrow, irregular band widening northward between Cairns and Torres Strait. It is a significant facies in that it represents the limit of quartz penetration across the shelf and, in the deeper areas of the shelf where present energy conditions are inadequate for sand transportation, it provides an indication of relict deposits and of earlier sea levels. In the southern region, quartzose sediment does not occur in the marginal reef zone of the Swains Complex, which is separated from the shallower western shelf and mainland by water depths in excess of 32 fm. However, in the Pompey Complex quartz is present in substantial amounts and this low to moderate quartz facies can be traced westward through depths less than 32 fm to the richly quartzose facies of the Mackay area. It is very probable that quartz entered the Pompey Complex at the time of the 32-fm sea level. In the northern region where the shelf is considerably narrower and shallower, the presence of quartz in the reef zone is less significant.

The *nonquartz facies* (less than 1%) is restricted to the Swains Complex and to most of the reef zone of the rest of the province. It also forms a comparatively large area in the deeper shelf zone northeast of Mackay where heavy mud deposition has occurred. Quartz is rarely found on reef surfaces except for those of fringing reefs in the near-shore zone and marginal to islands.

D. FORAMINIFERAL COMPONENT

Foraminiferal detritus is possibly the most abundant and most widespread of the organic components in the province. Its distribution is plotted on Fig. 4a,b,c, at levels of 5, 10, 20, 30, 40, and 60%. Its facies pattern differs from that of other organic components because both benthonic and pelagic species contribute. Strong trends parallel with the coastline reflect the main benthonic component, but marked cross-trends are also evident, particularly in the southern region where the broad shelf zone is swept by tidal currents and surface drift from the east, carrying pelagic foraminifera. In the northern and north-central regions, the transportation of pelagic forms across the narrow shelf has resulted in their concentration in shallow, near-shore muds (Swinchatt, 1967, 1968; Maxwell and Swinchatt, 1970).

The *nonforaminiferal facies* (less than 5%) is extensive (up to 60 miles wide) on the western half of the southern region except in the far south where cross-trends from the east (largely pelagic) tend to displace it. In the central and northern regions it is much narrower, being restricted to a band mostly less than 2 miles wide, just beyond the near-shore zone.

The *low foraminiferal facies* (5–10%) is also very narrow and represents the transition from the barren western sediment to the reef-dominated material to the east. Most of the foraminiferal component shows evidence of transportation and reworking; severely abraded tests of *Alveolinella, Marginopora, Operculina,* and *Elphidium* form the main fraction. Certain areas of interreef floor in the reef zone (e.g., Swains and Pompey Complex in the south, and the Blackwood Channel area in the northern region) develop this facies as a result of heavy dilution by other organic components. Interreef areas normally have high foraminiferal values.

The *moderate foraminiferal facies* (10–20%) and *high foraminiferal facies* (20–30%) lie to the east of the previous facies in most parts of the region and are generally wider than the previous one. Several expansions of the facies (related to westerly flowing currents) occur in the southern region (west of the Swains and the Northern Pompey Complex), and in the northern region (Princess Charlotte Bay and Cape York). In all such areas there is normally an increased pelagic fraction (mainly *Globigerina*) in the sediment.

In the southern region, the 10–30% range represents the most extensive foraminiferal facies. This contrasts strongly with the central and northern regions where higher ranges of concentration are more common, possibly as a result of the narrower shelf and lower rate of dilution by other components.

The *very high foraminiferal facies* (greater than 30%) is found in the southern region near the shelf margin, but is not extensive. Its main development occurs in the central region on the more open shelf between lat. 17°S and 20°30'S. Concentrations in excess of 60% are not uncommon in sediment from this region. Such strong development is related to the lower density of reefs and their smaller contribution of other components to the shelf sediment, the freer movement of oceanic water into the region, and thus the increased pelagic population and possibly the existence of favorable environments for benthonic faunas, e.g., offshore islands which are numerous in the region. The facies is also well represented in the northern region, although not on the same scale as

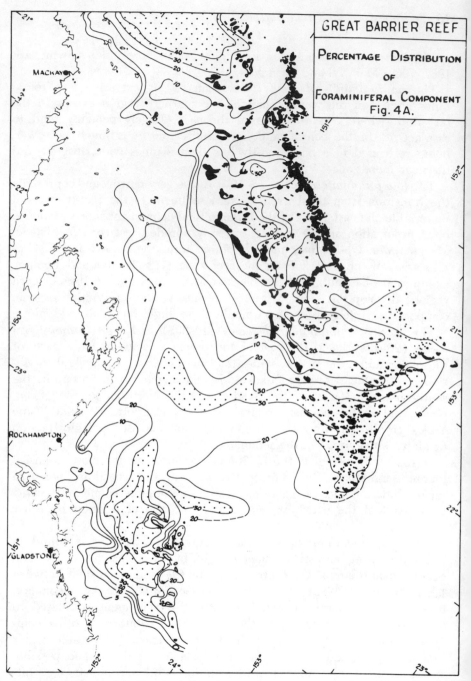

Fig. 4. (A, B, and C) Distribution of the foraminiferal component. (For part C, see foldout.)

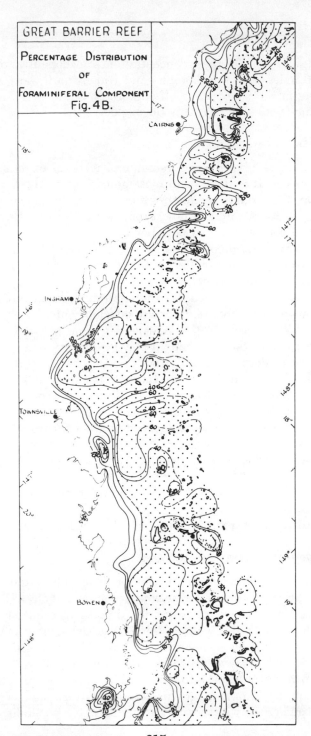

GREAT BARRIER REEF

PERCENTAGE DISTRIBUTION
OF
FORAMINIFERAL COMPONENT
Fig. 4B.

CAIRNS

INGHAM

TOWNSVILLE

BOWEN

in the central region. Highest concentrations develop in a zone which begins 5–10 miles from the shoreline and widens northward, its eastern margin lying in the lee of the shelf-edge reefs. South of Princess Charlotte Bay, the facies is not strongly developed.

E. Molluscan Component

Molluscan debris approaches foraminiferal material in terms of abundance and extent. It is the second most ubiquitous of the organic components. On Fig. 5a,b,c, its distribution pattern has been plotted at levels of 1, 10, 20, 30, 40, and 60%. From this pattern it is evident that the molluscan component does not show a marked correlation with any particular bathymetric or biological trend. The interval of 10–30% occupies more than two-thirds of the southern region; lower concentrations may be related to relict facies.

North of the southern region distribution and abundance of molluscan detritus change markedly. Widespread development of rich molluscan sediment ($>40\%$) occurs in the Whitsunday area, both in the passage and on the shelf to the east and in the Bowen off-shore area further north. The facies declines until Cairns is approached, and then the axial third of the shelf from Cairns to Cape York is virtually covered by molluscan sediment with values in excess of 40%.

The stronger development of the molluscan component in the central and northern regions is possibly influenced by three factors:

1. The smaller reef development of both regions compared with that of the southern region, and consequently less dilution of shelf sediment by reef-derived material

2. The smaller width of shelf and shallower bathymetry (except for the Whitsunday area)

3. The existence of relict deposits of shoreline shell gravels

F. *Halimeda* Component

Halimeda is a significant component of sediment in the reef zone, particularly on the southern and northern marginal shelves. Beyond the reef zone, its abundance declines rapidly so that approximately one-third of the shelf (mainly the western third) has sediment containing less than 1% *Halimeda*. This component appears to be a sensitive index of reef influence and wherever its concentration exceeds 1% reefs are found in close proximity (within 5–10 miles). The apparent exception is the belt north of the Capricorn group, but in this case submerged shoals of reefal character are widespread. Furthermore, the possibility of earlier reef development in this zone is suggested by the total facies character and the age of this sediment (Maxwell, 1969d,e).

On Fig. 6a,b,c, *Halimeda* concentrations have been plotted at levels of 1, 10, 20, 30, 60, and 80%. The reef zone of the southern region western shelf and of the central region is characterized by values exceeding 1% and in certain areas by values greater than 10%. In the southern region eastern shelf and in northern region, the values are greater than 10% and in large parts of the Swains Complex and in the lee of the shelf-edge reefs north of lat. 13°30′, *Halimeda* reaches levels of 60% and more.

G. BRYOZOAN COMPONENT

Bryozoan detritus has been plotted on Fig. 7a,b,c at concentrations of 1, 5, 15, 30, and 40%. The main developments, in excess of 15%, occur in the southern region as a widening tongue that extends northward from the Capricorn–Bunker Complex along the inner shelf almost to Mackay (more than 100 miles) and as a band of variable width in the back-reef zone of the Swains and Pompey Complexes. In the central and northern regions, this facies declines abruptly and values greater than 15% are found only in small, isolated zones. North of Townsville, the greater part of the shelf sediment contains less than 1% bryozoan detritus.

The strong concentration in the southern region and the virtual lack of bryozoa to the north may be a reflection of relict material and tidal current effects. In the south the wide marginal shelf with its abundant reef development may provide a suitable environment for the living organism and the swift tidal currents may serve to concentrate the detritus in the quieter axial part of the reef zone. The material further west appears to be located in a zone aligned with the 32-fm contour and the reefal shoals to the north of the Capricorn–Bunker Complex, thereby suggesting that relict material may be present in the bryozoan facies.

H. CORAL COMPONENT

As with *Halimeda*, the coral component in sediment is closely related to reef development and location and the distribution patterns of both are remarkably similar. On Fig. 8a,b,c, the coral values have been plotted at 1, 5, 10, 15, and 20%. The main coral facies (greater than 10% coral) is well developed on the inner shelf of the southern region (from Pine Peak Island to Lady Musgrave Reef, a distance of 170 miles) and on

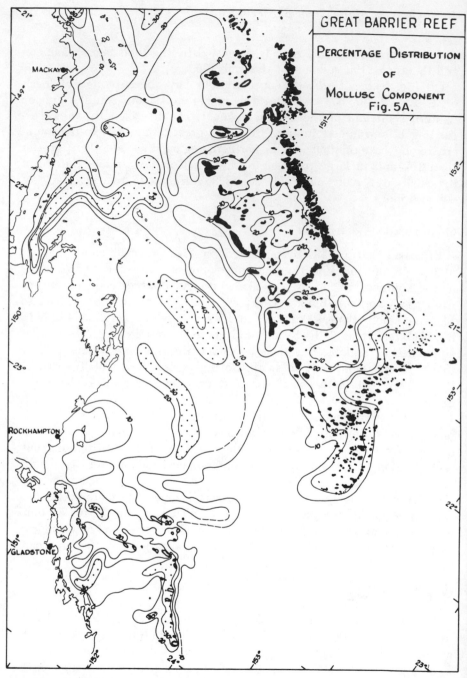

Fig. 5. (A, B, and C) Distribution of the molluscan component. (For part C, see foldout.)

318

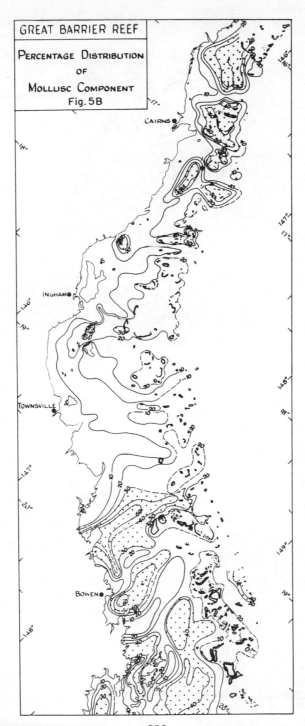

GREAT BARRIER REEF

PERCENTAGE DISTRIBUTION
OF
MOLLUSC COMPONENT
Fig. 5B

319

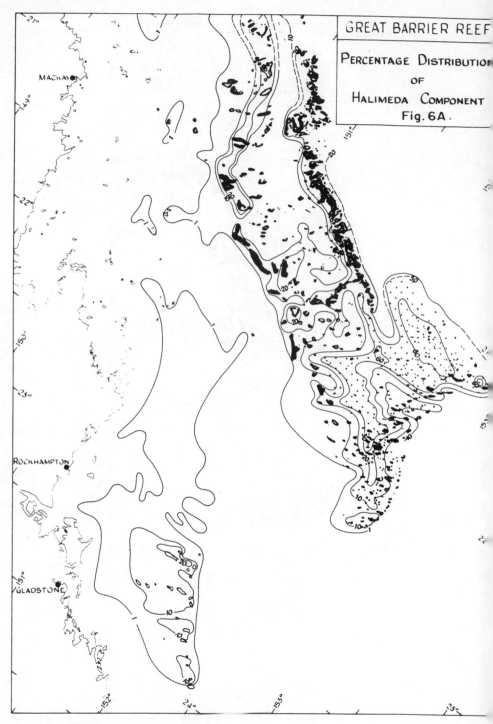

Fig. 6. (A, B, and C) Distribution of the *Halimeda* component. (For part C, see foldout.)

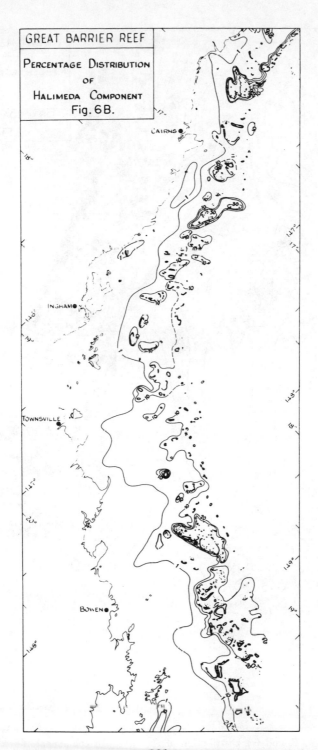

GREAT BARRIER REEF

PERCENTAGE DISTRIBUTION
OF
HALIMEDA COMPONENT
Fig. 6B.

CAIRNS●

INGHAM●

TOWNSVILLE●

BOWEN●

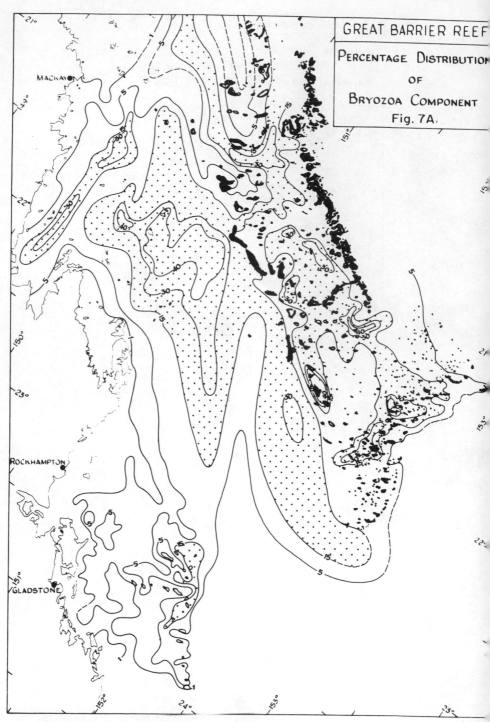

Fig. 7. (A, B, and C) Distribution of the bryozoan component. (For part C, see foldout.)

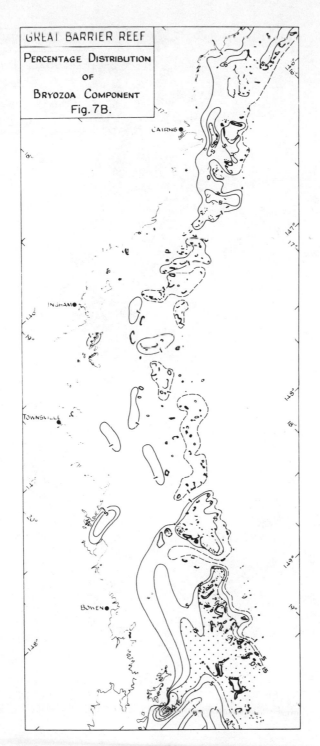

GREAT BARRIER REEF
PERCENTAGE DISTRIBUTION
OF
BRYOZOA COMPONENT
Fig. 7B.

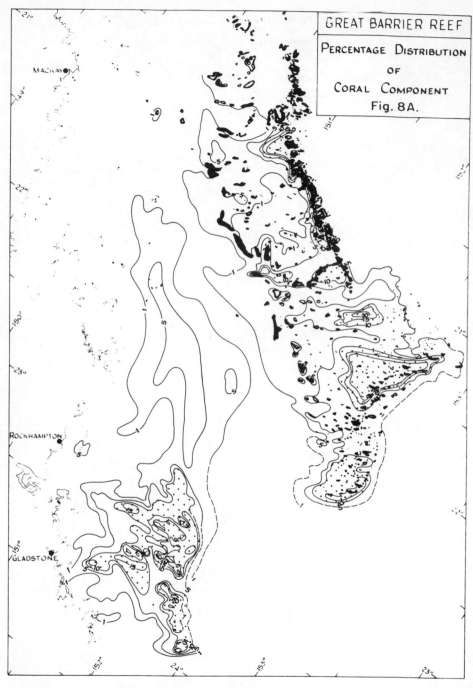

Fig. 8. (A, B, and C) Distribution of the coral component. (For part C, see foldout.)

324

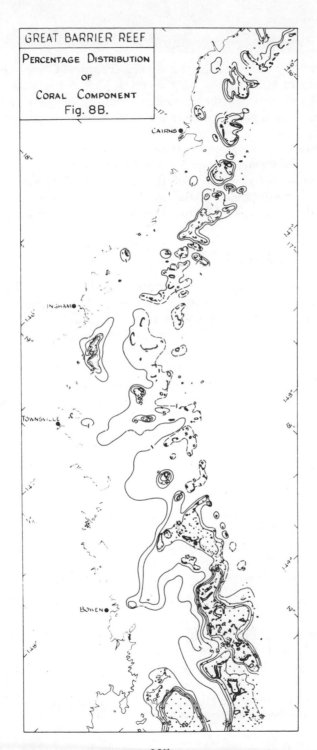

GREAT BARRIER REEF

PERCENTAGE DISTRIBUTION
OF
CORAL COMPONENT
Fig. 8B.

CAIRNS

INGHAM

TOWNSVILLE

BOWEN

the eastern marginal shelf in the Swains–Pompey Complexes. In the central and northern regions, this facies is only weakly developed in narrow isolated zones in the immediate vicinity of reefs. The greater part of these two regions is covered with sediment containing less than 5% of coral.

The difference in coral concentrations between the southern and the other two regions may be explained in terms of the factors similar to those that have been invoked for the molluscan distribution, viz. (1) stronger reef development in the south, (2) relict deposits, and (3) greater shelf width, thus inhibiting dilution of the reef zone by nonreefal material.

I. ECHINODERM COMPONENT

The echinoderm component has been plotted in Fig. 9a,b,c at levels of 1, 5, 10, 20, and 30%. This component is best represented in the northern and central regions, mainly in the mud and muddy sand facies of the inner shelf. In the southern region it rarely reaches concentrations greater than 5%, and in these instances the facies is associated with the reef zone. In the central region, the component is abundant in mud and muddy sand to the east of the Whitsunday Islands. It also occurs in high concentration to the east of Hinchinbrook and Palm Islands, in similar muddy sediment. Further north, it is found in small areas of the reef zone (south of Cairns) and to the leeward of reefs north of Cairns. From Port Douglas to Cape York, the high echinoderm facies forms a near-continuous belt, 3–9 miles wide, that occupies the inner shelf to the leeward of the main reef development. It expands considerably in the Torres Strait area.

The abundance of echinoderm material in the northern region is probably due in part to the extensive muddy sand facies which living echinoids favor, and to the unusually numerous ophiuroids that occur on the inner shelf and near-shore reefs. In the southern region, the muddy sand facies is not suitably located, nor is it strongly developed.

J. LITHOTHAMNIOID COMPONENT

Lithothamnion and related algae are essentially reef-controlled components, and as such their distribution patterns tend to follow those of coral and *Halimeda*. However, they rarely reach concentrations of the same order as the other two components and their distribution is more restricted, as is evident in Fig. 10a,b,c. This difference is probably due to the greater resistance of the encrusting algae to erosion by physical and organic processes and consequently to a much smaller supply of detritus. Only near the reef masses does the concentration in shelf-floor sediment exceed 5%. On the reef surface, particularly near the rim, the values may reach 40%. For the greater part of the province, lithothamnioid components are less than 1% of the sediment.

Lithothamnion values are particularly low in the central region, even in the reef zone. This is probably due to the low density of reefs in the region. In the northern region, quite high values exist in the thin band to the leeward of the shelf-edge reefs and also around the reefs of the inner shelf north of Princess Charlotte Bay.

III. Regional Facies Pattern

The broad facies trends within the province (Fig. 11) are controlled by the carbonate:noncarbonate ratio and the mud:sand ratio. The former is essentially a reflection of reef contribution to the shelf sediment, although components such as mollusc, echinoderm, and pelagic Foraminifera tend to confuse this relationship. Relict shell gravels as well as existing molluscan communities are important in this respect. The main components of the noncarbonate fraction are quartz and terrigenous mud. The mud:sand ratio is more significant in the noncarbonate material and effectively separates two natural facies—the quartzose sand which is restricted to the western part of the shelf near major stream outlets, and the mud which is typical of the deeper, axial zones of the shelf and the near-shore zone of the higher rainfall belt in the central region.

Transitional zones from carbonate to terrigenous and from mud to sand are generally narrow and abrupt. The exception occurs in the muddy sand facies of the central region, the northern extremity of the southern region and the northern part of the northern region. In the first two cases, the expansion of this facies is probably due to the excessive supply of mud to the shelf in the high rainfall belt and to the comparatively wide shelf, thus preventing dilution of the facies by carbonate sand from the reefs. In the northern region, the expansion of the facies in the Cape York–Torres Strait area is probably related to the effective scouring of the adjacent area by strong tidal current and the settling of transported mud in the more open, deeper areas to the east and south.

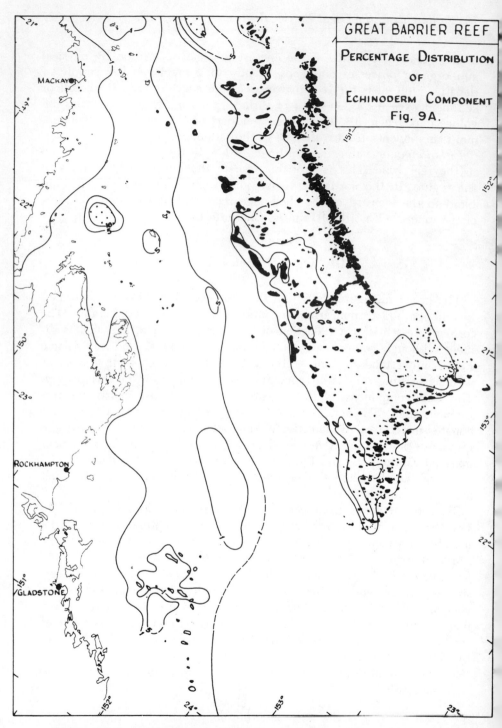

Fig. 9. (A, B, and C) Distribution of the echinoderm component. (For part C, see foldout.)

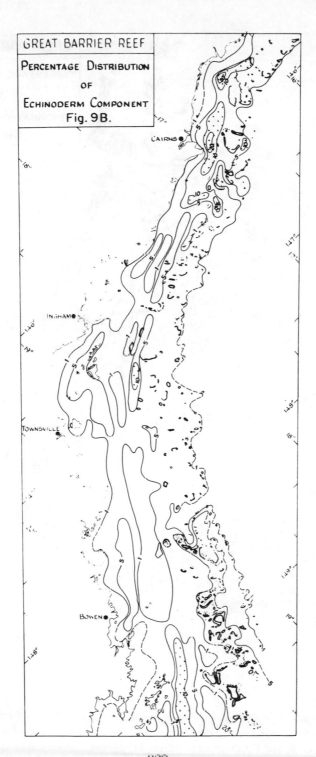

GREAT BARRIER REEF
PERCENTAGE DISTRIBUTION
OF
ECHINODERM COMPONENT
Fig. 9B.

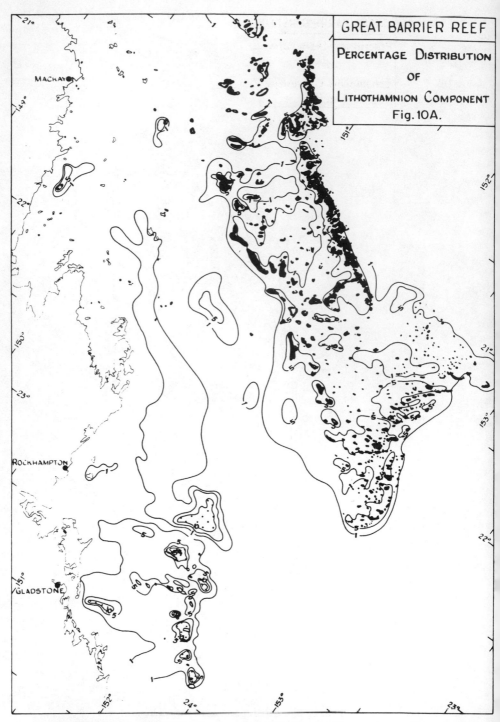

Fig. 10. (A, B, and C) Distribution of the lithothamnioid component. (For part C, see foldout.)

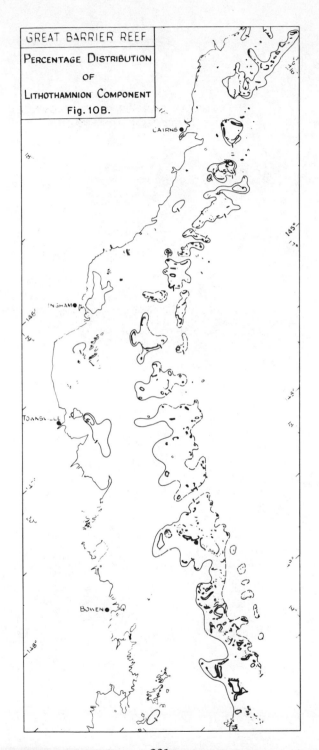

GREAT BARRIER REEF

PERCENTAGE DISTRIBUTION

OF

LITHOTHAMNION COMPONENT

Fig. 10B.

CAIRNS

INGHAM

TOWNSVILLE

BOWEN

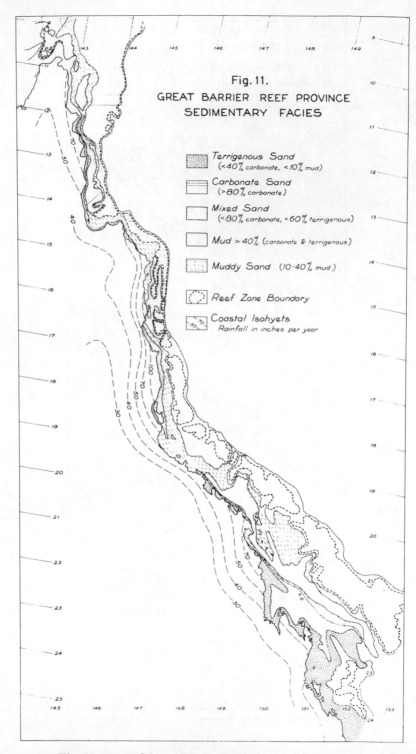

Fig. 11.
GREAT BARRIER REEF PROVINCE
SEDIMENTARY FACIES

Terrigenous Sand
(<40% carbonate, <10% mud)

Carbonate Sand
(>80% carbonate)

Mixed Sand
(<80% carbonate, <60% terrigenous)

Mud > 40% (carbonate & terrigenous)

Muddy Sand (10-40% mud)

Reef Zone Boundary

Coastal Isohyets
Rainfall in inches per year

Fig. 11. Regional facies pattern, Great Barrier Reef Province.

IV. Sediment of the Reef Mass

Detailed studies of sediment from reef surfaces have been carried out by Maxwell *et al.* (1961, 1964), Reid (1966), and Swinchatt (1967, 1968) on Heron, North-West, and Arlington Reefs, respectively. On Heron Reef (Figs. 12–16) the major components are *Halimeda, Lithothamnion,* and coral which all reach concentrations of more than 40%, and Mollusca and Foraminifera which reach values of 10% and, in isolated areas, values in excess of 40%. Minor components, echinoderm, bryozoan, sponge spicules, and crustacean, rarely aggregate more than 5%. The main *Halimeda* and coral concentrations occur on the sand flat and lagoon, i.e., on the deeper, more protected part of the reef, in the lee of the zone of living coral. *Lithothamnion* is dominant on the reef rim and in a central axial zone of the lagoon. Molluscan detritus has a relatively uniform distribution across the entire reef surface as has foraminiferal material. High concentrations of the latter are found in three areas in the lee of the reef rim and shingle banks. The distribution patterns of the five components, which account for more than 90% of the reef sediment, are controlled primarily by the location of the living organisms, and secondly by the movement of skeletal detritus from the growth areas by tidal currents and translatory wave action. The extent of dispersal is influenced by the type and size of detrital grain produced through breakdown of the parent organism. Thus, the more resistant *Lithothamnion,* which is dominant on the reef rim, produces a coarser detritus than does *Halimeda,* which is more prolific on the reef slope and reef flat. Coral also tends to produce finer grades of sand once it has been reduced from shingle. Thus coral sand and *Halimeda* sand tend to be more mobile and are transported further from their parent organisms than is *Lithothamnion.* The relationship between organic component and grain size is not a simple one and appears to be controlled by the physical parameters of the environment as well as by the microstructure of the particular organism. Measurements by Folk and Robles (1964) for modes of various skeletal materials from Isla Perez, Yucatan differ markedly from modes of similar materials from Heron Reef. Furthermore, the modes of interreef material differ from those of reef surface material.

The facies pattern that can be recognized on the Heron Reef corresponds broadly with the topographic and biological zonation of the reef. Five main facies types are present:

1. The *Lithothamnion*-dominated facies (*Lithothamnion* >40%, Mollusca >10%, Foraminifera >5%, *Halimeda*–coral <20%) is restricted to the reef rim and the axial zone of the lagoon.

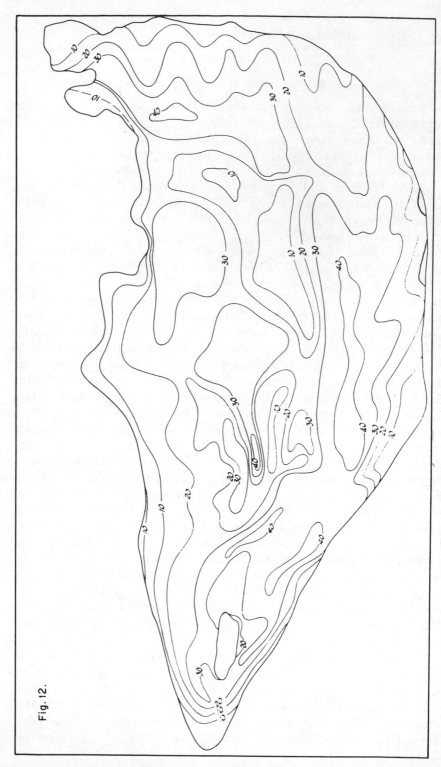

Fig. 12.

Fig. 12. Distribution of the *Halimeda* component, Heron Reef.

Fig. 13.

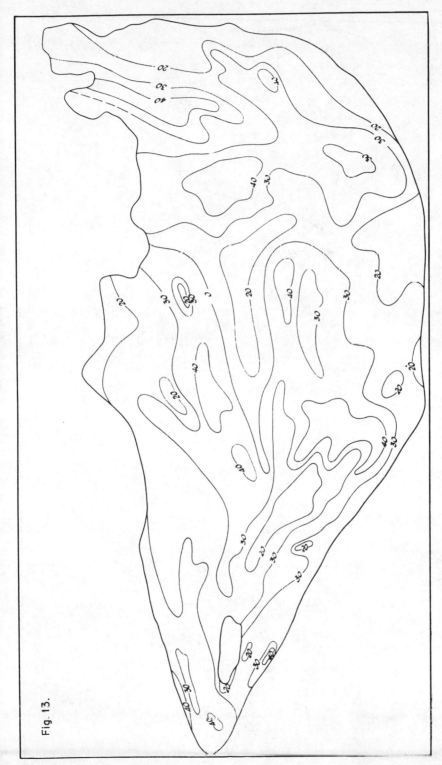

Fig. 13. Distribution of the coral component, Heron Reef.

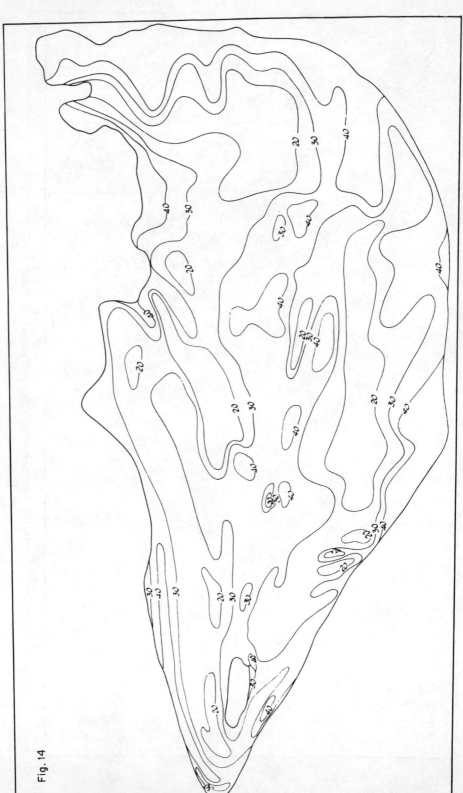

Fig. 14

Fig. 14. Distribution of the lithothamnioid component, Heron Reef.

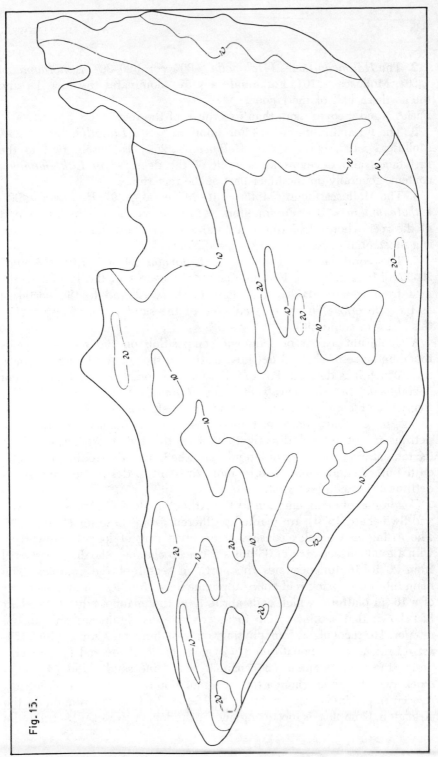

Fig. 15.

Fig. 15. Distribution of the mollusc component, Heron Reef.

2. The *Halimeda* facies (*Halimeda* >30%, coral 20–30%, *Lithothamnion* <20%, Mollusca <10%, Foraminifera <5%) dominates the reef flat and the northern half of the lagoon.

Facies 1 and 2 cover more than two-thirds of the reef.

3. The Foraminifera facies (Foraminifera >10%, *Lithothamnion* >25%, Mollusca >10%, coral <25%, *Halimeda* <25%) is closely tied to the first facies and develops as a result of the decrease in *Lithothamnion* content, generally on the inner part of the reef rim.

4. The Mollusca facies (Mollusca 10–20%, coral >30%, *Halimeda* <30%, *Lithothamnion* <10%, Foraminifera <5%) favors the northwestern part of the reef where algal rim and coral zone are both narrow and there is a substantial cover of dead coral rubble.

5. The coral facies (coral >30%, *Halimeda* 20–30%, *Lithothamnion* 20–30%; Mollusca <5%, Foraminifera <5%) occurs in the lagoon in the area from which tidal currents flow to the west and to the northeast as the tide ebbs. The beach sediment of the sand cay also approaches this facies in composition.

A significant aspect of sediment composition on the reef surface is the abundance of skeletal detritus relative to that of the living organism from which it is derived. Possibly the most abundant detrital component is *Halimeda*, but the density of living *Halimeda* appears to be far less than that of coral. In large areas of the reef it is difficult to find the living algae. Thus, one must infer that the abundance of *Halimeda* detritus is not related directly to that of the living organism, but to its rate of regrowth as the fronds are shed into the sediment. Coral, on the other hand, appears to be more resistant to destruction and sheds detritus at a much lower rate.

Conditions of sedimentation in the Arlington Reef Complex (Maxwell and Swinchatt, 1970) are markedly different from those on Heron Reef. The Arlington Complex consists of an open ring of long, narrow reefs with a total surface area of 35 miles^2. A deep channel, shoaling westward from 26 to 16 fm, occupies the northern sector of the complex. The main interior or back reef zone, approximately 45 miles^2 in area, consists of a 16-fm platform which ends abruptly at the channel edge. Swinchatt has shown that sedimentation in the complex is significant only in the interior. He recognized four physiographic–sedimentary zones (Fig. 17), viz. (1) zone of carbonate mud, (2) zone of coral-covered bottom bordering the deep channel, (3) channel floor (fine sand), and (4) hummocky western zone where a thin accumulation of coarse skeletal detritus is occurring. Unlike the situation on Heron Reef, sedimentation in the Arlington Complex is occurring in much deeper water (16 fm); the

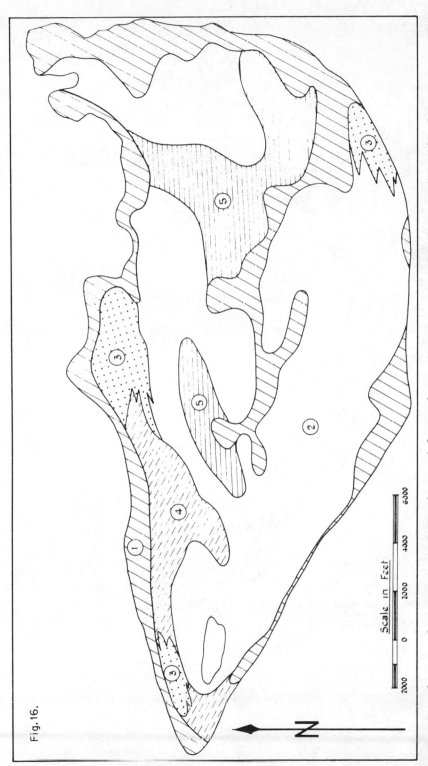

Fig. 16.

Fig. 16. Sedimentary facies, Heron Reef. Distribution of detritus from major groups: (1) Lithothamnion, (2) *Halimeda,* (3) Foraminifera, (4) Mollusc and (5) Coral.

Scale in Feet

N

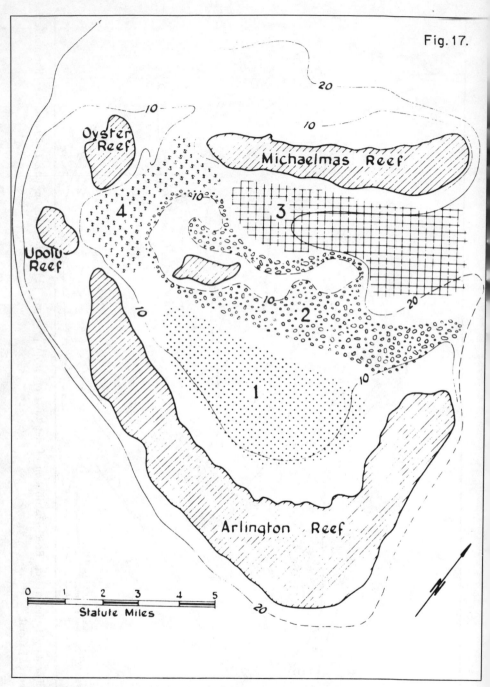

Fig. 17. Facies in the Arlington Reef Complex. [After Maxwell and Swinchatt (1970).]

TABLE I
RADIOCARBON DATINGS, GREAT BARRIER REEF

Sample no.	Location	Depth (fm)	Composition	Age (years)
194	23°46'S 151°56'E	20	Interreef "speckled" foraminiferal sand	4950
3003	22°18'S 151°17'E	29	Shell gravel	3520 ± 20
3268	21°52'S 150°42'E	32	Muddy "speckled" sand	2970 ± 90
3321	21°40'S 150°12'E	17	Shelly sand and gravel	1700 ± 90
3271	21°40'S 150°55'E	39	Bryozoan detritus from mud	0 ± 70
3277	21°33'S 150°21'E	22	Muddy shell sand	2150 ± 90
3043	21°12'S 152°42'E	24	Coarse interreef Foram- inifera–*Halimeda* sand	390 ± 80
8000	22°00'S 152°10'E	35 (base of 15 ft of core)	Pelagic Foraminifera in mud	5680 ± 260
3097	21°17'S 152°06'E		Intertidal beach rock (Twin Cay)	1110 ± 80
3082	21°18'S 152°03'E		Intertidal beach rock (Twin Cay)	630 ± 90
5595	14°40'S 145°18'E		Intertidal "reef rock" (Nymph Reef)	530 ± 80
5690	14°45'S 145°14'E		Intertidal "reef rock" (Turtle Island)	250 ± 80
5877	11°50'S 143°18'E		Intertidal beach rock (Magra Reef)	680 ± 80
5880	11°50'S 143°18'E		Intertidal coral head (Magra Reef)	770 ± 90
5961	10°42'S 142°09'E		Intertidal coral head (Prince of Wales Island)	5120 ± 120
5949	10°24'S 142°06'E		Intertidal coral head (Hawkesbury Island)	2810 ± 90

reef surface appears to retain only the coarse gravels. The dominant components are molluscan and foraminiferal, with smaller amounts of coral, *Halimeda*, and *Lithothamnion*. The mud fraction of the sediment contains an average of 20% noncarbonate material (i.e., terrigenous mud). The dominance of foraminiferal and molluscan components, as well as the presence of a substantial terrigenous mud fraction (in strong contrast with Heron Reef), are a reflection of the greater water depth

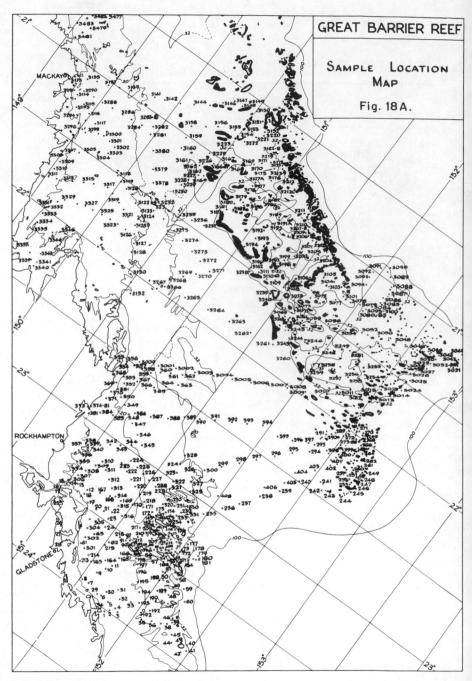

Fig. 18. (A, B, and C) Sample location map, Great Barrier Reef Province. (For Part C, see foldout.)

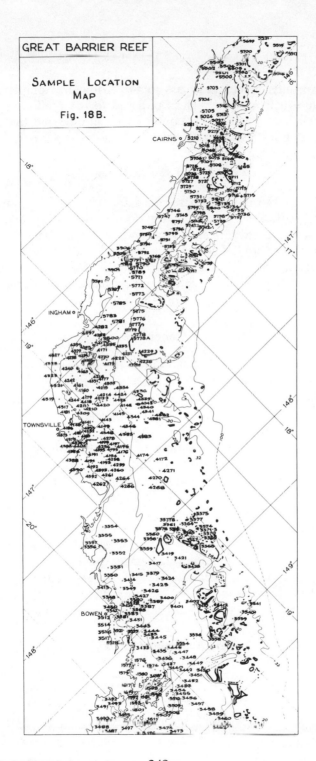

and the proximity of the land mass to the Arlington Complex. Furthermore, because it is a linear reef as opposed to the platform type (e.g., Heron Reef), less sediment is retained on its surface and the accumulation is occurring in the back-reef zone.

V. Age of the Sediment

Although the broad facies pattern of the province is comparatively simple and predictable, detailed mapping of sedimentary units in small areas has revealed great complexity and variation. The main causes of such complexity are the varied, but generally slow, rates of deposition and the survival of relict facies, particularly sands and shell gravels. Radiocarbon dating of a number of samples has confirmed the existence of relict sediment. These results can be found in Table I.

VI. Conclusion

The sedimentary character of the Great Barrier Reef Province appears to have been moulded in early post-Pleistocene time when much of the present shelf was still exposed and processes of erosion and dispersal were more effective than they are now. Tectonic events toward the close of the Tertiary also modified the source areas on land and later led to severe reduction in the supply of sediment. Rising sea level and stillstands resulted in the westward migration of the main zones of deposition (i.e., the near-shore zone) as well as to the isolation of the reef zone and its associated facies. Thus, strong near-shore sand and shell gravel deposits were developed and many have survived as relicts on the deeper shelf as the shoreline receded westward.

High-energy conditions in the present near-shore zone have led to progradation of the coastal mud facies in the high-rainfall belt, and to intensive erosion and dispersal of reef surface sediment. However, the great depths of the main part of the shelf inhibit the sedimentary process and little deposition is occurring at the present time.

VII. Acknowledgments

This paper is based on research originally sponsored by the American Petroleum Institute and the American Chemical Society (Petroleum Research Fund). Considerable information was provided by the Australian Gulf Oil Co., and in particular, by the late Dr. T. C. Wilson and Mr. E. Ericson. Work by P. J. Conaghan, W. R. Maiklem, K. O. Reid, B. M. Thomas, and E. Frankel, graduate students under the author's supervision, contributed to the results presented in this paper.

References

Australian National Tide Tables. (1970). Parts I, II, and III. Hydrographic Sect., R.A.N. Canberra. (E. R. Whitmore, Hydrographer).

Australia Pilot. (1962). Vol. 4. Hydrographic Dept., London.

Cloud, P. E. (1962). *U.S., Geol. Surv., Prof. Pap.* **350**, 1–138.

Conaghan, P. J. (1967). Unpublished Ph.D. Thesis, University of Queensland, Brisbane.

Ebanks, W. J. (1967). Ph.D. Thesis, Rice University, Houston, Texas.

Folk, R. L., and Robles, R. (1964). *J. Geol.* **72**, 255–292.

Ginsburg, R. N. (1956). *Bull. Amer. Ass. Petrol. Geol.* **40**, 2384–2427.

Ginsburg, R. N., Lloyd, R. M., Stockman, K. W., and McCallum, J. S. (1963). *In* "The Seas" (M. Hill, ed.), Vol. 3, pp. 554–582. John Wiley & Sons, New York.

Illing, L. V. (1954). *Bull. Amer. Ass. Petrol. Geol.* **38**, 1–94.

Kendall, C. G. St. C., and Skipwith, P. A. d'E. (1968a). *Amer. Ass. Petrol. Geol., Bull.* **53**, 841–870.

Kendall, C. G. St. C., and Skipwith, P. A. d'E. (1968b). *J. Sediment. Petrol.* **38**, 1040–1058.

Kendall, C. G. St. C., and Skipwith, P. A. d'E. (1969). *Bull. Geol. Soc. Amer.* **80**, 865–892.

Maiklem, W. R. (1970). *J. Sediment. Petrol.* **40**, 55–80.

Matthews, R. K. (1965). Ph.D. Thesis, Rice University, Houston, Texas.

Maxwell, W. C. H. (1968). "Atlas of the Great Barrier Reef." Elsevier, Amsterdam.

Maxwell, W. G. H. (1969a). *Aust. Conserv. Found., Spec. Publ.* **3**, 5–14.

Maxwell, W. G. H. (1969b). *In* "Stratigraphy and Palaeontology" (K. S. W. Campbell, ed.), pp. 353–374. Australian National University Press, Canberra.

Maxwell, W. G. H. (1969c). *Aust. Oil Gas Rev.* **15**, 15–22.

Maxwell, W. G. H. (1969d). *Aust. J. Sci.* **31**, 85.

Maxwell, W. G. H. (1969e). *Sediment. Geol.* **3**, 331–333.

Maxwell, W. G. H. (1972). *Deep-Sea Res.* **17**, 1005–1018.

Maxwell, W. G. H., and Maiklem, W. R. (1964). *Univ. Queensl. Pap., Dep. Geol.* **5**, 1–21.

Maxwell, W. G. H., and Swinchatt, J. P. (1970). *Bull. Geol. Soc. Amer.* **81**, 691–724.

Maxwell, W. G. H., Day, R. W., and Fleming, P. J. (1961). *J. Sediment. Petrol.* **31**, 215–230.

Maxwell, W. G. H., Jell, J. S., and McKellar, R. G. 1964. *J. Sediment. Petrol.* **34**, 294–308.

Neuman, A. C. (1963). Manuscript, Lehigh University, Bethlehem, Pennsylvania.

Newell, N. D., Purdy, E. G., and Imbrie, J. (1960). *J. Geol.* **68**, 481–497.

Purdy, E. G. (1963). *J. Geol.* **71**, 334–355 and 472–497.

Reid, K. O. (1966). Unpublished Manuscript, University of Queensland, Brisbane.

Sugden, W. (1963). *J. Sediment. Petrol.* **33**, 355–364.

Swinchatt, J. P. (1965). *J. Sediment. Petrol.* **35**, 71–90.

Swinchatt, J. P. (1967). *Annu. Meet. Program, Geol. Soc. Amer.* Abstract, p. 219.

Swinchatt, J. P. (1968). *Amer. Ass. Petrol. Geol., Bull.* **52**, 351 (abstr.).

Tebbutt, G. E. (1967). Ph.D. Thesis, Rice University, Houston, Texas.

Thomas, B. M. (1967). Honours' Thesis, University of Queensland, Brisbane.

Wyrtki, K. (1960). *Commonw. Sci. Ind. Res. Organ., Div. Fish. Oceanogr., Tech. Pap.* **8**, 1–44.

11

FORAMINIFERA OF THE GREAT BARRIER REEFS BORES

Alan R. Lloyd

I. Introduction

Four bores drilled along the Great Barrier Reefs, Queensland (Fig. 1) have provided foraminiferal faunas which form the subject of this paper. The first, drilled in 1926 on Michaelmas Cay, lat. 16°36′S, long. 145°59′E, 22 miles northeast of Cairns, was sponsored by the Great Barrier Reef Committee. The second bore, also by the Great Barrier Reef Committee, was drilled in 1937 on Heron Island, lat. 23°26′S, long.

Fig. 1. Locality map.

151°57′E, 51 miles northeast of Gladstone. Wreck Island bore, lat.
23°20′S, long. 151°57′E, 7 miles north of Heron Island, was drilled by
Mines Administration Pty. Ltd., on behalf of Humber Barrier Reef Oil
Pty. Ltd., in 1958. The latest well was drilled by Tenneco Australia
Inc., in 1969 on Anchor Cay on the south east of Bligh Entrance at
the northern end of the Reef, lat. 3°7′S, long. 144°6′E.

Wreck Island bore penetrated the first proven Tertiary marine sediments off the coast of Queensland and provided the first firm datings of sediments below the reef. Unfortunately no samples from the top 530 ft were available for examination, because of the lack of sample return over this interval during drilling. The other bores do little to fill this gap; their value lies in determining the nature and thickness of reef rock and other material. This paper deals primarily with the biostratigraphic value of the Foraminifera, the stratigraphic sequence, and the geological history of the reefs themselves.

II. Basis for Age Determinations

The ranges of the species and genera of Foraminifera I have selected to subdivide the Tertiary sequences in the Indo-Pacific region in terms of the European Ages are based on those published by Blow (1969), Clarke and Blow (1969), and Adams (1970). The planktonic Foraminifera zones of Blow (1969) are not easy to distinguish in subsurface work (when drill cuttings form the bulk of the samples available for study) because of the effects of contamination from caving which make it necessary to use the extinction point of species rather than their incoming. The European Ages are therefore used and subdivided into upper, middle, and lower units. Minor changes in the recognition of Pliocene to middle Miocene sections have been made and these are outlined below.

A. Recent to Upper Pliocene

The upper sections of off-shore wells in Indo-Pacific waters so far examined by myself have not been divisible on the basis of Foraminifera and, on the assumption that a complete sequence is present, the section above the middle Pliocene is designated Recent to upper Pliocene. The fauna commonly includes the benthonic species *Operculina bartschi* Cushman, *Cellanthus craticulatum* (Fichtel and Moll), *Amphistegina quoyi* d'Orbigny, *A. gibbosa* d'Orbigny, and *Pseudorotalia schroeteriana* (Carpenter, Parker and Jones), which, in the Indo-Pacific region, have not been found below the lower Pliocene as defined below.

B. Middle Pliocene

The highest recognizable extinction point in a well is that of *Globigerinoides obliquus extremus* Bolli and Bermudez and the extinction of this species is used to define the top of the middle Pliocene.

The association of *G. obliquus extremus* in cores or outcrop samples with all or any one of the benthonic species *Operculina bartschi, Cellanthus craticulatum, Amphistegina quoyi, A. gibbosa,* and *Pseudorotalia schroeteriana,* or with the planktonic species *Sphaeroidinella dehiscens* (Parker and Jones) and *Pulleniatina obliqueloculata obliqueloculata* (Parker and Jones), which range from early Pliocene to Recent, can be taken to indicate a lower to middle Pliocene age, a division being possible only when positive lower Pliocene is present lower in the sequence. In shallow-water facies *G. obliquus extremus* may be absent, and in these cases the middle Pliocene cannot be separated from the upper Pliocene to Recent.

C. LOWER PLIOCENE

The extinction of *Globoquadrina altispira altispira* (Cushman and Jarvis), *Sphaeroidinellopsis seminulina* (Schwager), or *S. subdehiscens* (Blow) is used to define the top of the lower Pliocene. In cores or outcrop samples, the association of these species with the Pliocene to Recent benthonic or planktonic species listed above, can be used to define the lower Pliocene.

In shallow-water facies where planktonic species are absent it is not possible to distinguish the lower Pliocene from the middle or upper Pliocene and only a general Pliocene to Recent age can be given on the basis of the benthonic species cited above.

D. UPPER MIOCENE

The top of the upper Miocene is defined by the extinction point of *Globoquadrina venezuelana* Hedberg. Blow (1969) shows the extinction of *G. venezuelana* as falling within the basal Pliocene so there is very little difference between the top of the upper Miocene as defined by Blow and as defined in this paper. The late upper Miocene can be defined in cores and outcrop samples by the association of *G. venezuelana* and *Globorotalia tumida* (Brady), while a general upper Miocene age can be defined on the association of *G. venezuelana* with *Turborotalia acostaensis* Blow. The extinction of *G. venezuelana* has been found to be a very reliable marker horizon.

E. MIDDLE MIOCENE

The top of the middle Miocene has proved to be difficult to recognize. *Globoquadrina dehiscens* (Chapman, Parr and Collins) has been found in a few cases to have its extinction point below that of *Globoquadrina venezuelana* and above the extinction of *Turborotalia mayeri* (Cushman and Ellison). For this reason the extinction point of *G. dehiscens* is taken tentatively to mark the top of the middle Miocene. The extinction of *T. mayeri* being earlier than that of *G. dehiscens* is taken to mark a horizon about the middle of the middle Miocene, which has proved to be a very good marker horizon.

The correlation of the Indonesian Letter Ages, based on planktonic Foraminifera, used in this paper is a combination of the proposals of Adams (1970) and Clarke and Blow (1969). Adams (1970, p. 120, fig. 30) tentatively shows *Miogypsina* spp. becoming extinct at the top of the "lower f" and *Lepidocyclina* (*Nephrolepidina*) spp. ranging only tentatively into the "upper f," while Clarke and Blow (1969, p. 89) shows these genera ranging positively into the "Tf3" or "upper f." Concerning *Lepidocyclina* (*Nephrolepidina*), Adams also states (1970, p. 117), " . . . it is almost certain that it does not extend to the top of *Tf*. However, since *Tg* is normally defined by the absence of *Nephrolepidina*, its occurrence high in *Tf* has at present to be assumed." My own work has not proven these genera to range above the "lower f" and the positive ranges of these genera as shown by Adams are therefore accepted. Adams equates his "lower f" with all of the middle Miocene while Clarke and Blow equate the "lower f" with the early middle Miocene. My own observations support the correlation of Clarke and Blow so *Miogypsina* spp. and *Lepidocyclina* (*Nephrolepidina*) spp. are considered to have become extinct at the end of the early middle Miocene in the Indo-Pacific region.

III. The Foraminiferal Faunas

A. MICHAELMAS CAY BORE

Chapman (1931) listed his determinations of foraminiferal species obtained from the samples from Michaelmas Cay. The sampling from this bore was poor so the species distributions shown by Chapman cannot be taken as reliable. Chapman was wise in not attempting any age determinations from the fossils.

The faunas are shown to be poor above 348 ft but better developed between 348 ft and total depth at 600 ft. The main species recorded over the bottom part of the hole belonged to the genera *Operculina*, *Elphidium*, *Cellanthus*, *Amphistegina*, and *Pseudorotalia*. Species included *Operculina bartschi, Cellanthus craticulatum, Amphistegina quoyi,* and *Pseudorotalia schroeteriana,* a fauna typical of a tropical shallow water back-reef environment of early Pliocene to Recent age in the Indo-Pacific region. The precise age of any part of the section cannot be determined, but no part of the section is, at least, older than Pliocene, based on the occurrence of *Cellanthus craticulatum, Operculina bartschi,* and *Pseudorotalia schroeteriana,* which do not range below the Pliocene.

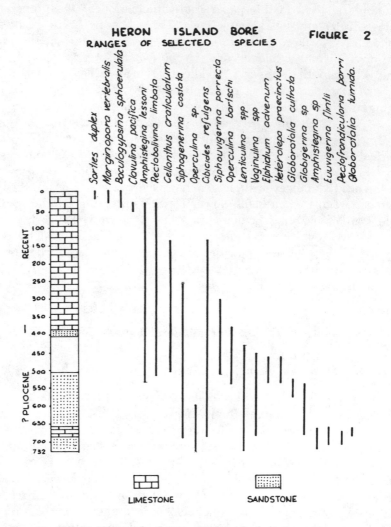

Fig. 2. Heron Island Bore, ranges of selected species.

B. HERON ISLAND BORE

Cushman (1942) made detailed lists of his determinations of the foraminiferal species extracted from the samples from the Heron Island bore, illustrating a few of the species. Lloyd (1967, 1970) described a part of the fauna. The ranges of selected species are set out on Fig. 2.

The assemblages are mostly poor in numbers of specimens but diverse in number of species. The preservation is generally poor, preventing positive identification of many of the species. The fauna is similar to that found in the Michaelmas Cay bore, but includes some deeper water genera such as *Brizalina* and *Euuvigerina* and rare planktonic species, *Globorotalia tumida, G. cultrata* (d'Orbigny), and *Globigerina* sp. The assemblages down to 678 ft included *Operculina bartschi, Cellanthus craticulatum*, and *Amphistegina quoyi* which do not occur below the Pliocene in the Indo-Pacific region. The faunas from 678 ft to final depth of 732 ft do not contain any age diagnostic species.

Cushman did not attribute any age to the section. Iredale (1942), on the basis of molluscs, considered that the section was entirely Recent in age. Jones and Endean (1967, p. 5), quoting a personal communication from myself, show an unconformity between the Recent/Pleistocene and Miocene at 131 m. Lloyd (1967, 1970) gave taxonomical notes on some of the Heron Island bore Foraminifera and tentative ages to the section—Recent to Pleistocene from the surface to 445 ft and Miocene from 445 ft to total depth of 732 ft. These ages are now considered to be incorrect, based on more recent findings. It is now considered that precise ages cannot be assigned to any part of the Heron Island bore section, but that most, if not all, of the section lies somewhere between Pliocene and Recent.

For details of Michaelmas Cay and Heron Island Bores reference can by made to Richards (1938) and Richards and Hill (1942).

C. WRECK ISLAND BORE

1. Previous Work

Details of the lithologies penetrated by the Wreck Island bore are set out in Maxwell (1962, p. 222). Derrington (1960) discussed the history of the well and briefly described the rock types, while Crespin (1960) gave a brief account of the foraminiferal faunas. Crespin ascribed tentative ages of lower Miocene to upper Pliocene (or younger) to the sequence between 1750 and 550 ft; the presence of lower Miocene was proven on the basis of *Lepidocyclina* found at 1150 ft, but the section above that depth was dated by Derrington on lithological criteria with the assumption that a continuous section is present. Lloyd (1961), in a preliminary study of the "smaller" Foraminifera, considered that the sequence from 530 to 1795 ft is Miocene and unconformably overlain by a Pleistocene to Recent sequence from 530 ft to surface. These ages were based mainly on the benthonic species; there was an absence of published guidelines using either planktonic or benthonic species at that time. Lloyd (1967, 1970) and Jones and Endean (1967) maintained this tentative dating. This dating can now be revised in the light of more recent publications, mainly Banner and Blow (1967), Blow (1969), Clarke and Blow (1969), and Adams (1970), and my own observations since the completion of those manuscripts and their subsequent publication. A comparison of these datings is presented on Fig. 3.

2. Nature and Age Significance of the Faunas

The faunal distribution of Wreck Island bore is set out in Fig. 4.

a. ?Upper Pliocene to Recent, 530 ft to Surface. The section between 530 ft and the surface could not be examined because of the absence of samples over this interval. It was previously thought that a disconformity was present at 530 ft, but because of the lack of positive evidence and in the light of the evidence from more recent drilling in the Gulf of Papua, it is thought that a continuous sequence possibly exists from 900 ft to surface. This sequence cannot be subdivided on the basis of the Foraminifera present.

b. ?Upper Pliocene, 900–530 ft. The faunas between 530 and 900 ft include *Globigerinoides quadrilobatus quadrilobatus* (d'Orbigny), *G. ruber* (d'Orbigny), *G. conglobatus* (Brady), *Turborotalia dutertrei* (d'Orbigny), *Globorotalia tumida, Pulleniatina obliqueloculata obliqueloculata, Elphidium parri* Cushman, *Cellanthus craticulatum, Amphistegina quoyi, Operculina bartschi, Cibicides refulgens* Montfort, *Parrelina* sp. A., *Parrelina* sp. B., *Plectofrondicularia parri* Finlay, *Vaginulina gippslandicus* Chapman and Crespin, *Heterolepa praecinctus* (Karrer), and *H. mediocris* (Finlay); benthonic species predominate. Apart from *Operculina bartschi, Cellanthus craticulatum,* and *Pulleniatina obliqueloculata obliqueloculata,* which give a general Pliocene to Recent age, no age diagnostic species are present which would permit an accurate age determination. The section between 900 and 530 ft is placed tentatively in the upper Pliocene by virtue of its position above positive middle Pliocene at 900 ft.

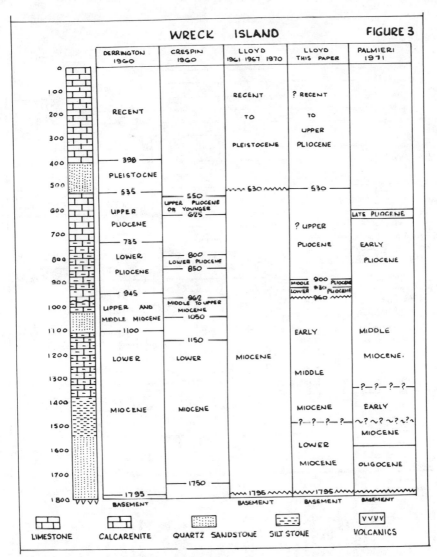

Fig. 3. Wreck Island Bore, correlation chart.

c. *Middle Pliocene, 930–900 ft.* The top of the middle Pliocene is placed at 900 ft, based on the highest occurrence of *Globigerinoides obliquus extremus* at that depth. Planktonic species are well developed at 900 ft; above this depth the number of planktonic species decrease considerably. The main species present are *Globigerinoides quadri-*

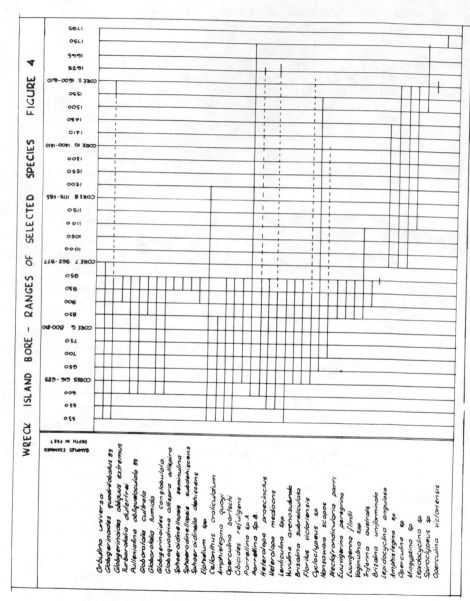

Fig. 4. Wreck Island Bore, ranges of selected species.

lobatus quadrilobatus, G. *quadrilobatus sacculifer* (Brady), G. *ruber,*
G. *obliquus extremus, Turborotalia dutertrei, Globorotalia cultrata,* G.
tumida, Orbulina universa d'Orbigny, *Amphistegina quoyi, Operculina
bartschi, Heterolepa praecinctus, Vulvulina arenasuturata* (LeRoy),
Euuvigerina peregrina (Cushman), and E. *flintii* (Cushman); all of
these forms being common to abundant.

d. *Lower Pliocene, 960–930 ft.* The lower Pliocene age is based on
the occurrence of *Globoquadrina altispira altispira, Sphaeroidinellopsis
seminulina,* and S. *subdehiscens,* which do not extend above the lower
Pliocene, and the species *Pulleniatina obliqueloculata obliqueloculata*
and *Sphaeroidinella dehiscens,* which range from early Pliocene to Recent; the last two species are common within this interval and only
rare above 930 ft. The main species present within the lower Pliocene
section are *Globigerinoides quadrilobatus quadrilobatus,* G. *quadrilobatus
sacculifer,* G. *obliquus extremus, Turborotalia dutertrei, Pulleniatina
obliqueloculata obliqueloculata, Globorotalia tumida,* G. *cultrata,
Globoquadrina altispira altispira, Orbulina universa, Sphaeroidinella
dehiscens, Sphaeroidinellopsis seminulina,* S. *subdehiscens, Operculina bartschi, Heterolepa praecinctus,* H. *mediocris, Vulvulina arenasuturata, Brizalina subreticulata* (Parr), B. *uniforminata* (Le Roy),
Trifarina australis (Heron-Allen and Earland), *Cibicides* n. sp., and
Vaginulina hirsuta (d'Orbigny); all except S. *subdehiscens* are common
to abundant. The faunas at 950 and 930 ft are well developed and
there is no disputing the age of the fauna.

e. *Early Middle Miocene, ?1500–960 ft.* The top of the early middle
Miocene is placed at 960 ft because of the marked faunal break at
this depth. The highest occurrence of *Lepidocyclina* (*Nephrolepidina*)
angulosa Provale in core 7 (963 ft) is the basis of the early middle
Miocene age. The L. (N.) *angulosa* occurs in abundance between 1175
and 1410 ft and is rare above 1175 ft. *Lepidocyclina* (N.) sp. is rare
to common from 1450 to 1500 ft.

Rare *Miogypsina* sp. occur from 1150 to 1500 ft. Poorly preserved
specimens of *Elphidium* sp., *Amphistegina* sp., and *Operculina* sp. occur
over most of the interval. *Parrellina* sp. A [a new species referred to
as a species of *Elphidium,* probably new but closely related to *Elphidium
reginum* (d'Orbigny) var. *caucasicum* Bogdanowicz by Crespin (1960)]
is rare throughout the interval.

f. *Lower Miocene, 1750–?1500 ft.* The occurrence of *Spiroclypeus* sp.,
Miogypsina sp., and *Lepidocyclina* (*Nephrolepidina*) sp. at 1625 ft indi-

cate an upper "Te" age which is equated with the lower Miocene. The top of the lower Miocene is placed tentatively at 1500 ft on the basis of a lithological change at this depth and the fact that most of the fossils below 1500 ft are completely replaced by or infilled with glauconite.

Lepidocyclina (*Nephrolepidina*) sp., is rare to common between 1500 and 1750 ft, while *Miogypsina* sp., is rare between 1500 and 1625 ft. The foraminiferal assemblage at 1625 ft is richly developed, and besides the genera listed above, contains numerous specimens of *Hanzawaia scopos* (Finlay), *Elphidium* sp., *Parrellina* sp. B, and *Operculina victoriensis* Chapman and Parr. *Operculina victoriensis* and *Parrellina* sp. A occur in abundance at 1750 ft. *Parrellina* sp. B and *Parrellina* sp. A both occur above 960 ft in the Pliocene section.

Numerous dwarfed specimens of planktonic Foraminifera occur in core 11 (1600–1610 ft) together with a few larger specimens. There appears to be only one species present which is closely comparable with *Globigerina* sp., cf. *G. trilocularis* of Bolli (1957, pl. 22, fig. 8, 9).

D. ANCHOR CAY NO. 1 WELL

The samples from Anchor Cay No. 1 Well were not very satisfactory for detailed stratigraphic control. Each sample represented a 100-ft interval and there were no samples from above 1500 ft available for examination. Some of the boundaries could therefore only be placed tentatively. The faunal distribution from Anchor Cay No. 1 Well is set out in Fig. 5.

1. Pliocene, 1500–?2800 ft

The occurrence of the benthonic species *Cellanthus craticulatum* between 1500 and 2200 ft indicates that this interval is not older than Pliocene. Planktonic species are rare over this interval and there is an absence of index species which would permit accurate age determinations and finer subdivisions. Lower Pliocene limestones lie disconformably on early middle Miocene limestones in the Wreck Island bore, and recent drilling adjacent to and in the Gulf of Papua has revealed lower Pliocene clastic sediments lying directly on early middle Miocene limestones; it is therefore considered that the section between 1500 and 2200 ft is Pliocene in age. The base of the Pliocene cannot be placed accurately, but it is put tentatively at 2800 ft because of the discovery of *Lepidocyclina* in the sample from 2800 to 2900 ft. The species between 2200 and 2800 ft are long-ranging and cannot be used for age determina-

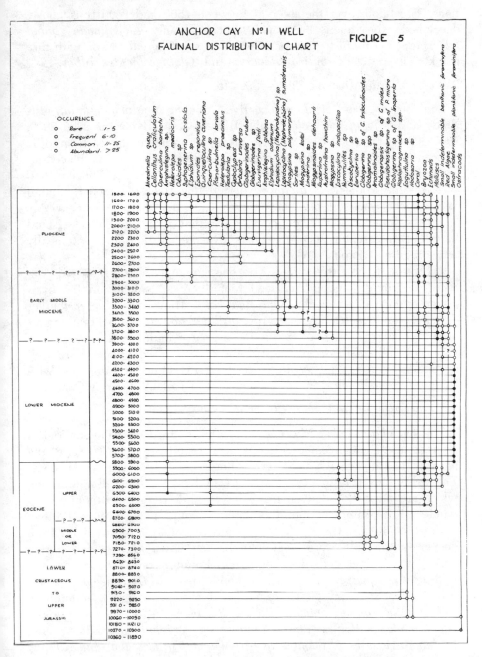

Fig. 5. Anchor Cay No. 1 Well, faunal distribution chart.

tions. The fauna over the interval between 1500 and 2800 ft is poorly developed, the main benthonic species being *Alveolinella quoyi* (d'Orbigny), *Amphistegina* sp., *Elphidium* sp., *Operculina* sp., with rare planktonic species *Orbulina universa* and *Globigerinoides ruber*.

2. Early Middle Miocene, ?2800–?3900 ft

The occurrences of *Lepidocyclina* (*Nephrolepidina*) *sumatrensis* Brady and *Miogypsina polymorpha* Rutten, respectively, at 2800–2900, 2900–3000, and 3300–3400 ft are taken to indicate an early middle Miocene age, and the top of the early middle Miocene is placed at 2800 ft on this evidence. The disconformity between the Pliocene and early middle Miocene is therefore placed tentatively at 2800 ft, giving an increased thickness of Pliocene to Recent limestones at Anchor Cay and thus showing a marked contrast to that found in Wreck Island bore; but this greater thickness can be expected in the more tectonically active region of Papua.

The base of the early middle Miocene cannot be defined on palaeontological grounds and it is placed tentatively at 3900 ft on the basis of a lithological change. *Miogypsina kotoi* Hanzawa, *Miogypsinoides dehaarti* van der Vlerk?, and *Austrotrillina howchini* (Schlumberger) occur within the interval placed in the early middle Miocene.

The overall fossil assemblages of the early middle Miocene are well developed with red algae often occurring in abundance together with coral, bryozoa, molluscan fragments, and "smaller" Foraminifera.

3. Lower Miocene to Upper Oligocene, ?3900–5850 ft

The top of the lower Miocene cannot be defined on palaeontological grounds and for convenience it is placed tentatively at 3900 ft on the basis of a lithological change from limestone to calcareous claystone. The calcareous claystone is characterized by planktonic Foraminifera, which are quite abundant at many levels, and by rare benthonic Foraminifera. The samples from this interval could not be broken down to free the fossils. Cavings of this calcareous claystone obtained from deeper in the hole were eventually broken down and yielded a rich fauna which included the benthonic Foraminifera *Lenticulina* sp., *Laticarinina pauperata* (Parker and Jones), *Hanzawaia scopos*, *Euuvigerina* sp., and *Cibicides* sp., and the planktonic Foraminifera *Turborotalia opima opima* (Bolli), *T. mayeri?*, *T. obesa* (Bolli), *Globorotalia variablis* Bolli, *Globigerina praebulloides* Blow, and *Globigerinoides quadrilobatus quadrilobatus*. *Turborotalia opima opima* has a short restricted range within the upper Oligocene and *G.*

quadrilobatus ranges from the early lower Miocene to Recent. This fauna therefore proves that at least a part of the calcareous claystone sequence is upper Oligocene and early Miocene in age.

4. Upper Eocene, 5850–?6800 ft

The occurrence of *Discocyclina indopacifica* Hanzawa in the sample from 5800 to 5900 ft is taken to define the top of the upper Eocene at about 5850 ft. The nature of the Oligocene/Eocene boundary cannot be determined, but is thought to be disconformable. *Discocyclina indopacifica* was found down to 6800 ft and the lower limit of the upper Eocene is placed tentatively at this depth.

The overall fossil assemblage is characterized by an abundance of *D. indopacifica* in some limestone chips, together with abundant bryozoa, red algae, echinoids, and the Foraminifera *Operculina* sp., *Amphistegina* sp., and *Nummulites* sp.

5. Middle or Lower Eocene, ?6800–?7300 ft

A number of small planktonic foraminiferal species and a single benthonic species were found between 7090 and 7300 ft. The specimens are mostly iron-stained similar to much of the accompanying sand grains, proving that the fossils were in place. Most of the specimens are poorly preserved, thus preventing positive identifications. The single benthonic species belongs to the genus *Anomalinoides*. The planktonic species include forms comparable with *Globigerina triloculinoides* Plummer, *Globigerapsis index* (Finlay), and *Pseudohastigerina micra* (Cole), the last being the best preserved. The absence of positive identifications does not permit an accurate age determination, but the genus *Pseudohastigerina* indicates an Eocene–early Oligocene age. The presence of upper Eocene limestones above suggests a possible lower or middle Eocene age.

Lithologically there does not appear to be any difference between the possible lower or middle Eocene and the underlying Mesozoic and the boundary between the two can be placed, but only tentatively, at 7300 ft. Indurated black shales between 6800 and 7003 ft were found to be barren, but are included in the possible lower or middle Eocene and the top of this section is placed tentatively at 6800 ft.

6. Lower Cretaceous to Upper Jurassic, ?7300–11890 ft

The Lower Cretaceous to Upper Jurassic age is based on the occurrence of a species of *Eoguttulina* between 9130 and 10,000 ft together with specimens of *Nodosaria, Haplophrugmoides*, and *Hyperammina;*

these forms are common in the Lower Cretaceous and Upper Jurassic of Papua and the Lower Cretaceous of Queensland.

IV. Geological History of the Great Barrier Reefs

Jones (1966) raised a number of geological questions about the Great Barrier Reefs to which he was seeking answers. These included the origin, nature, and history of the platform on which the reefs are built and movements of sea level relative to the land as have occurred since the growth of the reefs began. Answers to these questions have long been sought and it was to this end that the Great Barrier Reef Committee drilled Michaelmas Cay and Heron Island bores. The first answers to some of these questions, however, did not come until the drilling of Wreck Island bore.

Subsequent drilling of Aquarius No. 1 (22°37′13″S; 152°39′02″E) and Capricorn No. 1A (22°42′14″S; 152°16′55″E) Wells off-shore central Queensland, Anchor Cay No. 1 Well and other wells on-shore and off-shore in the Gulf of Papua coupled with great advances in studies of worldwide Tertiary stratigraphy now permit a better understanding of the geological history of the Great Barrier Reefs.

In this discussion the term northeastern Australian continental shelf is used as including a narrow coastal strip along the Queensland coast north of Gladstone, Torres Strait, and the southwest of Papua south of the Fly River, as well as the continental shelf proper; the term is used in this sense as it is considered that the whole was an entity throughout the Tertiary.

A. MESOZOIC

Anchor Cay No. 1 Well is the deepest well so far drilled in the Great Barrier Reefs province and the first to penetrate Eocene and Mesozoic marine sediments. The Mesozoic sediments belong to the Lower Cretaceous to Upper Jurassic and are closely related to sediments of this age occurring in Papua and Queensland. The volcanics penetrated at the base of Wreck Island bore (Maxwell, 1962) and volcanics and tuffaceous sediments penetrated at the bottom of Aquarius No. 1 and Capricorn No. 1A Wells (Palmieri, 1971) are also placed in the Lower Cretaceous. From this scattered information it seems likely that Mesozoic rocks are widespread, at depth, along and off the coast of Queensland and in the Gulf of Papua.

Upper Cretaceous sediments have not yet been proven in this area and it may well be that a hiatus existed at this time in this region.

B. EOCENE

Eocene strata so far have been penetrated only in the Gulf of Papua, with a possibly complete section at Anchor Cay No. 1 Well, represented by deep-water marine sediments of early or middle Eocene age and upper Eocene shallow-water shoal limestones. The early or middle Eocene sediments cannot be related to any other sediments of this age in Western Papua or Queensland. Upper Eocene shoal limestones are widespread in Western Papua, but have not yet been found to the south. It is possible therefore, that a hiatus existed along and off most of the coast of Queensland during the Eocene, with this area at a standstill while sedimentation took place to the north in the tectonically more active Papuan area.

C. OLIGOCENE

There was possibly a hiatus along the entire northeastern Australian continental shelf during the early and middle Oligocene with the possible exception of the Capricorn No. 1A and Aquarius No. 1 areas in the south where a sequence of paralic and deltaic, sparsely fossiliferous sandstones were deposited during this time. Sedimentation continued in the Aquarius No. 1, Capricorn No. 1A areas during the upper Oligocene with the deposition of marine clays and sands; a part of the paralic and deltaic sediments may also have been deposited during the early part of the upper Oligocene.

Subsidence and sedimentation recommenced within the upper Oligocene in the Anchor Cay area after a break during the early and middle Oligocene; the sediments are marine calcareous clays deposited under open oceanic conditions.

During the Oligocene, subsidence and sedimentation was apparently restricted to the tectonically active northern and southern ends of the northeastern Australian shelf; the intervening areas appeared to have been stable at this time and did not receive sediments.

D. LOWER MIOCENE

The Anchor Cay area continued to subside and receive calcareous clay sediments during the lower Miocene as it did during the upper Oligocene. Subsidence began at this time in the Wreck Island area with the deposition of shallow-water marine, glauconitic, fossiliferous quartzose sandstones. In the Aquarius No. 1 and Capricorn No. 1A areas, there was subsidence accompanied by the deposition of marine argillaceous sands, calcareous clays, and argillaceous limestones; this was a continuation of the upper Oligocene conditions.

The lower Miocene was the beginning of the widespread, but slow subsidence of the inner part of the northeastern Australian shelf which was accompanied by a slow rate of sedimentation. More rapid subsidence and sedimentation continued in the more tectonically active areas at the northern and southern ends and outer regions of the shelf.

E. Middle and Upper Miocene

Widespread subsidence along the inner and outer parts of the shelf continued during the early middle Miocene with the deposition of limestones which belong, in the main, to sheltered back-reef facies. These limestones are present in Wreck Island bore, Aquarius No. 1, Capricorn No. 1A, Anchor Cay No. 1, and in outcrop and subsurface in southwest Papua. These widespread limestones, belonging to reefal facies, may represent a Miocene reef development similar to the inner reefs present along the Queensland coast today. Fore-reef facies are rare in these limestones and it is considered that the equivalent of the present day outer Barrier Reefs has not yet been located off the coast of Queensland.

Late middle Miocene and upper Miocene sediments are absent in Wreck Island bore and Anchor Cay No. 1 Well and in outcrop and subsurface in southwest Papua, but are present as limestones in Aquarius No. 1 and Capricorn No. 1A Wells. This late middle and upper Miocene hiatus reflects a standstill along the entire length of the northeastern Australian shelf during this period, except for the Capricorn–Aquarius area which remained tectonically active at this time and continued to subside and receive sediments.

F. Pliocene to Recent

During the lower Pliocene there was a renewal of widespread subsidence similar to that of the early middle Miocene with predominantly limestone deposition along the northeastern Australian shelf. Lower Pliocene limestones are present in Anchor Cay No. 1, Wreck Island bore, Aquarius No. 1 Well, and Capricorn No. 1A Well, and possibly in Michaelmas Cay and Heron Island bores. The limestones in Anchor Cay No. 1 Well, Michaelmas Cay bore, and Heron Island bore contain only rare planktonic Foraminifera and reflect partially sheltered shoal reefal conditions, whereas the Wreck Island bore, Aquarius No. 1 Well, and Capricorn No. 1A Well limestones contain numerous planktonic Foraminifera and reflect fore-reef facies.

Except for brief periods of quartzose sandstone deposition in the Wreck Island and Heron Island areas during the late Pliocene, limestone or calcareous clays were deposited in the Wreck Island, Heron Island, Aquarius No. 1, Capricorn No. 1A, Michaelmas Cay, and Anchor Cay areas throughout middle Pliocene to Recent times. The southwest Papuan area was uplifted in the late Pliocene after a brief submergence during the Pliocene when shallow-water reef detrital sands and limestones were deposited.

G. Origin of the Great Barrier Reefs

Conditions similar to those existing today along the inner parts of the Great Barrier Reefs first came into existence along the northeastern Australian shelf during the early middle Miocene. There was a break in these conditions along most of the shelf during the late middle and upper Miocene, but reefal conditions appeared to have continued in the vicinity of Aquarius No. 1 and Capricorn No. 1A Wells during this time. In the lower Pliocene, conditions similar to the area of the outer Great Barrier Reefs came into existence and continued to the present time, possibly, however, with interruptions and even erosion at times in the Pleistocene.

Acknowledgments

Special thanks are extended to Dr. O. A. Jones of the Great Barrier Reef Committee for his kind invitation to write this paper and for his continued encouragement. Permission to publish the information on Anchor Cay No. 1 Well by Tenneco Australia Inc., is greatfully acknowledged. Thanks are also extended to my wife Beverley Lloyd for encouragement and assistance with the typing; to Dr. D. J. Belford and Mr. P. J. Jones of the Bureau of Mineral Resources, Geology and Geophysics, Canberra, for reading the original manuscript and their helpful criticisms; and to Mr. G. State of Sydney for the drafting of the figures.

References

Adams, C. G. (1970). *Bull. Brit. Mus.* (*Natur. Hist.*), *Geol.* 19(3), 85–137.
Banner, F. T., and Blow, W. H. (1967). *Micropaleontology* 13, 133–162.
Blow, W. H. (1969). *Proc. Int. Conf. Planktonic Microfossils, 1st, 1967* Vol. VI, pp. 199–421.
Bolli, H. M. (1957). *U.S., Nat. Mus., Bull.* 215, 97–123.

Chapman, F. K. (1931). *Rep. Gt. Barrier Reef Comm.* 3, 32–42.
Clarke, W. J., and Blow, W. H. (1969). *Proc. Int. Conf. Planktonic Microfossils, 1st, 1967* Vol. VI, pp. 82–97.
Crespin, I. (1960). *Bur. Miner. Resour. Aust. Petrol. Search Subs. Acts Publ.* 4, 12.
Cushman, J. A. (1942). *Rep. Gt. Barrier Reef Comm.* 5, 112–119.
Derrington, S. S. (1960). *Bur. Miner. Resour. Aust. Petrol. Search Subs. Acts Publ.* 4, 1–15.
Iredale, T. (1942). *Rep. Gt. Barrier Reef Comm.* 5, 120–122.
Jones, O. A. (1966). *Aust. Natur. Hist.* pp. 245–249.
Jones, O. A., and Endean, R. (1967). *Sci. J.* 2–9.
Lloyd, A. R. (1961). Unpublished Thesis, University of Adelaide.
Lloyd, A. R. (1967). *Bull. Bur. Miner. Miner. Resour. Aust.* 92, 69–112.
Lloyd, A. R. (1970). *Bull. Bur. Miner. Resour. Aust.* 108, 145–203.
Maxwell, W. G. H. (1962). *J. Geol. Soc. Aust.* 8, 217–238.
Palmieri, V. (1971). *Rep. Geol. Surv. Queensl.* 52.
Richards, H. C. (1938). *Rep. Gt. Barrier Reef Comm.* 2, 135–142.
Richards, H. C., and Hill, D. (1942). *Rep. Gt. Barrier Reef Comm.* 5, 1–111.
Traves, D. M. (1960). *J. Geol. Soc. Aust.* 7, 369–371.

AUTHOR INDEX

Numbers in italics refer to the pages on which the complete references are listed.

SUBJECT INDEX

Abbreviation Qd. for Queensland area.

A

Algae, see also Systematic Index
    calcareous, see Reef communities
    coralline, 1, 24–26, 31, 41–43
Algal mats, 38
Ancient conglomerate, see Coral
    conglomerate
Ancient reef, see Coral conglomerate
Andesite line, 113
Aragonite, 102, 123
Atolls, see Reef types, Locality Index
Australia–New Guinea massif, 170
Australian C.S.I.R.O., 189

B

Basalt, 275, 278, 280, 291, 294
Basins, see Tectonic features
Bathythermograph, 190, 199
Beach rock, see Reef features
Bedrock, older Pleistocene, 38
Bioherms, 1
Boreholes, bores, 122, 123, 129, 136, 139,
    146, 164–165
Bores, deep, Great Barrier Reef Province
    Anchor Cay 1, 348, 358–365
    Aquarius 1, 235, 264, 266, 362–365
    Capricorn 1A, 235, 264, 266, 362–365
    Heron Island, 265, 266, 267, 347,
        352–353, 362
    Michaelmas Cay, 347, 351–353, 362,
        364
    Wreck Island, 235, 266, 348, 353–356,
        358, 360, 362
Bouger anomaly, 282, 294
Boulder ramparts, see Reef features
Burdekin shelf, 290
Buttresses, see Reef features

C

Calcite, 102
Calcium carbonate saturation, 190, 220,
    221
Capricorn trough, see continental shelf,
    Qd., southern shelf embayment;
    troughs; Locality Index, Capricorn
    Channel
Carbon-14 dates, see Radiometric
    dating
Carbonate:non-carbonate ratio, 305, 306,
    327
Caribbean corals, 14–15
Cays
    carbonate, 36–40
    Caribbean, 36–40
    leeward, 40
    mud, 36, 38
    mud carbonate, 38
    Pacific, 38, 40
    pre-Holcene bedrock, 36, 38
    sand, 26, 36, 38, 40, 43
Chagos-Laccadive ridge, 59
Chesterfield trough, see Corridors;
    Locality Index, Townsville Trough
Climatic changes, effects on reefs, 41–42
Coastal plains, 240–243
Coastal ranges, 240, 242, 243
Community zones, see Reef communities
Continental drift, 291, 292, 296, 297
Continental run off, 196, 199, 200, 203,
    204, 206, 207, 216, 230
Continental shelf, Queensland
    axial zone, 327
    deeper shelf zone, 312
    eastern marginal shelf, 244, 263
    inner shelf, 243, 244, 294, 295, 304,
        308, 309, 317, 326, 327
    main reef zone, 263
    marginal reef zone, 312
    marginal shelf, 243, 244, 301

376

384

W

Water
  carbonate-enriched, 219, 220
  cellular motion, 227, 228
  circulation, 219, 220
  density, 224, 226
  isohaline, 206, 217, 218, 229, 230
  isothermal, 199, 218, 229, 230
Water masses, 208
  Antarctic deep, 224
  Antarctic intermediate, 222, 227
  boundary between, 204, 208, 212
  coastal (Australian), 200, 204
  Coral Sea, 198, 200, 203, 204, 206,
      207, 210, 211, 213, 216–231
  Coral Sea bottom, 224
  downsinking of, 222
  equatorial, 213
  Gulf of Carpentaria, 204, 212
  Gulf of Papua, 200, 207, 227
  high salinity, 200, 204, 227, 230
  low salinity, 206, 214, 227, 230, 231
  mixing of, 199, 203, 208, 211, 230
  Papuan, 200, 231
  Papuan–northern Torres Strait, 200
  phosphorus content of, 227
  seasonal, 200
  shelf, 200, 203, 204, 207, 211, 212,
      216, 218, 220, 230

south equatorial, 222, 224, 227
southern Papuan-Fly River coastal,
    231
Subtropical, 213
Torres Strait, 200, 212, 214
uplifting of, see Upwelling of
upper salinity maximum, 227
upwelling of, 214, 219, 222, 227, 229,
    231
west central, South Pacific, 224
West Irian, 200, 207, 231
Winds, 56, 57
  monsoon, 56
  monsoonal circulation, 56
  monsoonal winds, 301
  trade wind(s), 56, 57, 97, 115, 131,
      144, 155, 156, 158, 300, 301
  northwest monsoon, 190, 191, 192,
      199, 200, 203, 204, 206, 207, 213,
      214, 216, 217, 222, 230, 231
  southeast trade, 190, 191, 194, 198,
      199, 203, 206, 207, 213, 216, 222,
      230
  southeasterlies, 191
Wreck Island, 235, 264, 265, 266, see
    also Bores, deep, Great Barrier Reef
    Province; Locality Index

Z

Zonation of reefs, see Reef zones

LOCALITY INDEX

Abbreviations used:

Car.   Caribbean-Florida area
Fr. Pol.   Fench Polynesia
Ind.   Indian Ocean area
Mar.   Marshall Islands
N. Cal.   New Caledonia, Loyalty Islands
N.G.S.   New Guinea, Solomon Islands area
Qd.   Queensland, Great Barrier Reef Province
*   Denotes a figure or map

A

Abaco Reef, Bahamas, Car., 34
Addu Atoll, Ind., 53, 54, 56, 62, 64,
    67, 69, 71, 74, 75; 80, 82–85 (reef
    zonation)
Agalega (islands), Ind., 55, 56, 58, 63
Ailinginae Atoll, Mar., 95
Akiaki Atoll, Fr. Pol., 125
Alacran Reef, Car., 7, 13, 15; 20
    (geomorphic form), 24, 27, 29, 30;
    34, 35 (composition of sediments),
    41
Albuquerque Cays, Car., 18,* 20*
    (geomorphic form), 27; 34, 35, 39
    (composition of sediments)
Aldabra Atoll, Ind., 53, 54, 56, 58, 60,
    64, 67, 72, 73, 74, 75, 86
Amazon River, 1
Ambergris Cay, Car., 39 (composition
    of Cay sediments)
Ami Island, N. Cal., 156
Amirante Islands, Ind., 53, 55, 58, 60, 74
Anaa Atoll, Fr. Pol., 124, 125
Anchor Cay, Qd., 348 (oil exploration
    bore 1969)
Andaman Islands, Ind., 53, 55, 58, 61,
    67, 69, 74, 86
Andaman Sea., Ind., 60
Andros Islands, Bahamas, Car., 23, 34
    (composition of sediments), 36, 38
Annan River, Qd., 309 (possible source
    of sand)
Arabia, 61

Arabian Sea, 56, 58
Arafura Sea, 169, 172, 175, 185, 213,
    220
Archipelago see individual names
Arlington Reef, Qd., 333
Arlington Reef complex, Qd., 338; 340,*
    344
Arnhem Land, northern Australia, 173,
    185
Arno Atoll, Mar., 69
Aru Islands, N.G.S., 169, 185
Assumption Island, Ind., 56, 64
Astove Island, Ind., 64
Astrolabe Reefs (half-atolls), N. Cal., 160
Atoll das Racos, South Atlantic, 1
Atolls, see individual names
Austral Archipelago, Fr. Pol., 113, 115,
    136–138
Austral Chain (Guyots), South Pacific,
    182
Austral Islands see Austral Archipelago
Australia, 61, 74, 86

B

Badu Island, N.G.S., 172, 174
Bahama Banks, Car., 27, 65
Bahama Reefs, Car., 42
Bahamas (Islands), Car., 2, 4, 5, 10,
    15, 24, 30, 38; 39 (composition of
    sediments), 304 (contrast with
    Great Barrier Reef Province)
Baja Nuevo, Car., 20
Banc Farsan, Ind., 54
Barbados, Car., 7, 12, 15, 24, 30

385

SYSTEMATIC INDEX

396

References, Chapter 1

Adams, R. D. (1968). The leeward reefs of St. Vincent, West Indies. *J. Geol.* **76**, 587–595.

Agassiz, A. (1894). A reconnaissance of the Bahamas and of the elevated reefs of Cuba. *Bull. Mus. Comp. Zool. Harvard Univ.* **26**, 1–203.

Almy, C. C., Jr., and Carrion-Torres, C. (1963). Shallow water stony corals of Puerto Rico. *Carib. J. Sci.* **3**, 133–162.

Atwood, D. K., and Bubb, J. N. (1970). Distribution of dolomite in a tidal flat environment, Sugarloaf Key, Florida. *J. Geol.* **78**, 499–505.

Baars, D. L. (1963). Petrology of carbonate rocks. *Shelf Carbonates Paradox Basin, Symp., 4th Field Conf., 1963, Four Corners Geol. Soc.* pp. 101–129.

Ball, M. M. (1967a). Carbonate sand bodies of Florida and the Bahamas. *J. Sediment. Petrol.* **37**, 556–571.

Ball, M. M. (1967b). Tectonic control of the configuration of the Bahama Banks. *Trans., Gulf Coast Ass. Geol. Soc.* **17**, 265–267.

Ball, M. M., Shinn, E. A., and Stockman, K. W. (1967). The geologic effects of Hurricane Donna in South Florida. *J. Geol.* **75**, 583–597.

Bandy, O. (1964). Foraminiferal biofacies in sediments of Gulf of Batabano, Cuba, and their geological significance. *Bull. Amer. Ass. Petrol. Geol.* **48**, 1666–1679.

Banks, J. E. (1959). Limestone conglomerates (Recent and Cretaceous) in South Florida. *Bull. Amer. Ass. Petrol. Geol.* **43**, 2237–2243.

Bathurst, R. G. C. (1967a). Oolitic forms on low energy carbonate sand grains, Bimini Lagoon, Bahamas. *Mar. Geol.* **5**, 89–109.

Bathurst, R. G. C. (1967b). Subtidal gelatinous mat, sand stabilizer and food, Great Bahama Bank. *J. Geol.* **75**, 736–738.

Bavendamm, W. (1932). Die microbiologisch Kalkfällung in der tropischen See. *Arch. Mikrobiol.* **3**, 205–276.

Black, M. (1933a). The precipitation of calcium carbonate on the Great Bahama Bank. *Geol. Mag.* **70**, 455–466.

Black, M. (1933b). The algal sediments of Andros Island, Bahamas. *Phil. Trans. Roy. Soc. London, Ser. B* pp. 165–192.

Bock, W. D., and Moore, D. R. (1969). The foraminifera and micromollusks of Hogsty Reef and Serrana Bank and their paleoecological significance: *Carib. Geol. Congr., Prepr.*

Boyd, D. W., Kornicker, L. S., and Rezak, R. (1962). Recent algal bioherms near Cozumel Island, Mexico. *Geol. Soc. Amer., Spec. Pap.* **73**, 121–122 (abstr.).

Branner, J. C. (1904). The stone reefs of Brazil, their geological and geographical relations, with a chapter on the coral reefs. *Bull. Mus. Comp. Zool., Harvard Univ.* **44**, 1–285.

Broecker, W. S., and Takahashi, T. (1966). Calcium carbonate precipitation on the Bahama Banks. *J. Geophys. Res.* **71**, 1575–1602.

Brooks, H. K. (1962). Reefs and bioclastic sediments of the Dry Tortugas. *Geol. Soc. Amer., Spec. Pap.* **73**, 1–2 (abstr.).

Bryan, E. H., Jr. (1953). Check list of atolls. *Atoll Res. Bull.* No. 19, pp. 1–38.

Bubb, J. N., and Atwood, D. K. (1968). Recent dolomitization of Pleistocene limestone by hypersaline brines, Great Inagua Island, Bahamas. *Amer. Ass. Petrol. Geol., Bull.* **52**, 522 (abstr.).

Burkholder, P. R., and Burkholder, L. M. (1960). Photosynthesis in some Alcy-onacean corals. *Amer. J. Bot.* **47**, 866–872.

Cary, L. R. (1918). The Gorgonaceae as a factor in the formation of coral reefs. *Carnegie Inst. Wash., Dep. Mar. Biol. Pap., Publ.* No. 213, pp. 341–362.

Chevalier, J. P. (1966). Contribution a l'etude des Madreporaires des cotes occidentales de l'Afrique tropicale. *Bull. Inst. Fr. Afr. Noire, Ser. A* **28**, 912–975.

Clifton, H. E., Mahnken, C. V. W., van Derwalker, J. C., and Waller, R. A. (1970). Tektite 1, man-in-the-sea project: marine sciences program. *Science* **168**, 659–663.

Cloud, P. E., Jr. (1959). Geology of Saipan, Mariana Islands; Part 4, Submarine topography and shoal water ecology. *U.S., Geol. Surv., Prof. Pap.* **280-K**, 361–445.

Cloud, P. E., Jr. (1962). Environment of calcium carbonate deposition west of Andros Island, Bahamas. *U.S., Geol. Surv., Prof. Pap.* **350**, 1–138.

Curray, J. R. (1965). Late Quaternary history, continental shelves of the United States. *In* "The Quaternary of the United States" (H. E. Wright and D. G. Frey, eds.), pp. 723–735. Princeton Univ. Press, Princeton, New Jersey.

Daetwyler, C. C., and Kidwell, A. L. (1960). The Gulf of Batabano, a modern carbonate basin. *World Petrol. Congr., Proc., 5th, 1959* Sect. 1, Pap. 1, pp. 1–21.

Dalrymple, D. W. (1964). Recent sedimentary facies of Baffin Bay, Texas. Unpublished Thesis, Rice University, Houston, Texas.

Darwin, C. (1851). "The Structure and Distribution of Coral Reefs" (Reprinted by University of California Press, 1962).

Davis, J. H. (1940). The ecology and geologic role of mangroves in Florida. *Pap. Tortugas Lab.* **32**, 302–412.

De Buisonje, P. H., and Zonneveld, J. I. S. (1960). De kustvormen van Curacao, Aruba en Bonaire. *Natuurwetensch. Werkgroep Ned. Antillen* No. 11, pp. 121–144.

Deffeyes, K. W., Lucia, F. J., and Weyl, P. K. (1965). Dolomitization of Recent and Plio-Pleistocene sediments by marine evaporite waters on Bonaire, Netherlands Antilles: in Pray, L. C. and Murray, R. C. (eds.), Dolomitization and Limestone Diagenesis. *Soc. Econ. Paleontol. Mineral., Spec. Publ.* **13**, 71–88.

Doran, E. (1954). Land forms of Grand Cayman Island, British West Indies. *Tex. J. Sci.* **6**, 360–377.

Doran, E. (1955). Land forms of the southeast Bahamas. *Tex., Univ., Dep. Geogr., Publ.* **5509**, 1–38.

Drew, G. H. (1914). On the precipitation of calcium carbonate in the sea by marine bacteria, and the action of denitrifying bacteria in tropical and temperate seas. *Carnegie Inst. Wash. Publ.* **182**, 7–45.

Duane, D. B., and Meisburger, E. P. (1969). Geomorphology and sediments of the nearshore continental shelf, Miami to Palm Beach, Florida. *U.S. Army Corps Eng., Coastal Eng. Res. Cent., Tech. Memo* **29**, 1–47.

Duarte-Bello, P. P. (1961). Corales de los Arrecifes Cubanos. *Acuario Nac., Ser. Educ.* No. 2, 1–85.

Ebanks, W. J., Jr. (1967). Recent carbonate sedimentation and diagenesis, Ambergris Cay, British Honduras. Ph.D. Thesis, Rice University, Houston, Texas.

Ekman, S. (1953). "Zoogeography of the Sea." Sidgwick & Jackson, London.

Emery, K. O. (1962). Coral reefs off Veracruz, Mexico. *Geofis. Int.* **3**, 11–17.

Emery, K. O., Tracey, J. I., Jr., and Ladd, H. S. (1954). Geology of Bikini and nearby atolls. *U.S., Geol. Surv., Prof. Pap.* 260-A, 1–254.

Emiliani, C. (1954). Temperatures of Pacific bottom waters and polar superficial waters during the Tertiary. *Science* 119, 853–855.

Emiliani, C., and Flint, R. F. (1963). The Pleistocene record. *In* "The Sea" (M. N. Hill, ed.), Vol. 3, pp. 888–927. Wiley (Interscience), New York.

Environmental Science and Services Administration. (1968). "Surface Water Temperature and Density," C & GS Publ. No. 31-1. U.S. Department of Commerce.

Fleece, J. B. (1962). "The Carbonate Geochemistry and Sedimentology of the Keys of Florida Bay, Florida," Contrib. No. 5. Sediment. Res. Lab., Dep. Geol., Florida State University, Tallahassee.

Folk, R. L. (1962). Sorting in some carbonate beaches of Mexico. *Trans. N.Y. Acad. Sci.* [2] 25, 222–244.

Folk, R. L. (1967). The sand cays of Alacran Reef, Yucatan, Mexico; Morphology. *J. Geol.* 75, 412–437.

Folk, R. L., and Robles, R. (1964). Carbonate sands of Isla Perez, Alacran Reef Complex, Yucatan. *J. Geol.* 72, 255–292.

Folk, R. L., Hayes, M. O., and Shoji, R. (1962). "Carbonate Sediments of Isla Mujeres, Quintana Roo, Mexico and Vicinity." Guide Book to Field Trip to Peninsula of Yucatan, New Orleans Geol. Soc., Louisiana.

Fosberg, F. R. (1962). A brief study of the cays of Arrecife Alacran, a Mexican atoll. *Atoll Res. Bull.* No. 93, pp. 1–25.

Freeman, T. (1962). Quiet water oolites from Laguna Madre, Texas. *J. Sediment. Petrol.* 32, 475–483.

Garrett, P., Patriquin, D., Smith, D. L., and Wilson, A. O. (1972). Lagoon reefs of Bermuda: Their physiography, ecology and sediments. *J. Geol.* (in press).

Gebelein, C. D., and Hoffman, P. (1968). Intertidal stromatolites and associated facies from Lake Ingraham, Cape Sable, Florida. *Geol. Soc. Amer., Annu. Meet.* p. 109 (abstr.).

Ginsburg, R. N. (1956). Environmental relationships of grain size and constituent particles in some South Florida carbonate sediments. *Bull. Amer. Ass. Petrol. Geol.* 40, 2384–2427.

Ginsburg, R. N. (1957). Early diagenesis and lithification of shallow water carbonate sediments in southern Florida: in Regional Aspects of Carbonate Deposition. *Soc. Econ. Paleontol. Mineral., Spec. Publ.* 5, 80–100.

Ginsburg, R. N., ed. (1964). "South Florida Carbonate Sediments Guidebook." Geol. Soc. Amer. Conv.

Ginsburg, R. N., and Lowenstam, H. A. (1958). The influence of marine bottom communities on the depositional environment of sediments. *J. Geol.* 66, 310–318.

Ginsburg, R. N., Isham, L. B., Bein, S. J., and Kuperberg, J. (1954). "Laminated Algal Sediments of South Florida and their Recognition in the Fossil Record," unpublished rep. No. 54.21. Marine Laboratory, University of Miami, Coral Gables, Florida.

Glynn, P. W. (1963). Species composition of *Porites furcata* reefs in Puerto Rico with notes on habitat niches. *Ass. Isl. Mar. Lab.*, 5th Meet. pp. 6–9.

Goldman, M. I. (1926). Proportions of detrital organic calcareous constituents and their chemical alteration in a reef sand from the Bahamas. *Pap. Tortugas Lab.* 23, 37–66.

Goreau, T. F. (1959a). Buttressed reefs in Jamaica, British West Indies. *Proc. Int. Congr. Zool., 15th,* p. 250.

Goreau, T. F. (1959b). The ecology of Jamaican coral reefs, I. Species composition and zonation. *Ecology* **40,** 67–89.

Goreau, T. F. (1961). "The Structure of the Jamaican Reef Communities, Geological Aspects." Dep. Biochem. Ecol. N.Y. Zool. Soc., New York.

Goreau, T. F. (1963). Calcium carbonate deposition by coralline algae and corals in relation to their roles as reef builders. *Ann. N.Y. Acad. Sci.* **109,** 127–167.

Goreau, T. F. (1964). Mass expulsion of zooxanthellae from Jamaican reef communities after Hurricane Flora. *Science* **145,** 383–386.

Goreau, T., and Burke, K. (1966). Pleistocene and Holocene geology of the island shelf near Kingston, Jamaica. *Mar. Geol.* **4.**

Goreau, T. F., and Graham, E. A. (1967). A new species of *Halimeda* from Jamaica. *Bull. Mar. Sci.* **17,** 432–441.

Goreau, T. F., and Hartman, W. D. (1963). Boring sponges as controlling factors in the formation and maintenance of coral reefs. *In* "Mechanics of Hard Tissue Destruction," Publ. No. 75, pp. 25–54. Amer. Ass. Advance. Sci., Washington, D.C.

Goreau, T. F., and Wells, J. W. (1967). The shallow-water Scleractinia of Jamaica: Revised list of species and their vertical distribution range. *Bull. Mar. Sci.* **17,** 442–453.

Goreau, T. F., Llauger, V. T., Mas, E. L., and Seda, E. R. (1960). On the community structure, standing crop and oxygen balance of the lagoon at Cayo Turrumote. *Ass. Isl. Mar. Lab., 3rd Meet.* pp. 8–9 (abstr.).

Gorsline, D. S. (1963). Environments of carbonate deposition Florida Bay and the Florida Straits. *Shelf Carbonates Paradox Basin, Symp., 4th Field Conf., 1963, Four Corners Geol. Soc.* pp. 130–143.

Gould, H. H., and Stewart, R. H. (1953). Continental terrace sediments in the northeastern Gulf of Mexico: in Finding Ancient Shorelines. *Soc. Econ. Paleontol. Mineral., Spec. Publ.* **5,** 2–19.

Heilprin, A. (1890). The corals and coral reefs of the western waters of the Gulf of Mexico. *Proc. Acad. Natur. Sci., Philadelphia* **42,** 303–316.

High, L. R., Jr. (1969). Storms and sedimentary processes along the northern British Honduras coast. *J. Sediment. Petrol.* **39,** 235–245.

Hoffmeister, J. E., and Multer, H. G. (1964). Growth-rate estimates of a Pleistocene coral reef of Florida: *Geol. Soc. Amer., Bull.* **75,** 353–358.

Hoffmeister, J. E., and Multer, H. G. (1965). Fossil mangrove reef of Key Biscayne, Florida. *Geol. Soc. Amer., Bull.* **76,** 845–852.

Hoffmeister, J. E., Stockman, K. W., and Multer, H. G. (1967). Miami Limestone of Florida and its recent Bahamian counterpart. *Geol. Soc. Amer., Bull.* **78,** 175–190.

Hoskin, C. M. (1963). Recent carbonate sedimentation on Alacran Reef, Yucatan, Mexico. *Nat. Acad. Sci.—Nat. Res. Conc., Publ.* **1089,** 1–160.

Hoskin, C. M. (1964). Molluscan biofacies in calcareous sediments, Gulf of Batabano, Cuba. *Bull. Amer. Ass. Petrol. Geol.* **48,** 1680–1704.

Hoskin, C. M. (1966). Coral pinnacle sedimentation, Alacran Reef lagoon, Mexico. *J. Sediment. Petrol.* **36,** 1058–1074.

Hoskin, C. M. (1968). Magnesium and strontium in mud fraction of Recent carbonate sediment, Alacran Reef, Mexico. *Amer. Ass. Petrol. Geol., Bull.* **52,** 2170–2177.

Illing, L. V. (1954). Bahaman calcareous sands. *Bull. Amer. Ass. Petrol. Geol.* 38, 1–95.

Jindrich, V. (1969). Recent sedimentation by tidal currents in lower Florida Keys. *J. Sediment. Petrol.* 39, 531–553.

Jones, J. A. (1963). Ecological studies of the southeastern Florida patch reefs. Part I. Diurnal and seasonal changes in the environment. *Bull. Mar. Sci. Gulf Carib.* 13, 282–307.

Kaye, C. A. (1959). Shoreline features and Quaternary shoreline changes, Puerto Rico. *U.S., Geol. Surv., Prof. Pap.* 317-B, 49–140.

Kellerman, K. F., and Smith, N. R. (1914). Bacterial precipitation of calcium carbonate. *J. Wash. Acad. Sci.* 4, 400–402.

Kissling, D. L. (1965). Coral distribution on a shoal in Spanish Harbor, Florida Keys. *Bull. Mar. Sci.* 15, 599–611.

Koldewijn, B. W. (1958). Sediments of the Paria-Trinidad shelf. *Rep. Orinoco Shelf Exped.* 3, 1–109.

Kornicker, L. S. (1958). Bahamian limestone crusts. *Trans., Gulf Coast Ass. Geol. Soc.* 8, 167–170.

Kornicker, L. S. (1964). Form replica of a submerged barrier chain with lagoonal basin off South Cat Cay, Bahamas. *Bull. Mar. Sci. Gulf Carib.* 14, 168–171.

Kornicker, L. S., and Boyd, D. W. (1962). Shallow-water geology and environments of Alacran Reef complex, Campeche Bank, Mexico. *Bull. Amer. Ass. Petrol. Geol.* 46, 640–673.

Kornicker, L. S., and Bryant, W. R. (1969). Sedimentation of continental shelf of Guatemala and Honduras. *Amer. Ass. Petrol. Geol., Mem.* 11, 244–257.

Kornicker, L. S., and Purdy, E. G. (1957). A Bahamian fecal-pellet sediment. *J. Sediment. Petrol.* 27, 126–128.

Kornicker, L. S., Bonet, F., Cann, R., and Hoskin, C. M. (1959). Alacran Reef, Campeche Bank, Mexico. *Publ. Inst. Mar. Sci., Univ. Tex.* 6, 1–22.

Kumpf, H. E., and Randall, H. A. (1961). Charting the marine environments of St. John, U.S. Virgin Islands. *Bull. Mar. Sci.* 11, 543–551.

Land, L. S., and Goreau, T. F. (1970). Submarine lithification of Jamaican reefs. *J. Sediment. Petrol.* 40, 457–462.

Lewis, J. B. (1960). The coral reefs and coral communities of Barbados, W.I. *Can. J. Zool.* 38, 1133–1145.

Lewis, J. B. (1965). A preliminary description of some marine benthic communities from Barbados, West Indies. *Can. J. Zool.* 43, 1049–1074.

Lewis, J. B., Axelsen, F., Goodbody, I., Page, C., and Chislett, G. (1969). "Comparative Growth Rates of Some Reef Corals in the Caribbean," Mar. Sci. Manuscript, Rep. No. 10. McGill University, Montreal.

Lloyd, R. M. (1964). Variations in the oxygen and carbon isotope ratios of Florida Bay mollusks and their environmental significance. *J. Geol.* 72, 84–111.

Logan, B. W. (1969). Carbonate sediments and reefs, Yucatan Shelf, Mexico. Part 2. Coral reefs and banks. *Amer. Ass. Petrol. Geol., Mem.* 11, 129–198.

Logan, B. W., Harding, J. L., Ahr, W. M., Williams, J. D., and Snead, R. G. (1969). Carbonate sediments and reefs, Yucatan Shelf, Mexico. Part 1. Late Quaternary sediments. *Amer. Ass. Petrol. Geol., Mem.* 11, 1–128.

Logan, B. W., Davies, G. P., Read, J. F., and Cebulski, D. E. (1970). Carbonate sedimentation and environments, Shark Bay, Western Australia, *Amer. Ass. Petrol. Geol., Mem.* 13, 1–223.

Lowenstam, H. A. (1955). Aragonite needles secreted by algae and some sedimentary implications. *J. Sediment. Petrol.* **25**, 270–272.

Lowenstam, H. A., and Epstein, S. (1957). On the origin of sedimentary aragonite needles of the Great Bahama Bank. *J. Geol.* **65**, 364–375.

Lucia, F. J. (1968). Recent sediments and diagenesis of South Bonaire, Netherlands Antilles. *J. Sediment. Petrol.* **38**, 845–858.

Ludwick, J. C., and Walton, W. A. (1957). Shelf edge province in the northwestern Gulf of Mexico. *Bull. Amer. Ass. Petrol. Geol.* **41**, 2054–2101.

McCallum, J. S., and Stockman, K. W. (1964). Florida Bay. Water circulation. *In* "South Florida Carbonate Sediments Guidebook" (R. N. Ginsburg, ed.), pp. 11–15. Geol. Soc. Amer. Conv.

Macintyre, I. G. (1967a). Recent sediments off the west coast of Barbados, W.I. Unpublished Ph.D. Thesis, McGill University, Montreal.

Macintyre, I. G. (1967b). Submerged coral reefs off the west coast of Barbados, West Indies. *Can. J. Earth Sci.* **4**, 461–474.

Macintyre, I. G., and Milliman, J. D. (1970). Physiographic features on the outer shelf and upper slope, Atlantic Continental Margin, southeastern United States. *Geol. Soc. Amer., Bull.* **81**, 2577–2598.

Macintyre, I. G., and Pilkey, O. H. (1969). Tropical reef corals: Tolerance to low temperatures on the North Carolina Shelf. *Science* **166**, 374–375.

Macintyre, I. G., Mountjoy, E. W., and D'Anglejan, B. F. D. (1968). An occurrence of submarine cementation of carbonate sediments off the west coast of Barbados, W.I. *J. Sediment. Petrol.* **38**, 660–664.

MacKenzie, F. T., Kulm, L. D., Cooley, R. L., and Barnhart, J. T. (1965). *Homotrema rubrum* (Lamarck), a sediment transport indicator. *J. Sediment. Petrol.* **35**, 265–272.

Matthews, R. K. (1963). Continuous seismic profiles of a shelf-edge bathymetric prominence in northern Gulf of Mexico. *Trans., Gulf Coast Ass. Geol. Soc.* **13**, 49–58.

Matthews, R. K. (1966). Genesis of recent lime mud in southern British Honduras. *J. Sediment. Petrol.* **36**, 428–454.

Milliman, J. D. (1967a). Carbonate sedimentation on Hogsty Reef, a Bahamian atoll. *J. Sediment. Petrol.* **37**, 658–676.

Milliman, J. D. (1967b). The geomorphology and history of Hogsty Reef, a Bahamian atoll. *Bull. Mar. Sci.* **17**, 519–543.

Milliman, J. D. (1969a). Four southwestern Caribbean atolls: Courtown Cays, Albuquerque Cays, Roncador Bank and Serrana Bank. *Atoll Res. Bull.* No. 129, pp. 1–41.

Milliman, J. D. (1969b). Carbonate sedimentation on four southwestern Caribbean atolls and its relation to the "oolite problem." *Trans., Gulf Coast Ass. Geol. Sci.* **19**, 195–206.

Milliman, J. D. (1972a). Atlantic continental shelf and slope of the United States. Petrology of the sand fraction, northern New Jersey to southern Florida. *U.S., Geol. Surv., Prof. Pap.* (in press).

Milliman, J. D. (1972b). "Modern Marine Carbonates." Appleton, New York (in press).

Milliman, J. D., and Emergy, K. O. (1968). Sea levels during the past 35,000 years. *Science* **162**, 1121–1123.

Milliman, J. D., and Supko, P. R. (1968). On the geology of San Andres Island, western Caribbean. *Geol. Mijnbouw* **47**, 102–105.

Monty, C. L. V. (1967). Distribution and structure of recent stromatolitic algal mats, eastern Andros Island, Bahamas. *Ann. Soc. Geol. Belg.* **90**, 55–100.

Moore, C. H., Jr., and Billings, G. K. (1971). Preliminary model of beach rock cementation, Grand Cayman Island, B.W.I. *Carbonate Cements, Bermuda Conf. Carbonates, Bermuda Biol. Sta., Spec. Publ.* (in press).

Moore, D. R. (1958). Notes on Blanguilla Reef, the most northerly coral formation in the western Gulf of Mexico. *Publ. Inst. Mar. Sci., Univ. Tex.* **5**, 151–155.

Müller, G., and Müller, J. (1967). Mineralogisch-sedimentpetrographische und chemische Untersuchungen an einem Bank-Sediment (Cross Bank) der Florida Bay, USA. *Neues Jahrb. Mineral., Abh.* **106**, 257–286.

Multer, H. G. (1969). "Field Guide to Some Carbonate Rock Environments." Florida Keys and Western Bahamas.

Multer, H. G., and Hoffmeister, J. E. (1968). Subaerial laminated crusts of the Florida Keys. *Geol. Soc. Amer., Bull.* **79**, 183–192.

Munk, W. H., and Sargent, M. C. (1954). Adjustment of Bikini Atoll to ocean waves, Bikini and nearby atolls, Marshall Islands. *U.S., Geol. Surv., Prof. Pap.* **260**-C, 275–280.

Neumann, A. C., and Land, L. S. (1969). Algal production and lime mud deposition in the Bight of Abaco: A budget. *Geol. Soc. Amer., Spec. Pap.* **121**, 219 (abstr.).

Neumann, A. C., Gebelein, C. D., and Scoffin, T. P. (1970). The composition, structure and erodability of subtidal matts, Abaco, Bahamas. *J. Sediment. Petrol.* **40**, 274–297.

Newell, N. D. (1955). Bahamian platforms: in Poldervaart, A., (ed.), The Crust of the Earth. *Geol. Soc. Amer., Spec. Pap.* **62**, 303–316.

Newell, N. D. (1959). The biology of coral reefs. *Natur. Hist., N.Y.* **68**, 226–235.

Newell, N. D., and Rigby, J. K. (1957). Geological studies on the Great Bahama Bank: in Regional Aspects of Carbonate Deposition. *Soc. Econ. Paleontol. Mineral., Spec. Publ.* **5**, 13–72.

Newell, N. D., Rigby, J. K., Whiteman, A. J., and Bradley, J. S. (1951). Shoal-water geology and environments, Eastern Andros Island, Bahamas. *Bull. Amer. Mus. Natur. Hist.* **97**, 1–29.

Newell, N. D., Imbrie, J., Purdy, E. G., and Thurber, D. L. (1959). Organism communities and bottom facies, Great Bahama Bank. *Bull. Amer. Mus. Natur. Hist.* **117**, 177–228.

Newell, N. D., Purdy, E. G., and Imbrie, J. (1960). Bahamian oolitic sand. *J. Geol.* **68**, 481–497.

Nota, D. J. G. (1958). Sediments of the western Guiana shelf. *Meded. Landbouwhogesch., Wageningen* **58**, 1–98.

Odum, H. T., Burkholder, P. R., and Rivero, J. (1959). Measurements of productivity of turtle grass flats, reefs and the Bahia Fosforescente of southern Puerto Rico. *Publ. Inst. Mar. Sci., Univ. Tex.* **6**, 159–170.

Ottmann, F. (1963). "l'Atoll das Rocas" dans l'Atlantique sud tropical. *Rev. Geogr. Phys.* **5**, 101–107.

Perkins, R. D., and Enos, P. (1968). Hurricane Betsy in the Florida-Bahama area. Geologic effects and comparison with Hurricane Donna. *J. Geol.* **76**, 710–717.

Purdy, E. G. (1963). Recent calcium carbonate facies of the Great Bahama Bank.

1. Petrography and reaction groups; 2. Sedimentary facies. *J. Geol.* **71**, 334–353 and 472–497.

Purdy, E. G., and Imbrie, J. (1964). "Carbonate Sediments, Great Bahama Bank Guidebook." Geol. Soc. Amer. Conv.

Purdy, E. G., and Matthews, R. K. (1964). Structural control of Recent calcium carbonate deposition in British Honduras. *Geol. Soc. Amer., Spec. Pap.* **82**, 157 (abstr.).

Pusey, W. C., III. (1964). Recent calcium carbonate sedimentation in northern British Honduras. Ph.D. Thesis, Rice University, Houston, Texas.

Randall, J. E. (1965). Grazing effect on sea grasses by herbivorous reef fishes in the West Indies. *Ecology* **46**, 255–260.

Rice, W. H., and Kornicker, L. S. (1962). Mollusks of Alacran Reef, Campeche Bank, Mexico. *Publ. Inst. Mar. Sci., Univ. Tex.* **8**, 366–403.

Roberts, H. H., and Moore, C. H., Jr. (1972). Recently cemented aggregates (grape-stones) Grand Cayman Island, B.W.I. in *Carbonate Cements, Bermuda Conf. Carbonates, Bermuda Biol. Sta., Spec. Publ.* (in press).

Roos, P. J. (1964). The distribution of reef corals in Curaçao. *Stud. Fauna Curacao, Natuurwetensch. Werkgroep Ned. Antillen,* **20**, 1–51.

Rusnak, G. A. (1960). Some observations of recent oolites. *J. Sediment. Petrol.* **30**, 471–480.

Scholl, D. W. (1966). Florida Bay: a modern site of limestone formation. *In* "The Encyclopedia of Oceanography" (R. W. Fairbridge, ed.), pp. 282–288. Van Nostrand-Reinhold, Princeton, New Jersey.

Scoffin, T. P. (1970). Trapping and binding of subtidal carbonate sediments by marine vegetation in Bimini Lagoon, Bahamas. *J. Sediment. Petrol.* **40**, 249–273.

Shinn, E. A. (1963). Spur and groove formation on the Florida reef tract. *J. Sediment. Petrol.* **33**, 291–303.

Shinn, E. A. (1966). Coral growth-rate, an environmental indicator. *J. Paleontol.* **40**, 233–240.

Shinn, E. A. (1968). Burrowing in recent lime sediments of Florida and the Bahamas. *J. Paleontol.* **42**, 879–794.

Shinn, E. A., Ginsburg, R. N., and Lloyd, R. M. (1965). Recent supratidal dolomite from Andros Island, Bahamas: in Pray, L. C., and Murray, R. C. (eds.), Dolomitization and Limestone Diagenesis. *Soc. Econ. Paleontol. Mineral., Spec. Publ.* **13**, 112–123.

Shinn, E. A., Lloyd, R. M., and Ginsburg, R. N. (1969). Anatomy of a modern carbonate tidal-flat, Andros Island, Bahamas. *J. Sediment. Petrol.* **39**, 1202–1228.

Siegel, F. R. (1963). Variations of Sr/Ca ratio and Mg contents in Recent carbonate sediments of the northern Florida Keys area. *J. Sediment. Petrol.* **31**, 336–342.

Smith, C. L. (1940). The Great Bahama Bank. 1. General hydrographical and chemical features. 2, Calcium carbonate precipitation. *J. Mar. Res.* **3**, 147–189.

Smith, F. G. W. (1954). Gulf of Mexico Madreporaria. *U.S., Fish Wildl. Serv., Fish. Bull.* **55**, 291–295.

Smith, N. R. (1926). Report on a bacteriological examination of "chalky mud" and sea water from the Bahama Banks. *Carnegie Inst. Wash. Publ.* **344**, 69–72.

Squires, D. F. (1958). Stony corals from the vicinity of Bimini, Bahamas, British West Indies. *Bull. Amer. Mus. Natur. Hist.* **115**, 219–262.

Stanley, D. J., and Swift, D. J. P. (1967). Bermuda's southern aeolianite reef tract. *Science* **157**, 677–681.

Steers, J. A. (1940). The cays and Palisados, Port Royal, Jamaica. *Geogr. Rev.* **30**, 279–296.

Steers, J. A., Chapman, V. J., and Lofthouse, J. A. (1940). Sand cays and mangroves in Jamaica. *Georgr. J.* **96**, 305–328.

Stetson, H. C. (1953). The sediments of the western Gulf of Mexico. *Pap. Phys. Oceanogr. Meteorol.* **12**, 1–45.

Stockman, K. W., Ginsburg, R. N., and Shinn, E. A. (1967). The production of lime mud by algae in south Florida. *J. Sediment. Petrol.* **37**, 633–648.

Stoddard, D. R. (1962a). Three Caribbean atolls: Turneffe Islands, Lighthouse Reef and Glovers Reef, British Honduras. *Atoll Res. Bull.* No. 87, pp. 1–147.

Stoddart, D. R. (1962b). A short account on catastrophic storm effects on the British Honduras reefs and cays. *Nature (London)* **196**, 512–515.

Stoddart, D. R. (1963). Effects of hurricane Hattie on the British Honduras reefs and cays, October 30–31, 1961. *Atoll Res. Bull.* No. 95, pp. 1–142.

Stoddart, D. R. (1964). Carbonate sediments of Half Moon Cay, British Honduras. *Atoll Res. Bull.* No. 104, pp. 1–16.

Stoddart, D. R., and Cann, J. R. (1965). Nature and origin of beach rock. *J. Sediment. Petrol.* **35**, 243–247.

Storr, J. F. (1964). Ecology and oceanography of the coral-reef tract, Abaco Island, Bahamas. *Geol. Soc. Amer., Spec. Pap.* **79**, 1–98.

Sverdrup, H. V., Johnson, M. W., and Fleming, R. H. (1942). "The Oceans, Their Physics, Chemistry and General Biology." Prentice-Hall, Englewood Cliffs, New Jersey.

Swinchatt, J. P. (1965). Significance of constituent composition, texture and skeletal breakdown in some Recent carbonate sediments. *J. Sediment. Petrol.* **35**, 71–90.

Taft, W. H., and Harbaugh, J. W. (1964). Modern carbonate sediments of southern Florida, Bahamas, and Espiritu Santo Island, Baja California: a comparison of their mineralogy and chemistry. *Stanford Univ. Publ., Univ. Ser., Geol. Sci.* **8**, 1–133.

Taft, W. H., Arrington, F., Haimowitz, A., MacDonald, C., and Woolheater, C. (1968). Lithification of modern marine carbonate sediments at Yellow Bank, Bahamas. *Bull. Mar. Sci.* **18**, 762–828.

Thiel, M. E. (1928). *In* "Meersfauna Westafrika" (D. D. Michaelsen, ed.), Vol. 3, pp. 253–350.

Thomas, L. P., Moore, D. R., and Work, R. C. (1961). Effects of Hurricane Donna on the turtle grass beds of Biscayne Bay, Florida. *Bull. Mar. Sci. Gulf Carib.* **11**, 191–197.

Thorp, E. M. (1936). Calcareous shallow-water marine deposits of Florida and the Bahamas. *Pap. Tortugas Lab.* **29**, 37–119.

Tracey, J. F., Jr., Ladd, H. S., and Hoffmeister, J. E. (1948). Reefs of Bikini, Marshall Islands. *Geol. Soc. Amer., Bull.* **59**, 861–878.

Traganza, E. D. (1967). Dynamics of the carbon dioxide system on the Great Bahama Bank. *Bull. Mar. Sci.* **17**, 348–366.

Uchupi, E., Milliman, J. D., Luyendyk, B. P., Bowin, C. O., and Emery, K. O. (1972). Structure and origin of the southeastern Bahamas. *Amer. Ass. Petrol. Geol., Bull.* (in press).

Vaughan, T. W. (1916). Some littoral and sublittoral physiographic features of the Virgin Islands and northern Leeward Islands and their bearing on the coral reef problem. *J. Wash. Acad. Sci.* **6**, 53–66.

Vaughan, T. W. (1918). Some shoal-water bottom samples from Murray Island, Australia, and comparison of them with samples from Florida and the Bahamas. *Carnegie Inst. Wash. Publ.* **213**, 235–288.

Vaughan, T. W. (1919). Fossil corals from Central America, Cuba, and Puerto Rico, with an account of the American Tertiary, Pleistocene, and Recent coral reefs. *U.S., Nat. Mus., Bull.* **103**, 189–524.

Vermeer, D. E. (1959). "The Cays of British Honduras." Dep. Geogr., University of California, Berkeley.

Vermeer, D. E. (1963). Effects of Hurricane Hattie, 1961, on the cays of British Honduras. *Z. Geomorphol.* **7**, 332–354.

von Arx, W. S. (1954). Circulation systems of Bikini and Rongelap lagoons. *U.S., Geol. Surv., Prof. Pap.* **260-B**, 265–273.

Voss, G. L., and Voss, N. A. (1955). An ecological survey of Soldier Key, Biscayne Bay, Florida. *Bull. Mar. Sci. Gulf Carib.* **5**, 203–229.

Wantland, K. F. (1967). Recent benthonic foraminifera of the British Honduras shelf. Unpublished Ph.D. Thesis, Rice University, Houston, Texas.

Ward, W. C., Folk, R. L., and Wilson, J. L. (1970). Blackening of eolianite and caliche adjacent to saline lakes, Isla Mujeres, Quinta Roo, Mexico. *J. Sediment. Petrol.* **40**, 548–555.

Wells, J. W. (1954). Recent corals of the Marshall Islands, Bikini and nearby atolls. *U.S., Geol. Surv., Prof. Pap.* **260-I**, 385, 486.

Wells, J. W. (1957). Coral reefs: in J. Hedgpeth (ed.), Treatise on Marine Ecology and Paleoecology. *Geol. Soc. Amer., Mem.* **67**, 609–631.

Wiens. H. J. (1962). "Atoll Environment and Ecology." Yale Univ. Press, New Haven, Connecticut.

Work, R. C. (1969). Systematics, ecology, and distribution of the mollusks of Los Roques, Venezuela. *Bull. Mar. Sci.* **19**, 614–711.

Yonge, C. M. (1963). The biology of coral reefs. in F. Russell (ed.), *Advan. Mar. Biol.* **1**, 209–260.

Zaneveld, J. S. (1957). Micro-atolls in the Netherlands Antilles. *Rep. Inter-Isl. Mar. Biol. Conf., Inst. Mar. Biol., Univ. Puerto Rico* pp. 18–19.

Zaneveld, J. S. (1958). A lithothamnion bank at Bonaire (Netherlands Antilles). *Blumea, Suppl.* **4**, 206–219.

Zans, V. A. (1958a). Recent coral reefs and reef environments of Jamaica. *Geonotes* **1**, 18–25.

Zans, V. A. (1958b). The Pedro Cays and Pedro Bank. Report on the survey of the cays, 1955–57. *Jam., Geol. Surv. Dep., Bull.* No. 3, pp. 1–47.